Ecological Studies

Analysis and Synthesis

Volume 80

Ecological Studies

Volume 63
Long-Term Forest Dynamics of the Temperate Zone (1987)
By P.A. Delcourt and H.R. Delcourt

Volume 64
Landscape Heterogeneity and Disturbance (1987)
Edited by M. Goigel Turner

Volume 65
Community Ecology of Sea Otters (1987)
Edited by G.R. van Blaricom and J.A. Estes

Volume 66
Forest Hydrology and Ecology at Coweeta (1987)
Edited by W.T. Swank and D.A. Crossley, Jr.

Volume 67
Concepts of Ecosystem Ecology: A Comparative View (1988)
Edited by L.R. Pomeroy and J.J. Alberts

Volume 68
Stable Isotopes in Ecological Research (1989)
Edited by P.W. Rundel, J.R. Ehleringer, and K.A. Nagy

Volume 69
Vertebrates in Complex Tropical Systems (1989)
Edited by M.L. Harmelin-Vivien and F. Bourliere

Volume 70
The Northern Forest Border in Canada and Alaska (1989)
By J.A. Larsen

Volume 71
Tidal Flat Estuaries: Simulation and Analysis of the Ems Estuary (1988)
Edited by J. Baretta and P. Ruardij

Volume 72
Acidic Deposition and Forest Soils (1989)
By D. Binkley, C.T. Driscoll, H.L. Allen, P. Schoeneberger, and D. McAvoy

Volume 73
Toxic Organic Chemicals in Porous Media (1989)
Edited by Z. Gerstl, Y. Chen, U. Mingelgrin, and B. Yaron

Volume 74
Inorganic Contaminants in the Vadose Zone (1989)
Edited by B. Bar-Yosef, N.J. Barnow, and J. Goldshmid

Volume 75
The Grazing Land Ecosystems of the African Sahel (1989)
By H.N. Le Houérou

Volume 76
Vascular Plants as Epiphytes: Evolution and Ecophysiology (1989)
Edited by U. Lüttge

Volume 77
Air Pollution and Forest Decline: A Study of Spruce *(Picea abies)* on Acid Soils (1989)
Edited by E.-D. Schulze, O.L. Lange, and R. Oren

Volume 78
Agroecology: Researching the Ecological Basis for Sustainable Agriculture (1990)
Edited by S.R. Gliessman

Volume 79
Remote Sensing of Biosphere Functioning (1990)
Edited by R.J. Hobbs and H.A. Mooney

Volume 80
Plant Biology of the Basin and Range (1990)
Edited by C.B. Osmond, G.M. Hidy, and L. Pitelka

C.B. Osmond L.F. Pitelka G.M. Hidy

Plant Biology of the Basin and Range

With 117 Illustrations, 29 in Full Color

Springer-Verlag
Berlin Heidelberg New York
London Paris Tokyo Hong Kong

C. Barry Osmond
Botany Department
Duke University
Durham, NC 27706
USA

Louis F. Pitelka
Environment Division
Electric Power Research Institute
Palo Alto, CA 94303
USA

George M. Hidy
Environment Division
Electric Power Research Institution
Palo Alto, CA 94303
USA

Cover illustration: A typical contact between altered and unaltered andesite, Virginia Mountains, western Nevada. Sierran vegetation of *Pinus ponderosa* and *Pinus jeffreyi* on altered rock at the left; Great Basin vegetation of *Pinus monophylla, Juniperus osteosperma,* and *Artemisia tridentata* on unaltered rock on the right. Elevation about 1,500 meters.

ISBN-13:978-3-642-74801-1 e-ISBN-13:978-3-642-74799-1
DOI: 10.1007/978-3-642-74799-1

Library of Congress Cataloging-in-Publication Data
Plant biology of the basin and range / C. Barry Osmond, Louis F. Pitelka,
George M. Hidy, editors.
p. cm.—(Ecological studies ; v. 80)
Includes bibliographical references.
1. Plant physiological ecology—Great Basin. I. Osmond, C. Barry.
II. Pitelka, Louis F. III. Hidy, George M. IV. Series.
QK141.P58 1990
581.5'0979—dc20 90-9730
CIP

Softcover reprint of the hardcover 1st edition 1990

Camera-ready copy provided by the editors.
2131/543210 Printed on acid-free paper

Preface

In a very real sense, much of North American physiological plant ecology began in the Basin and Range and has been researched there over the last four decades. However, we believe that this book may be the first attempt to bring together the full range of contemporary research into the fascinating plant biology of the Basin and Range Province. We have invited contributions from researchers presently working in and around the Province and asked them to review the major vegetation zones and distinctive environmental issues from a predominantly plant ecophysiological perspective. As researchers interested in plant physiological and ecological processes, and in atmospheric processes affecting vegetation, we have tended to emphasize the atmosphere, plant, soil continuum in structuring this book. After an introduction to the geography of the region, we deal with atmospheric processes and climates of the Great Basin, follow with chapters on the different vegetational zones, treated from ecophysiological perspectives, and then place emphasis on plant-soil relations. We have not treated plant-animal interactions in the detail that the impacts of man and his domesticated animals on the desert ecosystem deserve. However we have included a review of a very promising technique (analysis of stable isotopes at natural abundance) for integration of these processes. We close with a compelling statement of the case for the Great Basin as a laboratory for climatic change research, prepared by a multidisciplinary team from the Desert Research Institute.

Our treatment could have been more comprehensive but for a very special time constraint, arising from events which gave rise to this book. In April 1989, the Desert Research Institute and the State of Nevada awarded the second Nevada Medal to W. Dwight Billings in recognition of his services to science and technology in the State. With the connivance of many of his colleagues in research, and with the confidence of his colleague editors of the series Ecological Studies with whom Billings has served since the inception of the series, we have striven to prepare this volume to celebrate the above recognition of Dwight Billings' achievements.

Dwight Billings moved from the University of Nevada, Reno to Duke University in 1952, but his research never really left the Basin and Range Province and its boundaries. As is amply illustrated by the work of Billings and his students since then, research in the Great Basin has had a large impact on plant ecophysiology, and plant biology in general. In 1981-82 developments at the Desert Research Institute, University of Nevada, Reno rekindled local research in this field. Dwight Billings and Frits Went were persuaded to lead a group of ecophysiologists on a memorable foray into some of the more interesting vegetation zones and environments. This reconnaissance in the company of experience more than achieved its purpose. These pioneers of desert plant physiology and ecology also introduced the newcomers to colleagues and friends with real enthusiasm for plants of the region. Until his "second retirement" from the Desert Research Institute in October 1985, Frits Went remained a stimulating member of the Biological Sciences Center. Billings himself subsequently became an Adjunct Professor in the Institute, researching a problem which he had defined 30 years previously. He also presented broad issues of mountain forest plant ecology in the Institute and prepared a review of this topic which is published here.

As editors, we are especially grateful to the publishers for facilitating the project, and to the Environment Division, Electric Power Research Institute, Palo Alto for a contribution which makes it a distinctive volume in a distinguished series. We are also grateful to Susan Gerbeth-Jones for preparation of the layout and typesetting of the camera ready copy using Microsoft Word 5.0, and to Richard Thomas for patient reconstruction of the Figures.

C.B. Osmond
L.F. Pitelka
G.M. Hidy

Contents

Contributors

ARCHER, S.	Department of Range Science College of Agriculture Texas A & M University College Station, TX 77843
BALL, J.T.	Biological Sciences Center Desert Research Institute and Department of Plant Science Agricultural Experiment Station University of Nevada Reno, NV 89557
BILLINGS, W.D.	Department of Botany Duke University Durham, NC 27706
CALDWELL, M.M.	Department of Range Science and Ecology Center Utah State University Logan, UT 84322
DANSIE, A.J.	Anthropology Department Nevada State Museum Carson City, NV 89710

DAVIS, J.O.
Quaternary Science Center
Desert Research Institute
University of Nevada
Reno, NV 89506

DELUCIA, E.H.
Department of Plant Biology
University of Illinois
Urbana, IL 61801

DOBROWOLSKI, J.P.
Department of Range Science and Watershed Science Unit
Utah State University
Logan, UT 84322

FOX, C.A.
Biological Sciences Center
Desert Research Institute
University of Nevada
Reno, NV 89506

HIDY, G.M.
Environment Division
Electric Power Research Institute
Palo Alto, CA 94303

INGRAHAM, N.L.
Water Resources Center
Desert Research Institute
University of Nevada
Las Vegas, NV 89120

IRWIN-WILLIAMS, C.C.
Quaternary Science Center
Desert Research Institute
University of Nevada
Reno, NV 89506

KLIEFORTH, H.E.
Atmospheric Sciences Center
Desert Research Institute
University of Nevada
Reno, NV 89506

KNAPP, A.K.
Division of Biology
Kansas State University
Manhattan, KS 66506

MOUAT, D.A. — Biological Sciences Center
Desert Research Institute
University of Nevada
Reno, NV 89506

NOWAK, R.S. — Department of Range, Wildlife and Forestry
University of Nevada
Reno, NV 89512

OSMOND, C.B. — Department of Botany
Duke University
Durham, NC 27706

PITELKA, L.F. — Environment Division
Electric Power Research Institute
Palo Alto, CA 94303

REINHARDT, R.L. — Atmospheric Sciences Center
Desert Research Institute
University of Nevada
Reno, NV 89506

RICHARDS, J.H. — Department of Range Science and Ecology Center
Utah State University
Logan, UT 84322

ROSE, M.R. — Biological Sciences Center
Desert Research Institute
University of Nevada
Reno, NV 89506

SCHLESINGER, W.H. — Department of Botany
Duke University
Durham, NC 27706

SMITH, S.D. — Department of Biological Sciences
University of Nevada
Las Vegas, NV 89154

SMITH, W.K.
Department of Botany
University of Wyoming
Laramie, WY 82071

TIESZEN, L.L.
Department of Biology
Augustana College
Sioux Falls, SD 57197

WHARTON, R.A.
NASA Headquarters
EBR, Washington DC 20546
and Biological Sciences Center
Desert Research Institute
University of Nevada
Reno, NV 89506

WIGAND, P.E.
Quaternary Science Center
Desert Research Institute
University of Nevada
Reno, NV 89506

Plate 1. Dwight Billings on altered andesite in the Virginia Mountains of the western Great Basin in May 1982. Also shown is Park S Nobel.

Plate 2. Thematic mapper image of the White Mountains, on the southwestern margin of the Great Basin, viewed on 7 July 1984, showing the major landscapes of the region. The image was formed from data collected in a visible channel (band 1) and two near infra red channels (bands 4 and 5) of the mapper. This combination shows irrigated agriculture and riparian vegetation as bright green, mountain forests olive green, with snow in blue, water in blue-black, clouds white, and the sparsely vegetated fans and playas in a range of hues. The spatial resolution of the primary data is 30 meters. (Photograph provided by C. Elvidge).

Plate 3.1. (upper) Panamint Mountains, eastern California, showing typical Basin and Range topography of the region, with extensive bajadas (alluvial fans). Death Valley is to the immediate east (right) of the picture.

Plate 3.2. (lower) Enclosed drainage basin, Eureka Valley, CA, with a large depositional playa in the center of the basin.

Plate 4.1. (upper) Aerial view of pebble-mounds at the Peg Wheat site, Hazen, NV, showing orientation in relation to natural drainage of the Lahontan terrace.

Plate 4.2. (lower) Tools of the Western Pluvial Lakes Tradition found in Carson Desert pebble-mounds. Scale bars, 1 cm.

Chapter 1

Man and Plants in the Great Basin

C.C. Irwin-Williams, C.B. Osmond, A.J. Dansie, and L.F. Pitelka

Approaching by air from the southwest of Utah, Edward Abbey wrote, "...we can plainly see the transition. Instead of canyons, mesas, plateaus and the sculptured forms of sandstone we see long, forested mountain ranges trending north and south, each range isolated from the next by an intervening broad valley or basin of dun colored desert true waste-lands of sparse shrubby vegetation, wide beds of waterless drainages, huge alluvial fans spreading out from the base of each mountain into the valley below. The mountains stand half buried in their own debris, the valleys marked by the winding subsurface rivers seeking an outlet, which most will never reach, to the sea. The aspect below is one of un-relieved bleakness and barrenness — tawny sands, dull clays and gravels, gray rocky peaks and craggy breaks thinly covered with the olive drab of juniper and pinyon pine — but the scale of things is awesome..." (Muench and Abbey 1979).

Crossing Nevada by automobile, John McPhee describes an equally graphic picture of the Basin and Range; "Basin, Fault. Range. Basin. Fault. Range. A mile of relief between basin and range... Each range here is like a warship standing on its own, and the Great Basin is an ocean of loose sediment with these mountain ranges standing in it as if they were members of a fleet without precedent... Some of the ranges are forty miles long, others a hundred, a hundred and fifty. They point generally north. The basins that separate them — ten and fifteen miles wide — will run on for fifty, a hundred, two hundred and fifty miles with lone, daisy-petalled windmills standing over sage and wild rye" (McPhee 1980).

Few of the Earth's landscapes evoke so consistently for an observer the mixed emotions of boredom, fear and awe in the manner of the Great Basin. Few landscapes provide such a contrast of geological or biological textures (Plate 2). One of the last regions of the western United States to be explored, this huge tract of mountains and sinks, with its extremes of weather conditions, was a formidable barrier to the westward migrations of the middle 19th century. It was interposed inconveniently between the Mormon outpost in Utah, on the west face of the Wasatch Range, and the goldfields in the western foothills of the Sierra Nevada of California. Penetrated and described by Fremont in 1843-44, the heart of the Great Basin, the slender artery of the Humboldt River and the grandeur of Pyramid Lake were exposed to traffic that the fragile environment scarcely could sustain. With only California gold in mind, pioneers quickly discovered the reality of "one of that species of deserts whose concentrated hideousness shames the diffused and diluted horrors of Sahara" (Twain 1872). Even discounting his propensity for hyperbole, one cannot but respond to Twain's assessment of the western Great Basin in 1861. "The desert was one prodigious graveyard. And the log chains, wagon tires, and rotting wrecks of vehicles were almost as thick as the bones." Yet in another breath he could exclaim that "the view from this lonely and segregated mountain peak...is the loveliest picture with which the hand of the Creator has adorned the earth, according to the best of my knowledge and belief, so help me God." So it remains in the Great Basin today. Although the landscape carries the scars of failed mining, ranching and residential enterprise, it remains one of great beauty and color (Plates 3.1, 3.2).

Fremont (1845), the first to describe the Great Basin, recognized sagebrush (*Artemisia tridentata*) as the essence of life in the region. He recorded that, on leaving the forests of what is now northeastern California, and "looking forward to the east scarce a tree was to be seen. Viewed from our elevation, the face of the country exhibited only rocks and grass, and presented a region in which the artemisia became the principal wood, furnishing to its scattered inhabitants fuel for their fires, building material for their huts, and shelter for the small game which ministers to their hunger and nakedness." This introduction to the Basin and Range is probably that most familiar to the traveler today. Entering the region on Interstate 80, the transition from Sierran forests to desert remains startlingly abrupt. Fremont's first description of the Great Basin is distinguished for its extensive reporting of plant biology, and is characterized by a sensitivity to the fragile environment which was soon lost among the pioneers. Lurching from prose to pragmatism,

Twain (1872) saw sagebrush as "an imposing monarch of the forest in exquisite miniature...which is very fair fuel, but as a vegetable it is a distinguished failure."

The precise boundaries of the Great Basin of North America are not agreed. To the hydrologist they can be defined quite pragmatically in terms of present drainage (Chapter 7). To plant ecologists, the boundaries tend to become more diffuse because plant distributions reflect past and present interactions of biotic and abiotic environments (Chapter 6). Centered approximately 116° W and 39° N, settled largely on the modern state of Nevada, and extending into substantial portions of Utah and California with smaller excursions into Oregon, Idaho and Wyoming, (Figure 1.1) the Great Basin, or the Basin and Range Province, or the Intermountain Region, is one of the most distinctive parts of North America from the viewpoint of almost all realms of natural science. Bounded by the Wasatch Range to the east, the Sierra Nevada to the west, the watersheds of the Snake and Columbia Rivers to the north, and the Colorado River drainage to the south the Great Basin occupies about 650,000 km^2 of the most diverse desert terrain.

The Basin and Range is a high-elevation, mountainous, cool desert. Nevada alone claims more than 400 mountains, only 10% of which are shown in Figure 1.1. Oriented predominantly north-south, these characteristic fault block mountains are thought to have arisen as recently as 17 million years ago following the stretching and uplifting of a low-elevation, relatively featureless, volcanic plain deposited some 20 million years earlier (McPhee 1980; Fiero 1986), although some believe the mountains to be only about 10 million years old and the volcanic plain to be much younger. The unusual brown- to- grey lava crust which spread throughout central Nevada, was stretched by as much as 50-100% in some locations. As the crust thinned and cracked, the fault blocks slid down or were tilted as the crust was pulled apart. About two thirds of the fault blocks dropped down to the west and about one third dropped down to the east (Fiero 1986).

Although there is uncertainty as to the causes of the Great Basin extension, it is agreed widely that the thinning of the crust was related to the fault-controlled uplift of mountains and the drop down of basins. Evidently the uplifting followed closely the after the Sierran uplift which has continued over the last 10 million years. Not surprisingly, there is much evidence that these transformations of the landscape have been accompanied by marked changes in vegetation.

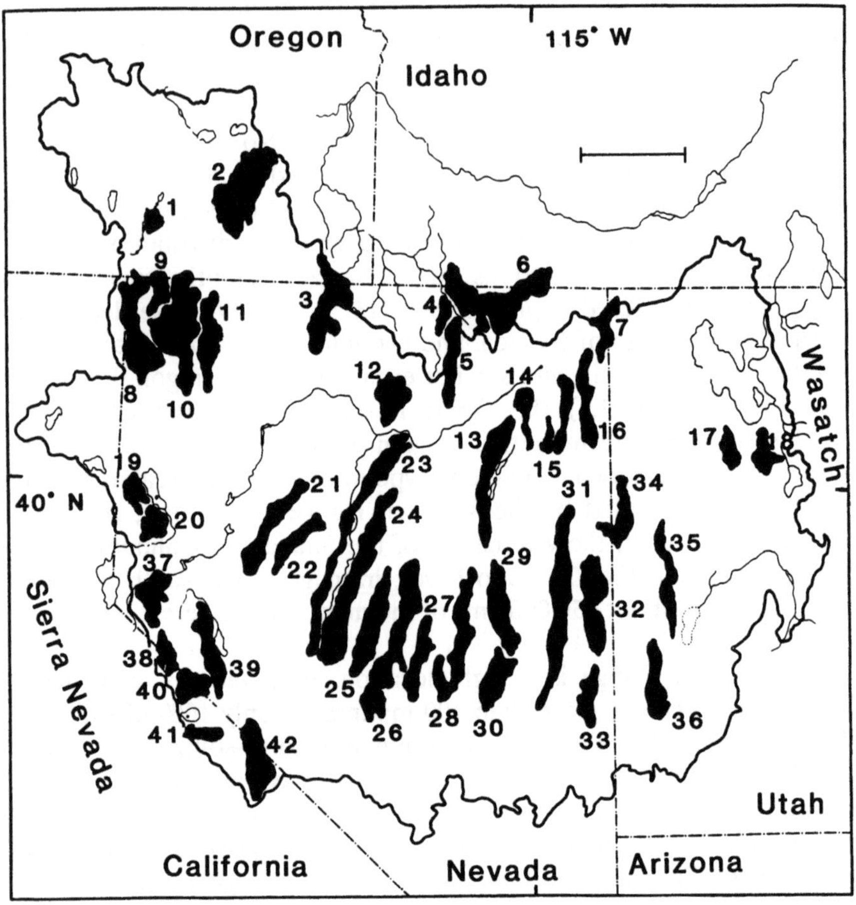

Figure 1.1. Principal mountain ranges of the Great Basin (redrawn from Lanner 1983 with permission). 1, Hart Mtn; 2, Steens Mtn; 3, Santa Rosa Mtns; 4, Bull Run Mtns; 5, Independence Mtns; 6, Jarbridge Mtns; 7, Goose Creek Range; 8, East Warner Mtns; 9, Bald Mtn; 10, Calico Mtns; 11, Black Rock Range; 12, Sheep Range; 13, Ruby Mtns; 14, East Humboldt Mtns; 15, Spruce Mtn and Pequop Mtns; 16, Toana Mtns; 17, Stansbury Mtns; 18, Oquirrh Mtns; 19, Virginia Mtns; 20, Pah Rah Mtns; 21, Stillwater Range; 22, Clan Alpine Range; 23, Shoshone Range; 24, Toiyabe Range; 25, Toquima Range; 26, Monitor Range; 27, Hot Creek Range; 28, Pancake Range; 29, White Pine Mtns; 30, Grant Range; 31, Schell Creek Range; 32, Snake Range; 33, Wilsons Creek Range 34, Deep Creek Mtns; 35, House Range; 36, Wah Wah Range; 37, Pine Nut Range; 38, Sweetwater Mtns; 39, Wassuk Mtns; 40, Bodie Mtns; 41, Glass Mtn; 42, White Mtns.

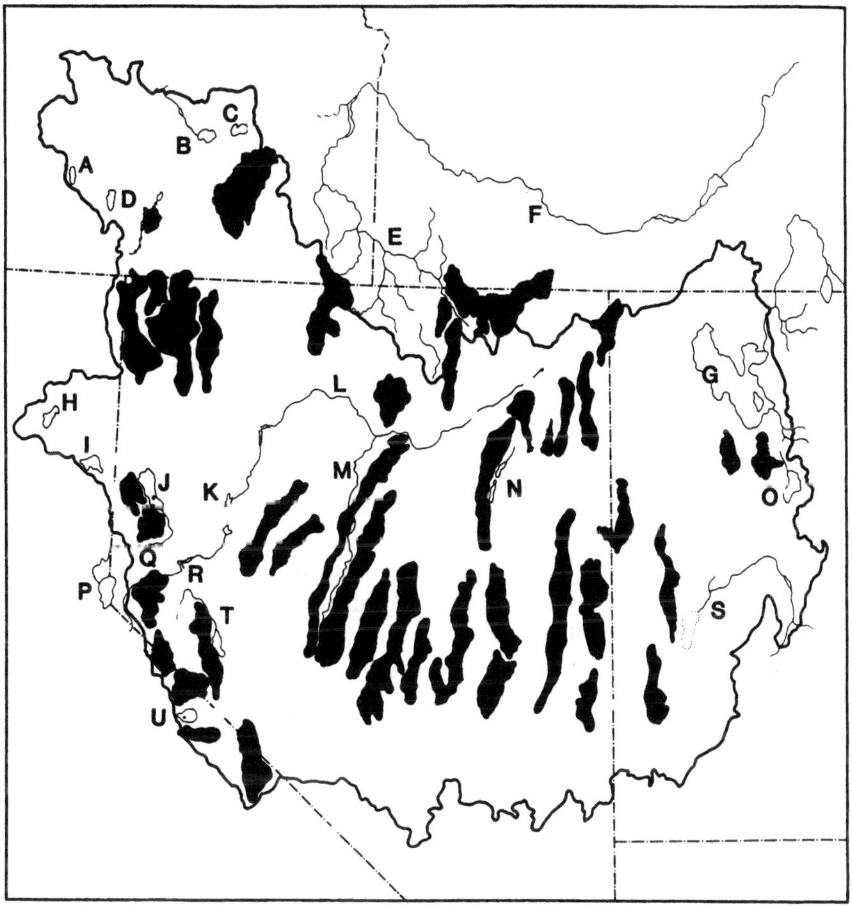

Figure 1.2. Principal rivers and lakes of the Great Basin (redrawn from Lanner 1983 with permission). A, Summer L; B, Harvey L; C, Malheur L; D, Lake Albert; E, Owyhee R; F, Snake R; G, Great Salt Lake; H, Eagle L; I, Honey L; J, Pyramid L; K, Humboldt R; M, Reese R; N, Franklin (upper) and Ruby (lower) Lakes; O, Utah L; P, Lake Tahoe; Q, Truckee R; R, Lahontan Reservoir; S, Sevier Lake and River; T, Walker Lake and River; U, Mono L. Major mountain ranges are shown in black and are identified in Figure 1.1.

All the margins of the Great Basin eventually drain to the Pacific Ocean, but the Basin itself is a complex of more than one hundred interior basins which have no outlet at all (Figure 1.2). In the west, many were linked in the late Quaternary as arms of the highly dissected pluvial Lake Lahontan, the terraces of which are still clearly visible over much of the Province. Remnants of this lake are now confined to Walker and Pyramid Lakes, the terraced shores testifying to their peripatetic status in more recent times. These terminal lakes became model systems in modern limnology (Hutchinson 1937) and today are sensitive indicators of man's activities in the western Great Basin (Galat et al. 1981). Water draining to these internal basins does not "sink mysteriously into the earth" as Twain speculated, but evaporates for the most part, often leaving extensive salt flats in the bottom of the basin. These are most spectacularly developed as the remnants of the larger lake to the east, Lake Bonneville, which has retreated to the Great Salt Lake and Utah Lake. From these playas in the basins, to the rocky outcrops on the tops of nearby ranges, a fascinating array of zoned, water-limited desert vegetation is displayed (Chapter 3).

Like many deserts the Basin and Range is pervaded by a misleading aura of antiquity. Yet as emphasized by Axelrod (1979) most semideserts of the world seem to have originated in the early Pliocene, and in North America, to have expanded rapidly in area over the last 10,000 years. A succession of rich fossil floras in the Great Basin provide an astonishing account of vegetational change since the early Oligocene (Figures 1.3, 1.4). Perhaps no change has been more dramatic than the retreat of *Sequoiadendron* to a few groves in the Sierra Nevada (Axelrod 1986). Areas now occupied by pinyon-juniper woodland display a fossil montane flora of the Miocene-Pliocene, and present-day sagebrush was formerly the site of the Tertiary woodlands.

In general terms, the fossil floras indicate a long-term trend from mesic to arid conditions that was accompanied by a shift from summer to winter precipitation and a shift to higher summer temperature and higher evaporation (Axelrod 1979). A late Eocene swamp flora near Elko is dominated by *Taxodium* and *Alnus*, with *Metasequoia*, *Acer*, and other species on nearby slopes. Further west the flora at Sutro Tuff in the Virginia Range region, which contains abundant conifers, a palm, and dominant evergreens such as *Lithocarpus* and *Quercus*, has been dated at 19 million years. It is thought that between 10 and 8 million years

Figure 1.3. (right) Leaves of fossil plants from forest vegetation in the Great Basin. Upper left, *Sequoiadendron* branchlet; upper right, *Sequoiadendron* cone; center left, *Lithocarpus* leaf; center right, *Prunus* leaf; lower left, *Mahonia* leaf; lower right, *Quercus* leaf. (Photographs provided by D.I. Axelrod, all x 0.75).

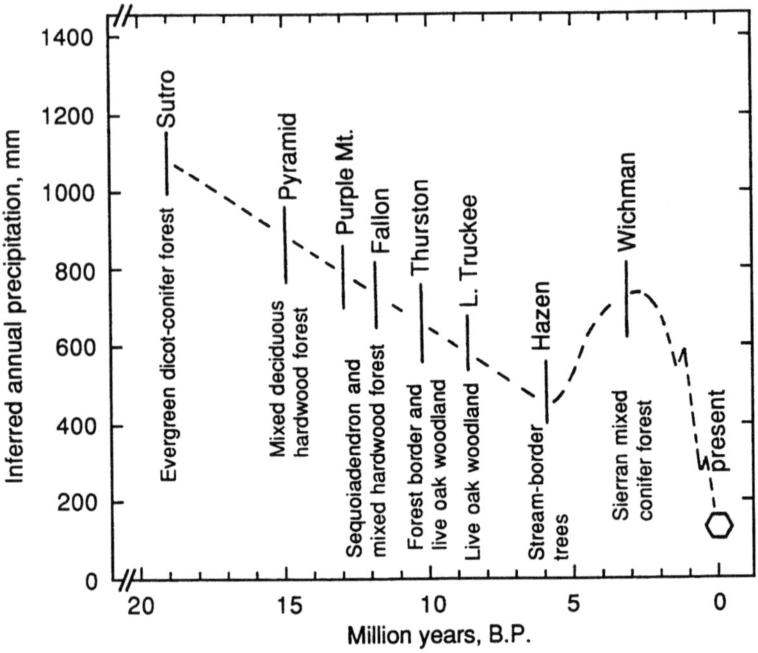

Figure 1.4. Reconstruction of past climatic change in the Great Basin using evidence from fossil floras. Redrawn from Axelrod (1979).

ago the mixed conifer and broad leaf schlerophyll forests in the Basins were replaced by live oak-juniper woodlands as displayed in the Esmeralda and Truckee floras. A subsequent increase in precipitation in the late Pliocene (3.5 million years ago) is indicated by the presence of an impoverished Sierra Nevada mixed conifer forest found in the western basins of the region. The double rainshadow cast over the Great Basin by the rise of the Sierra Nevada to the west and the Rocky Mountains to the east, led to the demise of basin forest floras and their replacement by semi-shrubs from south-western regions, especially after the last glaciation (Axelrod and Raven 1985). Excellent records of the recent history of forest and shrublands of the region are found in woodrat middens throughout the Great Basin ranging back 20,000 years in the south, to about 5,000 years in the north (Chapter 9).

The rapidly expanding desertification of the Great Basin over the last 10,000 years corresponds with human occupation of the region, but until very recently the "dominant mammal" was responsible for little impact on the landscape. Historic hunter-gatherer cultures practiced some environmental and vegetational management in the Great Basin, including sowing of wild seeds, burning to increase plant productivity and, more rarely, stream irrigation to increase crops of wild seeds (Pierce 1984). Although the region produces many useable native forage plants, and abundant small faunal resources, the absence of surface water during most of the year renders them effectively inaccessible to larger mammals and human beings. Whether for the purpose of obtaining potable water, increasing wild crops, or irrigating domestic crops, it is increasingly clear that water management of one variety or another was an important element of Great Basin hunter-gatherer subsistence technology. One of the more interesting and widespread of these practices was associated with complexes of surface features made up of hundreds and thousands of pebble-mounds.

First encountered by amateur archaeologists in the mid-60s, six of these pebble-mound complexes have been investigated in the western Great Basin (Tuohy 1981; Dansie 1981). Four of these occur on the coarse desert pavement of the south-facing slope of the 1,200 m beach of Lake Lahontan in the Carson Desert. A fifth is located in a similar but north-facing setting on the south edge of the Black Rock Desert and a sixth is found along a tributary to the Carson Sink. Characteristically, the pebble-mounds are about a meter across, less than half a meter high, and are composed of rounded pebbles and cobbles that have desert varnish on one side. There is little or no soil in the mounds, and there is nothing under them except unmodified desert pavement and hard silt. The mounds are surrounded by cleared areas from which the desert pavement has been completely or largely removed, exposing the hard impermeable soil formed on the old Lake Lahontan sediments. In areas in which the natural pavement is composed only of small flat pebbles, the scraped elements do not present mounds but only low areas of increased pebble density that are nearly invisible on the ground, but highly visible from the air (Plate 4.1).

The pebble-mounds occur in clusters covering from one half to about fifty hectares (from scores to many hundreds of mounds), and are associated uniformly with erosional drainage features that cut through the Lahontan beach bar (Plate 4.1). The terrain underlying the mound complexes slopes so as to drain towards an erosional gully or arroyo. Slope varies from very slight to near 30 °. Some of the mound clusters exhibit a very well-organized, linear form and may reflect a method of collecting the pebbles into linear strips before subdividing them into piles. In others, the mounds are not arranged so regularly, which seems to indicate that the mounds were built one at a time. Although the vegetation around the mounds has not been examined in detail, the mounds seem to have no direct effect on the viability or distribution of native vegetation. Shrubs sometimes grow in mounds, but they grow just as often elsewhere, on cleared areas or on unmodified desert pavement.

Speculations as to the function of the pebble-mounds have varied widely. They range from nonfunctional use of excess civilian construction corps labor in the 1930s to collection of gravel by Chinese workers for use in railroad construction in the late 1800s. Some have also invoked the favorite archaeological proposition of ceremonial purposes involving unspecified cultures. The fact that the original Sadmat site lay across a branch of the Southern Pacific Railroad initially led to confusion about the age and function of the mound complexes. The association with the railroad line is considered fortuitous, especially since other mound complexes are quite distant from it. Indeed, recent historic artifacts are rare or absent at the mound sites, and there is no evidence of the use of heavy mechanical or horse-drawn equipment.

In 1975-76, A. Dansie and S.L. Taylor became interested in the pebble-mound phenomenon. Departing markedly from existing interpretations, Dansie noted the physical, topographic, and environmental similarities between the mound complexes of the northern Great Basin and similar pebble-mound fields half a world away in the Negev Desert of Israel. Although the Negev examples had been known as archaeological curiosities since the 1900s, it was only in the 1950s that Evenari and his associates correctly interpreted the function of these large areas of modified desert pavement, pebble-mounds, and strips (Evenari et al. 1961, 1971). They concluded that the clearing of the desert pavement from the relatively impermeable underlying desert soils resulted in the creation of moderate to large-scale catchments for very efficient water-harvesting systems. These systems were of critical economic significance to the Nabatean and other indigenous agricultural and semi-agricultural inhabitants of the area.

The estimated age of the northwestern Great Basin pebble-mounds tends to covary with speculations as to function. Davis (1982) pointed out that the patination on the pebbles contained in the mounds is irregular, unlike that of the surrounding desert pavement, and possibly indicating a younger age. However, the irregular patination may be the result of frost heaving or turbation related to the relatively steep slope of the mounds, compared to the even quality of the undisturbed desert pavement. The best known and first of the mound localities discovered is the Sadmat site. This locality also produced a large quantity of "San Dieguito" surface artifacts belonging to what is now called the Western Pluvial Lakes Tradition, probably dating about 9,000 years before the present (BP). All four of the Carson Desert Mound complex sites studied to date have consistently yielded surface collections characterized by large weathered flake tools, and two have produced numerous stemmed points, steep edged side scrapers and gravers, characteristic of the Western Pluvial Lakes Tradition (Plate 4.2). Rock cairns and other surface rock alignments are features common to the Paleo-Indian/Archaic periods of the Great Basin and California desert.

Although there are alternative interpretations about age, there is little doubt about function. The pebble-mounds constitute a technique for acquiring potable water in an arid environment which lacks drinking water but still produces other useful resources, such as medium and small game and a variety of edible wild plants. This artificially "harvested" water could serve both to improve the range and productivity of the wild resources and to provide the potable water needed by the hunter-gatherers themselves. There is evidence of the Western Pluvial Lakes Tradition at the same beach line-lake elevation without pebble-mounds, these latter localities are characterized by a sandy substrate rather than by the hard impermeable silty clay found in association with the mound complexes, which would be a prerequisite for water harvesting.

The water harvesting functions of the pebble-mound complexes were verified experimentally by C.C. Irwin-Williams, A. Dansie, J.O. Davis and R. Jacobson. They compared a partially regenerated archeological plot (the Peg Wheat Site), an experimental site, artificially cleared of stones and arranged as the archeological site, and an adjacent control of untouched desert pavement. After a typical 25 mm rainfall event, the control plot collected no runoff, the archaeological plot collected about 10 liters, and the newly cleared experimental plot collected 84 liters. Similar results were obtained after several precipitation events of various magnitudes, with the experimental plot regularly producing the greatest runoff per square meter; the archaeological plot the second greatest; and the control plot the least (often none). A second set of experiments was conducted using a Mewig Micro-Rainfall Simulator, that allows the application of a known volume of water over an exact area at a carefully controlled rate. The simulator experiments showed cumulative rainfall capture was substantially more for the experimental plot than for the control plot. In addition, runoff occured earlier in the experimental plot, a fact of significance in view of the short duration of the infrequent summer showers characteristic of the northern Great Basin. These experiments demonstrate that clearing the catchment area of the Great Basin desert pavement substantially retards infiltration and substantially enhances runoff. The pebble-mounds which are the most obvious archaeological feature of the system, are by-products of the surface clearing operation.

The Carson Desert pebble-mound complexes are remarkably similar in form, size, composition and environmental setting to those in the Negev. The differences are related to the specific context of the Lahontan beach terraces on which they occur, with the Nevada mound fields being oriented toward erosional features which breach the gravel beaches. They evidently reflect the use of a very simple yet extremely effective (and very possibly early) water harvesting technology developed by desert hunters and gatherers in the Great Basin. Exactly how this technology was integrated into indigenous patterns of activity is not entirely understood. The most likely possibilities relate to the creation of potable, relatively non-saline water sources which would extend the range of the hunter-gatherer activities. The harvested water may have been used to encourage the growth of edible natural plants, such as wild rye. The larger mound fields, capable of producing thousands of liters of water, could have served as magnets for wild game (e.g., big horn sheep, antelope, etc.) which could have been hunted quickly and efficiently in this clustered and relatively immobile situation. The smaller, less well-organized but more numerous groups of mounds, along streams with impermeable substrate and along beds or banks, would be attractive to small mammals which could be taken in passive snares or traps and collected on regular sweeps of the area.

Water harvesting technology of a different sort came to the margins of the Great Basin in 1915 with the building of the first reservoir to support irrigated agriculture in the western U.S. It was named, appropriately, Lahontan Reservoir. With an impact on the landscape which is more obvious than the pebble-mounds of the hunter-gatherers, less than a century after implementation, these technologies have created major problems for sustainable water management in the western U.S. (Englebert and Scheuring 1984). Water management on the Truckee River, which led to the demise of Winnemucca Lake adjacent to

Pyramid Lake in the 1930s, continues to fuel long-standing litigation between the Paiute communities and the urban and agricultural interests that seems to ebb and flow with the snow pack in the Sierra Nevada.

Despite the arresting impression of irrigated agriculture in the Great Basin, whether viewed from the roadside or from space, research into plant biology in the Basin and Range has not been driven by concerns for water, nutrients, and salinity during cultivation, as it may have been in nearby California. Rather, it has been driven by the astonishing array of vegetational zonation, life-forms, and phenologies displayed in the flora of a diverse landscape that shows evidence of recent change. This source of interest, and its effect on the emerging discipline of plant physiological ecology, is evident in some of the early discussions of fossil floras. Shreve (1936) emphasized the need to keep in mind the evidence of past floras when seeking understanding of physiological and environmental determinants of present day plant distribution. Axelrod (1940) noted that "the fields of plant physiology and paleobotany overlap when an attempt is made to explain the later floristic evolution of these woodland and montane floras." Many aspects of the impacts of environment upon plant performance, survival, and distribution implied from the fossil record can be deduced from studies of present-day taxa throughout the Great Basin.

As argued by Chabot and Mooney (1985), much of North American physiological plant ecology can be traced to the definitive studies of Great Basin vegetation by Billings (1949,1951). Like the surveys of Fremont, these studies were based primarily on research funded by the U.S. Army. Accompanied by a substantial body of detailed soil and climatic data, Billings' papers began to develop a major theme around the environmental control of plant growth in habitats dominated by a small number of limiting factors. As instruments for the measurement of plant physiological functions improved, Billings pioneered the application of these instruments under field conditions. These activities led naturally to a niche in plant biology that became known as physiological plant ecology (Billings 1957).

The Basin and Range today is renowned for enthusiasm among its inhabitants, from all walks of life, for the preservation and propagation of its unique plant life. The Northern Nevada Native Plant Society publishes a regular newsletter, a journal (Mentzelia) and has sponsored a treatment of endangered species (Mozingo and Williams 1980). The University of Nevada Press has published well-illustrated natural histories dealing with the trees (Lanner 1984) and shrubs (Mozingo 1987) of the Great Basin. A regional journal, the Great Basin Naturalist, and a modern flora (Welsh et al. 1987) are further testimony to long-standing regional botanical activity. The newest National Park in the United States, The Great Basin National Park, was established in the eastern part of the Province in 1987.

We hope this volume may promote further pioneering research into the fascinating plant biology of the Basin and Range and a wider appreciation of the fragile but daunting beauty of the Province. In the words of Edward Abbey, "For some of those who have learned not only to live in, but also to love the desert, it offers rewards greater than its visual appeal to the sense of beauty — the promise implicit in all that rugged wildness, that open, unfenced, untraveled space — the sense of adventure, the reality of freedom, the hope of a refuge..." (Muench and Abbey 1979).

References

Axelrod DI (1940) Late Tertiary floras of the Great Basin and border areas. Bull. Tor. Bot. Club 67:477-487

Axelrod DI (1979) Desert vegetation, its age and origin. pp 1-72. In Goodin JR and Northington DK (editors) Arid Land Plant Resources. Texas Tech. Univ., Lubbock.

Axelrod DI (1986) The Sierra redwood (*Sequoiadendron*) forest: end of a dynasty. Geophytology 16:25-36.

Axelrod DI and Raven PR (1985) Origins of the Cordilleran flora. J. Biogeog. 12:21-47.

Billings WD (1949) The shadscale vegetation zone of Nevada and eastern California in relation to climate and soils. Am. Midl. Naturalist. 42:87-109.

Billings WD (1951) Vegetational zonation in the Great Basin of western North America. pp 102-122. In Comp. Rend. du Colloq. Ecologiques de la Regeneration de la Vegetation des Zones Arides, Union Intern. Soc. Biol. Paris.

Billings WD (1957) Physiological ecology. Annu. Rev. Plant Physiol. 8:375-392.

Chabot BF and Mooney, HA (1985) (editors) Physiological Ecology of North American Plant Communities. Chapman and Hall, NY.

Dansie A (1981) Pebble mound complexes of northwestern Nevada. Nevada Archaeologist 3(1):16-28.

Davis JO (1982) Comments on the pebble mound study. Nevada Archaeologist 3(2):5-6.

Englebert EA and Scheuring AF (1984) Water Scarcity: Impacts on Western Agriculture. Univ. California Press, Berkeley.

Evenari M, Shanon L, Tadmor N, and Aharoni Y (1961) Ancient agriculture in the Negev. Science 133:979-96.

Evenari M, Shanon L, and Tadmor N (1971) The Negev: the challenge of a desert. Cambridge Univ. Press.

Fiero B (1986) Geology of the Great Basin: A natural history. Univ Nevada Press, Reno.

Fremont JC (1845) The Exploring Expedition to the Rocky Mountains, Oregon and California.

Galat DL, Lider EL, Vigg S, and Robertson SR (1981) Limnology of a large, deep, North American terminal lake, Pyramid Lake, Nevada USA. Hydrobiologia 82:281-317.

Hutchinson GE (1937) A contribution to the limnology of arid regions. Trans. Conn. Acad. Sci. 33:47-132.

Lanner RM (1984) Trees of the Great Basin, Univ. Nevada Press, Reno, NV.

Mozingo HN (1987) Shrubs of the Great Basin, Univ. Nevada Press, Reno, NV.

Mozingo HN and Williams M (1980) Threatened and Endangered Plants of Nevada. U.S. Bureau of Land Management, Reno, NV.

McPhee J (1980) Basin and Range, Farrar Straus Giroux, NY.

Muench D and Abbey E (1979) Desert Images. Harcourt Brace Janovich, NY.

Pierce C (1984) Aboriginal modifications of native vegetation in the Great Basin. Unpublished manuscript. In Symposium on Prehistoric Desert Water Management, Society for American Archaeology, Denver, CO.

Shreve F (1936) The transition from desert to chaparral in Baja California. Madroño 3:257-264.

Tuohy DR (1981) A brief history of the discovery and explanation of pebble mounds, boulder cairns and other rock features at the Sadmat Site, Churchill County, Nevada. Nevada Archaeologist 3(1):4-15.

Twain, M (1872) Roughing It. New American Library Edition, NY.

Welsh SL, Atwood ND, Goodrich S, and Higgins LC (editors) (1987) A Utah Flora. Brigham Young Univ. Press, Provo, UT.

Chapter 2

Atmospheric Processes Affecting the Climate of the Great Basin

G. M. Hidy and H. E. Klieforth

Introduction

The Great Basin flora has experienced huge variations in climate and in soil in the course of its history. Today it is exposed to one of the most extreme and variable climates on earth. This chapter deals with the factors that affect present day climate of the Basin and Range, with emphasis on the climates of montane and sub-alpine forest regions that are home to some of the oldest living plants, the bristlecone pines, as well as economically important species. The chapter also deals with the difficulties of evaluating climatic interactions and measurement in such a topographically diverse environment. It also presents a preliminary assessment of research dealing with the local and remote impacts of modern man upon air quality and visibility, one of the more controversial and less tangible aspects of environmental change in the desert.

The climate of a region is basically the long-term average state of its weather elements and their extremes. The semantics of long-term averaging vary according to the scientific phenomenon of interest. However, generally averaging is taken over the period of record which, in the Great Basin, may vary from 10 to 130 years depending on location. Climatic variability is determined directly by available observations of the atmosphere, or indirectly from surrogate geological or biological records. Measured changes in climate that are relevant to plant growth at high elevations have time constants ranging from hours to months, to those that must be inferred over thousands of years, covering both cyclical and nonperiodic warm and cool periods of the earth. Spatial variability is important especially in mountain regions. Vegetation has responded to climate change in different ways depending on the nature and resilience of the species. Climate change and the associated disturbances of the lithosphere have sent some species into extinction, while creating a favorable environment for the migration and evolution of others (Chapter 1).

Major changes in climate are believed to have occurred in the past with minimum response times of hundreds of years, but there is evidence of shifts in the climate over smaller time intervals. Large increases of particulates in the atmosphere following volcanic eruptions can have cooling effects, and increased solar heat retention by increases in the atmospheric concentration of "greenhouse" gases, such as CO_2, N_2O, CH_4 and chlorofluorohydrocarbons (CFCs) may lead to warming. The theory of radiative heating of planets indicates that these gases, in combination with water vapor, create conditions for heat retention near the earth's surface. This takes place because the atmosphere is transparent to incoming solar radiation at wavelengths outside the spectral absorption bands of the greenhouse gases, whereas the outgoing radiation from the earth's surface is in the energy absorption range of the greenhouse gases. Increases in absorbing gas concentration will lead to increased heat retention in the absence of cooling processes (Hansen et al. 1981). The significance of a relatively rapid shift in climate over a century or two, due to increases in absorbing gases, is unknown. However, the potential of these components of the atmosphere to disrupt climate has increased the need to incorporate knowledge of both physical and chemical characteristics of the atmosphere into the study of the responses of forest systems, especially in the western United States (C.G.C. 1988).

The weather and climate of Nevada are discussed specifically by Houghton, et al., (1975), and the climate of Utah is discussed by Brough, et al., (1987). A good deal of climatic information relevant to vegetation was included by contributors to the recent comprehensive treatment of North American plant ecology (Barbour and Billings 1988). Other useful sources of climatic data are given in Houghton (1986), who provides an account of the geography and history of this region, McLane (1978), who describes 314 mountain ranges in Nevada alone, and Trimble (1989), who treats the natural history and biogeography of the Great Basin. A major problem in studying the historic climate of this region is the relative dearth of information from the higher elevations. For various reasons related to the settlement and development of the western United States, the earliest weather records derive from Army posts or forts and towns along the railroads. Later, with the growth of aviation and its reliance on meteorological information, weather stations were located at airports which, of course, were built in flat valley areas. Recently, however, there has been a notable increase in climatic data from higher altitude stations located at ranches, mines, state and national parks, and mountain research stations. Thus, for future monitoring of climatic change it is important to maintain measurements of weather elements at long-established sites while continuing to expand the network of high-altitude sites.

Climatic effects are discussed conventionally in terms of physical factors, including temperature, sunlight and cloudiness, precipitation, humidity, and winds. Recent research also has examined the influence of exposure to airborne chemicals, such as sulfur and nitrogen oxides, and atmospheric acidity (NAPAP 1987). These gases derive from natural sources as well as from air pollution associated with fossil fuel combustion or industrial processing. Sulfur and nitrogen oxides, along with heavy metals, are cited often as pollution components. The former are chemically reactive in the atmosphere, where as the metals are inert, in the form of oxides and salts. Sulfur oxides form salts with cations, or sulfuric acid, and the nitrogen oxides form nitric acid. These chemical aspects of climate can have marked biological and aesthetic effects on both local (urban) scales and regional scales. Vegetation may respond to these chemicals via direct contact with atmospheric gases and aerosol particles, or through the action of dissolved materials in precipitation (Winner and Atkinson 1986). Chemicals in precipitation also may affect vegetation by modifying the soil chemical characteristics or biological interactions in the root zone (Schulze et al. 1989). At the same time, vegetation itself is a source of both volatile organics (Rasmussen and Went 1964) and particulates (Went et al. 1967).

Physical Factors

The regional climates of the United States are quite varied and are determined by latitude, the proximity to oceans, major freshwater bodies, such as the Great Lakes, and on the influence of major weather-producing regions such as the northern Pacific Ocean, the intertropical convergence zone, and the western Atlantic region, centering on a region of high barometric pressure near Bermuda. The mid-continent patterns are strongly modulated by the southward surge of arctic air along the north side of meanders in the high-altitude jet stream, and northward flow of tropical air from the eastern Pacific and the Gulfs of California and Mexico. The great mountain ranges exert further influence on climate by modifying the storms, by blocking or steering a westerly flow across the continent, and by exerting a strong influence on precipitation patterns.

In contrast with much of the Midwest and the East, the western climate is quite heterogeneous, composed of major arid and wet regions with distinctive characteristics. Within each region, the climate may differ markedly within tens of kilometers. Much of this localized change can be attributed to the strong forcing by the topography of atmospheric motion (vertical rise and fall), causing clouds to form and precipitation to occur in the areas of rising and cooling air. This orographic influence is well illustrated going inland from the Pacific Coast of Southern California where temperature, humidity, and precipitation change sharply from a Mediterranean-like regime to a semi-desert environment over distances of less than 100 km. Another of these abrupt spatial transitions, with marked consequences for vegetation, is to be found along the Columbia River dividing Washington and Oregon.

The Great Basin is unique in its "basin and range" character, and its climatic extremes. Like other arid regions, it is characterized by an average net loss of surface moisture, and by an excess of evapotranspiration over precipitation. In the Great Basin region, the high mountain ranges force moist westerly air to rise and deposit precipitation along the western slopes and at ridge crests, thus leaving the eastern slopes relatively dry. The complex terrain, large differences in altitude and consequent distortions of air currents create high variability in local precipitation. Vegetation may receive intense rainfall for short times followed by very long periods without precipitation. This sporadic water supply creates widely varied seasonal and interannual growing conditions for Great Basin vegetation.

The mountain-valley environment is layered strongly with respect to temperature. The mountains tend to be cooler and more ventilated than the valley floors during most of the year. Surface air heating during the day can lead to very high valley temperatures, which are often accompanied by strong local turbulence (wind intensity and gustiness) and formation of localized microtornadoes called dust devils. At night, in contrast, the valleys lose heat rapidly by radiation, causing cool air to

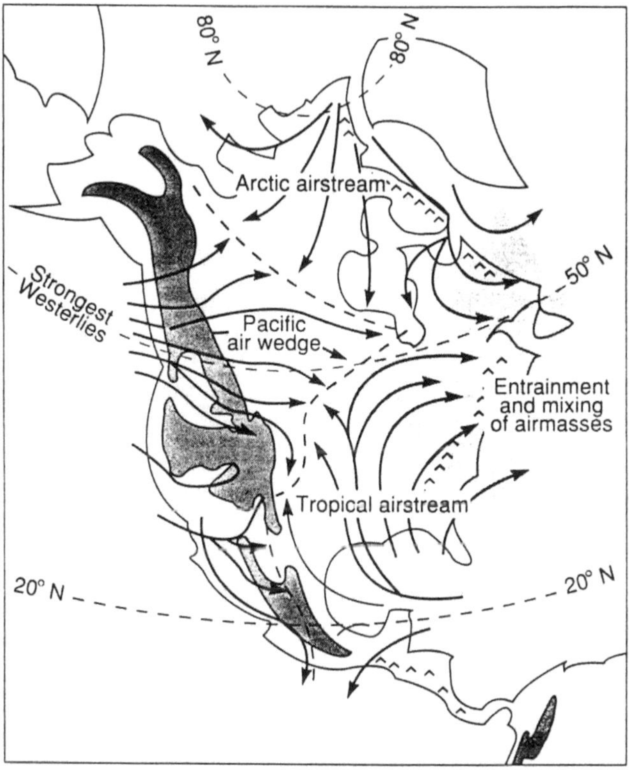

Figure 2.1. Schematic representation of the surface air flow across North America, illustrating the wedge of Pacific air east of the Rocky Mountains, and the downstream entrainment and mixing of air streams, based on July resultant surface winds. Redrawn from Bryson and Hare (1974).

pool in stagnant layers that are not mixed with the warmer air aloft (Billings 1954; Barry 1981). In heavily populated basins, such as Reno and Salt Lake City, very stable inversions form in winter, which accumulate pollutant gases and particulates in the lowest air layers.

Wind

On the synoptic or continental scale, wind systems are related to the contrasting thermal properties and movement of large air masses across the North American continent. Air mass transport and mixing to a great extent establish the broad-scale patterns of temperature, humidity and cloudiness prevailing over the United States. The large-scale wind

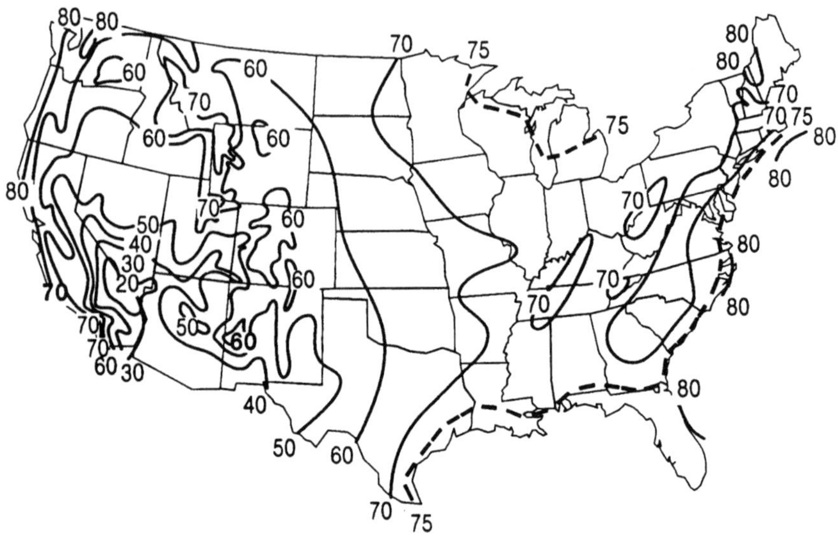

Figure 2.2. Annual mean relative humidity (%) over the continental United States. Redrawn from NOAA (1974).

patterns of interest are indicated in Figure 2.1. The West is affected by a band of westerlies that blow inland with strongest intensity in the Pacific Northwest. This moist, cool air is mixed continuously along the northern border of the U.S. with cold and dry arctic air streams. Along the mid- to southern reaches of the Pacific Coast, the westerlies are weaker and the air moving eastward is warmer and usually somewhat drier. Continental penetration of these westerlies in summer tends to be suppressed by the strong warming of air over the southwestern deserts, creating a region of high barometric pressure. In fact, this cell of high pressure (the Great Basin Anticyclone) induces its own air circulation that often brings warm, moist air northward from Mexico.

Pacific air blends farther inland over the Great Plains with a strong southerly flow of warm, moist air from the Gulf of Mexico. The region where this tropical airstream mixes with arctic and Pacific air is a regime of highly variable weather over the Plains, giving rise to occasional intense snowstorms in winter and strong thunderstorm and tornado activity in spring and early summer. With the blending of air masses over the central continent, the air moves eastward across the Atlantic Coast. Weather in this region is warm and humid in summer from the Gulf air, and can be characterized in winter by cold, dry, arctic outbreaks and

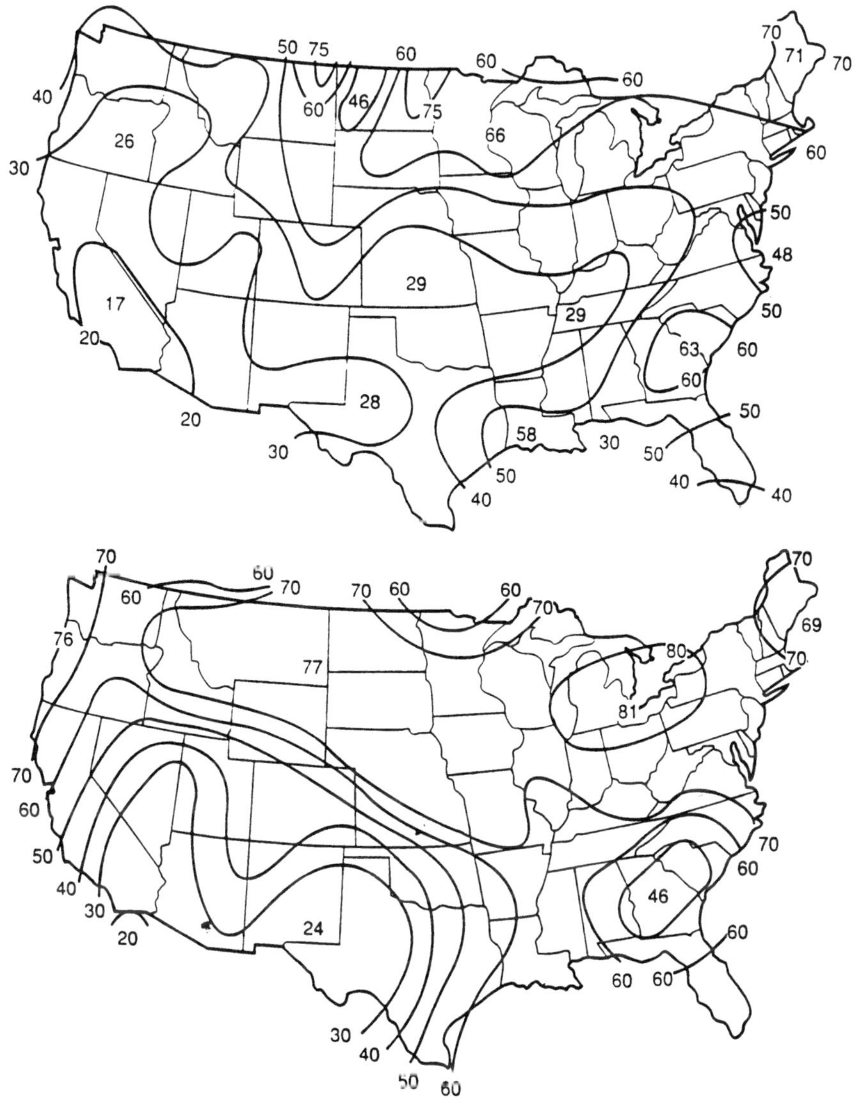

Figure 2.3. Summer (upper) and winter (lower) cloud cover (%) for July-August 1962 and December-January 1963 as calculated from TIROS II satellite data. Redrawn from Clapp (1964).

warmer, moist periods. The air mass blending essentially creates the average patterns of humidity and cloudiness in Figures 2.2 and 2.3. Air flows over the Great Basin predominantly from the west. The strongest winds aloft, of the order of 50 m s^{-1}, are associated with large

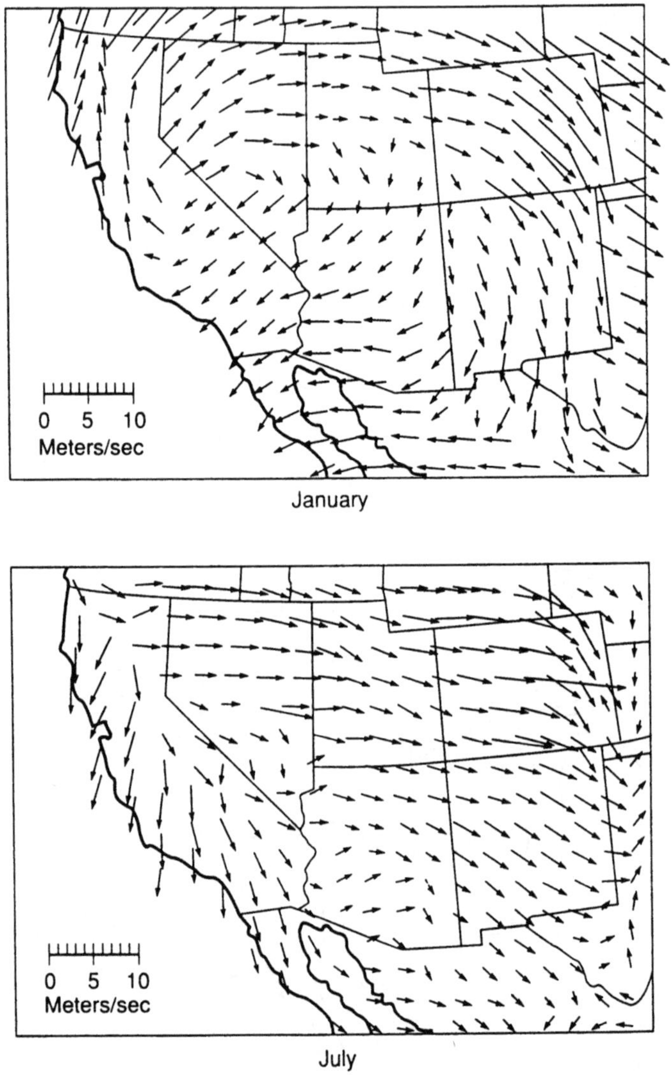

Figure 2.4. Monthly afternoon mixed-layer mean resultant wind fields in winter (upper) and summer (lower) 1981, calculated from the Limited Fine Mesh model of NOAA. Redrawn from Latimer (1985).

temperature gradients along the polar front. Strong southerly surface winds of 15 to 30 m s^{-1} often precede the passage of a cold front, and equally strong northerly surface winds follow the passage of a vigorous frontal trough or cyclone. In summer, as a thermally-induced anticyclone develops over the Southwest, a convergence zone develops between the southeasterly flow of moist subtropical air and the dry southwesterly flow from the mid-latitudes of the eastern Pacific.

On the mesoscale, a diurnal circulation pattern in summer is induced by differential heating on eastern and western slopes of the mountains. It begins in the morning with an easterly upslope breeze in response to solar heating of the east-facing slopes. During the afternoon the wind shifts to westerly in response to the increased heating and resultant convection on the west-facing slopes. This diurnal pattern is enhanced in the Sierra Nevada region by a prevailing westerly flow aloft and an apparent resonance with a diurnal summer sea-breeze pattern across California. In the Truckee Meadows, and Washoe, Eagle and Carson Valleys of the western Great Basin, summer breezes are often near 10 m s^{-1} during afternoons and early evenings.

Low-level wind patterns in the afternoon in the Great Basin region and the Southwest are shown for January and July in Figure 2.4. These maps give more detail of the interaction with the westerlies and southwestern high-pressure circulation for the year 1981. This was an unusual year with lower summer temperatures and a warmer winter, but the wind patterns for this year qualitatively illustrate the summer and winter extremes for the region. To the north, westerlies prevailed across the Great Basin, and brought periods of strong winds in winter, and generally westerly flow in summer. To the west, the Sierra Nevada experienced southerly flow in winter superseded by northerly flow in summer as the (clockwise) high-pressure circulation centered in southern Nevada strengthened and weakened. Winds tend to weaken in the southern Great Basin, but are strong in a band of westerlies to the north that reach maximum speed entering the western Rockies in Colorado. Winds are often relatively weak across the eastern Great Basin in winter, but southerly flow strengthens in summer as the high-pressure circulation expands to shift over Arizona and Utah.

Among the long-term toposcale effects of prevailing winds is their role in shaping the geomorphology of the region. In some of the valleys and basins, sand dunes have been formed by the large eddies within the confines of the surrounding mountains. In some areas, such as Eureka Valley, the dunes rise over 200 m above the valley floor. Snow blowing over mountain ridges in strong westerly winds is deposited on the upper leeward slopes creating deep drifts in some winters, thus enhancing the

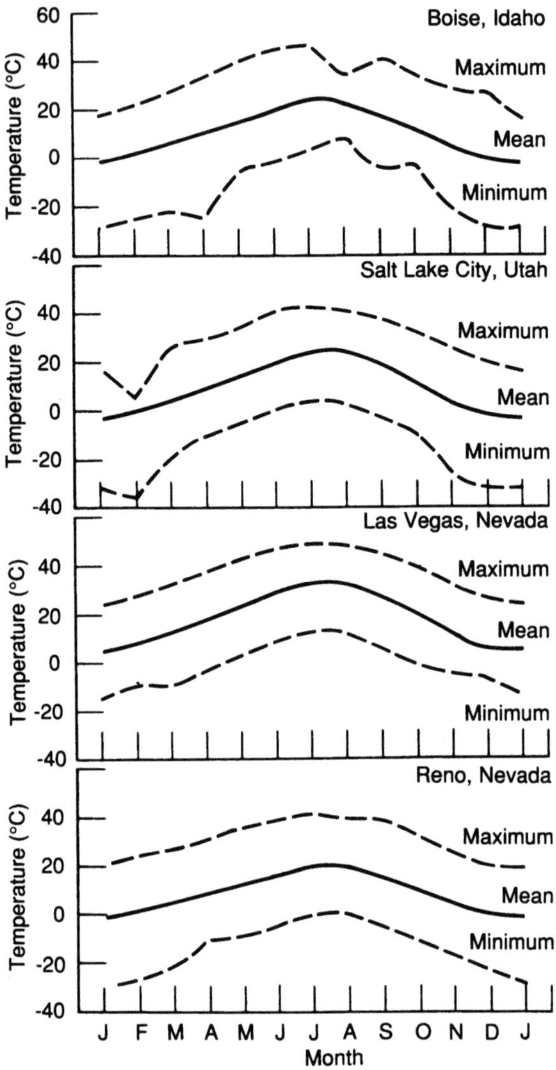

Figure 2.5. Monthly mean surface temperature and extreme values on the northern (Boise, ID), eastern (Salt Lake City, UT), southern (Las Vegas, NV) and western (Reno, NV) margins of the Great Basin. Redrawn from NOAA (1974).

runoff from that area. In the Pleistocene epoch of geologic history such drifts deepened to form permanent ice fields and glaciers, which together with reduced insolation on east slopes, carved cirques of varying size on the upper eastern slopes of the highest ranges. Heald (1956) describes the glacial topography on Wheeler Peak in the Snake Range, a feature that is now included in the Great Basin National Park.

Embedded in the larger-scale wind patterns are the local air flows that create the wind stress on vegetation. These are coupled to the large scale, but are modulated by orographic forcing, surface friction and thermal forcing. At the higher elevations, the ridge tops rise into the high winds of the upper layers. In the valleys and mountain slopes, winds will be highly variable, strong at times in stormy weather, during conditions of intense downslope flow driven by strong wind shear and thermal stratification, or in conditions of intense thermal mixing from surface heating. One of the more damaging factors of climate for plants is the intensity of winds. In alpine areas, winds can be very high in speed and very persistent. At the smallest scale of leaf and needle size, winds can cause mechanical damage, as well as controlling boundary layers for heat and moisture transfer, and chemical transfer, to and from the plant (Nobel 1981). However, the combination of very high winds and extremes in temperature (Chapter 4) suppresses growth to a point of establishing a zone above which no trees or woody vegetation can survive, or where vegetation is reduced to stunted alpine Krummholz (Arno and Hammerly 1984; Mutel and Emerick 1984). In studies of trees flagged by strong winds, it has been shown that the resultant asymmetry of tree ring growth is proportional to the strength of the prevailing wind. Other ground-level effects include the abrasion of vegetation and structures by wind-blown sand, the development of small hummocks about plant communities, and the shifting topography of sand dunes which trap and hold moisture a few centimeters below their surface thus supporting a unique biota. Timberline is generally found above 3000 m in the Great Basin, and it increases in altitude from north to south (Chapter 3).

Temperature

Conditions of temperature in valleys of the Great Basin vary widely each day, as well as seasonally. Daily temperatures can vary over 15 to 20° C, and seasonal averages may range from well below 0° C in winter to 40° C and above in summer. Typical seasonal patterns for boundaries of the Great Basin are shown at four locations in Figure 2.5. The extremes are generally larger than observed to the west across the Sierra Nevada, but are similar to continental conditions extending eastward into the grasslands of the Great Plains. They are greater in range than those observed along much of the eastern seaboard, where extremes are moderated by the Atlantic Ocean influence, as well as the northward flow from the Gulf of Mexico.

The seasonal variation of air temperature within the Great Basin is large because of its relative distance from moderating oceanic influences, the strong insolation during long summer days, and great heat loss by terrestrial radiation during long winter nights. Spatial differences are equally large because of great variation of altitude. During all months of the year under predominantly clear skies the diurnal range at most sites is of the order of 15 to 20° C. The average lapse rate (decrease of temperatures with altitude) is 6.5° C km^{-1} during most of the year but at mid-day in summer with strong thermal activity the lapse rate on sunny mountain slopes approaches the adiabatic value of about 10° C km^{-1}. For example, on days when the air temperature is 32° C in the Owens Valley (1,225 m), it may be 7 to 13° C on Mt. Whitney or White Mountain Peak (both above 4,320 m). As noted by Powell and Klieforth (1989), temperatures in the high elevations of the White Mountains of California-Nevada are generally warmer in winter and cooler in summer than at comparable altitudes in Utah and Colorado.

On clear, calm nights, especially in winter, cold-air-drainage and differential radiation effects greatly reduce or reverse this gradient forming inversions in which the temperature increases with height in the lowest hundred meters or so. Even in summer small mountain valleys often have minimum air temperatures of 10° C or more colder than slopes a short distance away, a phenomenon that has important implications for ecosystems, and for researchers who camp near study sites! Billings (1954) correlated these cold air drainage patterns with the boundaries of pinyon-juniper vegetation.

Temperature changes of 10 to 15° C over short periods of time during the day are frequently the result of air flow patterns in mesocale weather systems such as rapid warming in the downslope (Foehn or chinook) winds of mountain lee waves, or in equally rapid cooling during downbursts of cold air from thunderstorms. Air temperature at official weather stations is measured in an instrument shelter at a standard height of about 1.5 m above ground level. Thus for more accurate measurements of microclimates and vegetation near the ground, the researcher must establish appropriate instrumentation at the study site (Chapter 4).

Humidity

The wide range of temperature in the Great Basin is accompanied by considerable variability in humidity throughout the region. The Great Basin, including the eastern slope of the Sierra Nevada, is relatively dry in general, but parts of its interior are the driest in North America. Contours of annual mean relative humidity in the north of the region are similar to those of Arizona and New Mexico (Figure 2.2). However, the climate becomes extremely dry in southern Nevada, where the maximum aridity in the United States prevails. The spatial gradients in average relative humidity are quite strong in the western region relative to other parts of the United States. Local gradients are found within the mountains where the higher elevations will be somewhat more humid with more frequent cloud formation and precipitation. Diurnally, relative humidity will be lowest at midday and highest at night; seasonally relative humidity tends to be higher in winter, but absolute humidity is generally lower because the air retains less water vapor mass at colder temperatures.

Relative humidity is a function of both air temperature and the water vapor content of the air. Thus on calm days without a change of air mass, the diurnal pattern of relative humidity resembles a mirror image of the temperature trace in a hygrothermogram. Extremes of relative humidity occur during precipitation, in cap clouds or fog, and with the formation of dew or frost, in all of which it is near 100%, and in Foehn winds in which it may be as low as 2%. Extremes of absolute humidity occur in cold, dry arctic air masses during winter where dew points may be several degrees below freezing, and in warm moist tropical air masses of summer in which dew points may exceed 10° C. The latter incursions occur only during relatively infrequent invasions of tropical air. Dryness is the predominant feature of the Great Basin atmosphere, a characteristic which influences vegetation type, growth rates, and longevity. At the air-ground interface the occurrence of frost affects soil structure and vegetation processes. Throughout most of the Great Basin, the frost-free season is short (Chapter 6).

Cloud Cover, Fog, and Haze

Generally, high elevations in combination with relatively low cloud cover provide intense sunlight for forests in the West, particularly in the Great Basin region. Cloud cover patterns over the U.S., as deduced from satellite images, are shown in Figure 2.3. The Great Basin region is seen to encompass some of the lowest daytime cloud cover observed both in summer and winter. Summer cloud cover is much less across the Great Basin region than in eastern regions, by a factor of two or more. However, in winter, cloud cover, at least in the northern Great Basin, is similar to the continental average. In the southern Great Basin, winter cloudiness decreases, as it does in the southern Rockies. The daytime sunlight availability (the inverse of cloud cover) is somewhat less in winter than summer. In winter, particularly, there are relatively strong north-south gradients in cloudiness. Interestingly, the north-south gradients in cloudiness prevail in a zonal band across the continent, broken only on the northern shore of the Gulf of Mexico. This zonal pattern is not as strong in summer, but is still present.

An increase in cloud cover usually accompanies cyclonic air flow patterns in which ascending air cools and condenses. This may be manifest as layer type (stratiform) high- and middle-level clouds if the air mass is thermally stable, or as convective (cumuliform) clouds if the air mass is unstable. Distance from marine influence results in a general lack of low clouds and fog during most of the year. Consequently, visibility is greater than in other regions of the U.S. In advance of Pacific storms, air flow over the many mountain ranges forms a complex pattern of up- and down-drafts and spectacular "wave" clouds throughout a great depth of the atmosphere, cloud formations which may be visible for many hours from great distances.

In anticyclonic ("fair weather") circulation patterns, the air is most often nearly cloudless but in many cases there may be extensive high cirrus or cirrostratus clouds. In winter anticyclones, subsiding air results in the formation of low-level temperature inversions which, if they persist, may lead to the formation of fog or smog. If the temperature is above freezing these remain as "warm" fogs composed of small water droplets. The presence of water droplets at air temperatures below freezing is known as a "supercooled" fog or pogonip, which, on contact with solid objects such as vegetation, forms rime ice. The freezing of supercooled water, which leads to riming and ice formation on leaves, needles, and branches, can be particularly damaging to trees, especially in combination with high winds (Chapter 4). Nevertheless, when such ice thaws and melts, it makes a significant contribution to the annual precipitation on forested mountain ridges. Fog increases substantially with elevation, even in the western mountains. The subalpine forests of the Great Basin region are more often "in-cloud" at ridge tops relative to montane and valley conditions. Such fogs add some moisture to the alpine ecosystem, which is not well accounted for in most climate assessments.

The low humidity combined with generally low levels of airborne particle concentrations creates the finest visibility prevailing in North America. One of the great aesthetic pleasures of the region is the magnificent viewing conditions, in which vistas can extend for more than 200 km. These conditions of high visibility are susceptible to change from natural sources, such as dust storms, and to haze sources associated with cities, power plants, and other industrial activities. Haze and reduction of visibility may also be caused by smoke from forest fires or athropogenic sources, from fine dust transported from dry lake beds, or from distant volcanic eruptions (Klieforth 1987). If long-lasting, these elements could affect the local or regional climate. Even when these sources are absent, the excellent visibility can give way to haziness, which is believed to be related to hydrocarbon vapor emissions from vegetation, such as conifers and sagebrush (Rasmussen and Went 1964). These hydrocarbons may react in the air to produce condensed materials that forms aerosol particles. Evaluation of natural versus anthropogenic contributions to desert haze remains problematic (Hidy 1984).

Precipitation

The combination of air flow, temperature, and atmospheric moisture ultimately produces the precipitation that provides surface water supply for the maintenance of soil moisture. As one might expect, the heterogeneity observed in other climatic variables is reflected also in precipitation. The mean annual precipitation isohyets for the United States are shown in Figure 2.6. The distribution of precipitation in this figure shows a strong parallel with relative humidity (Figure 2.2) and cloudiness (Figure 2.3) as expected. Pacific coastal precipitation is generally heavy in the Northwest, decreasing with latitude to southern California, and is comparable with that found in the U.S. east of the Mississippi River. In parts of the Olympic Peninsula precipitation exceeds that of anywhere else in the continental U.S. On the east side of the Sierra Nevada and the Cascades, there are strong local gradients in precipitation.

These gradients derive from the orographic modification of air flow. When air is forced to rise over mountains, water vapor cools and condenses forming clouds and precipitation. Generally, the windward

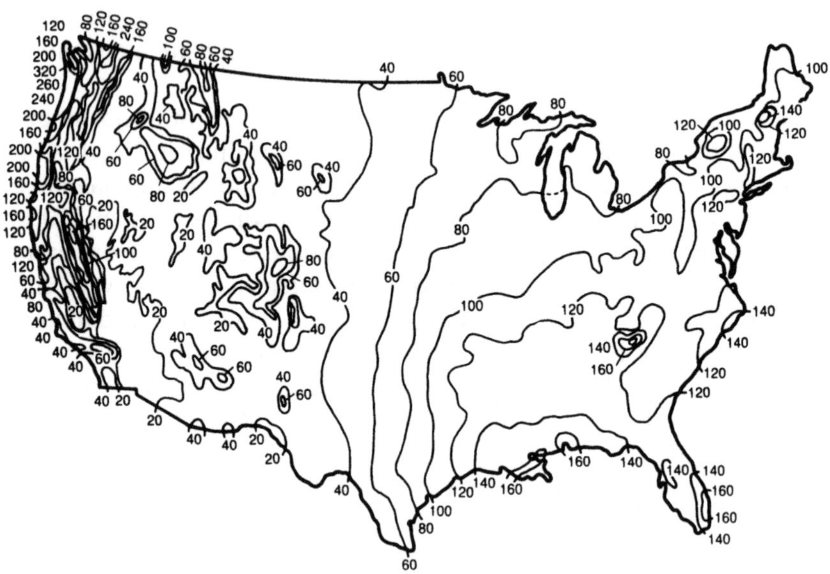

Figure 2.6. Mean annual precipitation isohyets (1931-60) in centimeters. Redrawn from NOAA (1974).

side of the mountains (prevalently the west side) and the ridge tops receive the most precipitation. The leeward side receives precipitation at the higher elevations, but the valleys and lowlands between ranges receive lesser amounts and are often very dry. The southwestern Great Basin mountain ranges generally receive the least mean annual precipitation and the valleys near them are the driest in the region.

Precipitation patterns vary seasonally in distinct ways. In the eastern U.S., precipitation falls more uniformly by season than in the West. In some Eastern locations, precipitation is actually greatest in spring and summer. In the West, precipitation is generally heavier in late winter and early spring than in the remainder of the year. This pattern is particularly true of the Great Basin, where most of the precipitation comes in winter and spring. The random or stochastic variability in precipitation also is a major factor in the Western climate. This is illustrated on a large scale in Figure 2.7, which shows the geographical patterns of the coefficient of variation in annual precipitation for the U.S. The variability of precipitation patterns east of the Mississippi is much smaller than to the west. The highest variability is in the southern

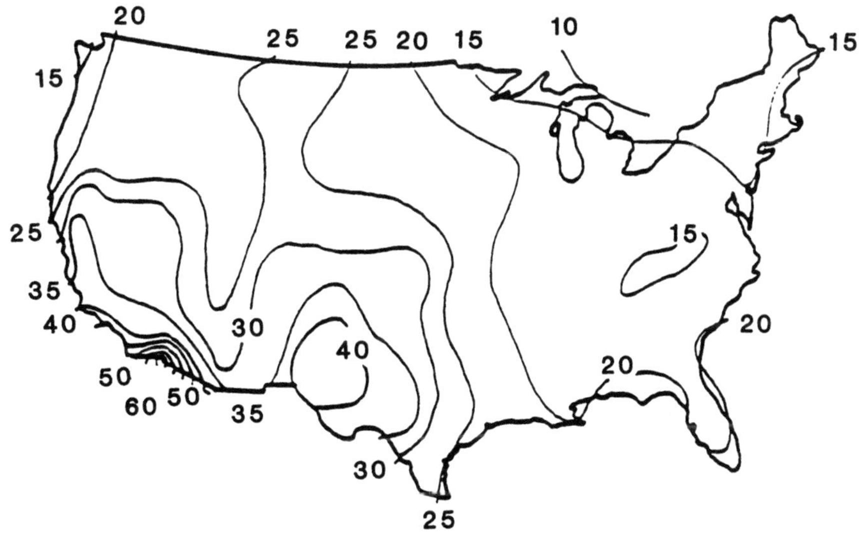

Figure 2.7. Coefficient of variation (%) in annual precipitation. The coefficient was calculated as the standard deviation of precipitation quantity divided by the annual mean value over the period 1931-1960. Redrawn from Hirschfield (1962).

desert region along the Mexican border, north of the Gulf of California. This variability declines northward across the Great Basin. On a small scale, orographic influences combine with thermal circulations and moisture to create sporadic disturbances that produce scattered and brief precipitation across the desert. Conditions can develop in which precipitation is seen as a veil (known as "virga") from the cloud base, but fails to survive the dry air below cloud base.

In his analysis of the temporal and spatial distribution of precipitation in the Great Basin, Houghton (1979) recognized three principal regimes:

1. Pacific: frontal cyclones from the Pacific Ocean, which cause winter to be the wettest season in the western and northern Great Basin.

2. Continental: cold cyclones involving mainly Pacific air, which develop east of the Sierra most often in spring and autumn, causing a spring maximum in Utah and central and eastern Nevada.

3. Gulf: summer thunderstorms in subtropical air masses, which produce a primary or secondary summer maximum in the southeastern Great Basin.

The storms associated with these weather regimes have been described by Powell and Klieforth (1989). "Warm" storms from the vicinity of Hawaii occur during some winters bringing a high freezing level and heavy rainfall to altitudes as high as 3,000 m, and over a large geographical area. A particularly intense series of such storms in February 1986 caused extensive avalanches in the Sierra Nevada, Carson Range, Ruby Mountains, and the Wasatch. Orographic effects in frontal storms are most pronounced in the Sierra Nevada and nearby ranges in the western Great Basin. In the Carson Range, for example, it is common for the ratio of storm precipitation at the summit of the Mt. Rose Highway at 2,700 m to that 16 km distant in the leeward valley at 1,510 m to be 10:1 or even 20:1. Such local patterns in winter storms are reflected in similarly large gradients in annual precipitation. Reno in the Truckee Meadows between Lake Tahoe and Pyramid Lake has an average annual precipitation of about 180 mm, where as nearby areas in the Sierra Nevada receive an average of about 1,800 mm.

The largest snowstorms in most Great Basin ranges occur with cold cyclones involving moist Pacific air. This is especially true of the southern Sierra Nevada and southern Nevada where the winter maximum precipitation in Las Vegas and the Spring Mountains results from these storms. In the western Great Basin such storms are known as "Tonopah Lows" and are associated with low clouds with rain or snow on the eastern slope of the Sierra Nevada and ranges farther east, while the western slope of the Sierra Nevada may be sunny. In such cyclonic storms wind speeds are less and the orographic effect is reduced, leading to a ratio between precipitation amounts at the highest elevations to that in adjacent valleys of about 2:1.

In some years the remnants of former tropical storms or hurricanes bring general precipitation to large areas of the Great Basin in late summer and early fall. However, most summer precipitation is associated with intense but very local thunderstorms. Often heavy rain will fall for a brief period over one canyon while no rain will fall a short distance away. Large, long-lived thunderstorms cause localized flash floods, which wash out roads and often destroy trees and cover other vegetation with boulders, sand, and silt especially on the fans and playas of the basins. Such storms also bring frequent lightning which causes many forest and range fires. The hail which often accompanies thunderstorms is not as large or as destructive as that experienced in the Rocky Mountains and Great Plains. Convective storms have a pronounced diurnal pattern in western Nevada as discussed by Klieforth (1987). Also it should be noted that, in contrast to winter storms, the heaviest rain in thunderstorms often occurs in the valleys.

In a 20-year study of precipitation in the Lake Tahoe-Truckee River-Pyramid Lake watershed (Klieforth, unpublished), it has been found that, on average, at elevations near 1,980 m about half the annual precipitation falls as rain and half as snow. Near 1,600 m about 25% of the annual precipitation falls as snow where as at 2,600 m snowfall accounts for 85 to 90% of the annual total. At one station in this network (Tahoe Meadows, 2,600 m), the average annual precipitation is 1,352 mm; extremes in annual amounts have been 565 mm in 1975-76 and 2,462 in 1982-83. During the latter year the total snowfall was 1,735 cm and the greatest snow depth was 5.0 m with a water content of 243 cm. During two months (January 1969 and February 1986) of the 20-year period of record, the total monthly precipitation was 940 mm of water.

The extremes of precipitation, snowfall and snowpack described in the last paragraph are from a site on the western edge of the Great Basin and are more typical of precipitation in the central Sierra Nevada; it is not surprising that these are record amounts for Nevada, the driest of the 50 states. A site typical of the eastern Great Basin is Lehman Caves in the Great Basin National Park. At that location (2,080 m) the average annual precipitation is 335 mm with eight months of the year in which the average precipitation exceeds 25 mm and the difference between the wettest month (March) and the driest month (June) is only 20 mm. Although the annual precipitation in central and eastern Great Basin ranges is less than that in the Sierra Nevada, there is perhaps less susceptibility to long periods of drought because, with all three precipitation regimes (Pacific, Continental, and Gulf) represented, there is less probability that all three will be deficient in any one year. In fact it is frequently noted that a decrease in precipitation during one season is often compensated by an increase in precipitation in another season of the same year.

Although it was necessary in the past for scientists studying mountain or desert ecosystems to use climatological data from the nearest weather station often many kilometers distant, there is a growing tendency for the establishment of electronic weather stations at biological field stations and specific study areas. In addition to standard measurements of wind velocity, air temperature, humidity, and precipitation, other parameters, such as solar radiation and soil temperatures are also included. Other useful climatic information that is becoming increasingly available includes LANDSAT photography in various spectral bands, airborne snow surveys, and remote-recording data from SNOTEL (snow telemetry), ALERT (automatic local evaluation in real time), and RAWS (remote automatic weather station) sites. Most of this information is now available by computer.

Chemical Factors

The chemical climate of the atmosphere in which primary plant processes occur can be defined in terms of trace constituents in the air, and certain chemicals that reach the terrestrial ecosystem through precipitation. Air contains both gases and suspended particles that affect vegetation. These components derive from natural sources, and from air pollution. They include naturally occurring gases, such as CO_2, H_2S, NH_3, N_2O, and dimethyl sulfide (DMS). Air pollution accounts for additional CO_2 as well as SO_2, NO, NO_2 and nitric acid. Particles in the air are a complex mixture of soil dust, sulfate (mainly as ammonium salts and sulfuric acid), ammonium nitrate, heavy metal salts or oxides and organics. Organic gases are also present, ranging from methane to volatile aliphatic, olefinic and oxygenated material. Vegetation contributes to the natural hydrocarbons in the air through emissions of pinene, isoprene and other volatiles (Monson and Fall 1989).

Table 2.1 Representative gas and particulate concentrations in rural western North America (Data of Young et al. 1988)

Constituent	Concentration
NO_2	0.2 - 11.0 ppb
HNO_3	0.03 - 1.0 ppb
NH_4	0.8 - 1.1 ppb
O_3	20 - 58 ppb
particulate SO_4^{2-}	0.8 - 5.0 $\mu g\ m^{-3}$
particulate NO_3^-	0.05 - 1.0 $\mu g\ m^{-3}$

Although atmospheric chemists have come to appreciate the diversity of chemicals in the air, they have examined few of them in relation to the impact on plants and ecosystems. However, the results of much U.S. and European research into the effects of air pollution on plants have been summarized by Winner et al. (1985) and Schulze et al. (1989), respectively. The interest in atmospheric sulfur and nitrogen from anthropogenic air pollution and as airborne acidifying materials has emphasized studies of consequences of these gases or particle components. Thus, SO_2 is oxidized to H_2SO_4 and NO and NO_2 are oxidized to HNO_3. Sulfur and nitrogen are nutrients in forests (Winner and Atkinson 1986), but high concentrations of SO_2 and NO_2 (above a few ppm) in the air are known to damage vegetation (Winner et al. 1985). Ozone, which is a product of photochemical reactions involving NO, NO_2 and certain organic vapors, can damage vegetation at even lower concentrations (above 0.1 ppm).

Exposure to gases and particles at low humidities results in dry deposition by direct contact, followed by absorption or assimilation of the chemicals. Deposition of dissolved chemicals also takes place in precipitation, or by contact and capture of fog or cloud particles. Precipitation deposition is called wet deposition. If acidity develops from scavenging of acid-forming materials the deposition is referred to as "acid rain". Capture of cloud or fog particles is sometimes termed "occult deposition". Acidity in deposition is measured in terms of hydronium ion concentration (H_3O^+ or H^+) which arises from the dissolution of acidifiers, or the fallout of acidic species, ammonium bisulfate, sulfuric acid or nitric acid. In the last decade, considerable attention has been given to atmospheric acidity and its potential for adverse effects on terrestrial and aquatic ecosystems.

Concentration Patterns and Dry Deposition

There exist relatively small bodies of data that characterize ambient concentrations of gases and particle constituents, especially in the rural and thinly populated areas of the West. Information prior to 1986 has been summarized by Young et al. (1988). Ambient concentrations of gases of interest in alpine and subalpine areas of the West are as shown in Table 2.1. The low range of concentrations is probably typical for the central Great Basin and, except for ozone, the concentrations are far below those measured in western urban areas, or eastern rural and urban areas. In general, they all approach known background levels at remote sites of the world. In terms of concentrations that produce adverse effects on vegetation, ozone is by far the most significant. For example, damage to pines has been observed in southern and central California at persistent ozone levels at approximately 0.1 ppm.

Atmospheric concentration of gases or particles tells an incomplete story of exposure. A more important measure of exposure is the rate of contact and absorption of these materials into a plant (Winner and Atkinson 1986). The rate of gas or particle transfer depends on factors such as the surface moisture, the solubility of materials, and biochemical reactions at the leaf or stem surface. The overall rate of accumulation is given in terms of the product of atmospheric concentration and a mass transfer coefficient (a dry deposition velocity). For soluble, biologically active gases, such as SO_2 or O_3, the mass transfer coefficient is about 1 cm s^{-1}. For particle migration to surfaces, the mass transfer coefficient is 0.1 to 0.5 cm s^{-1}.

Direct measurements of dry deposition rate are very difficult to make. Results from careful experiments indicate that this rate is highly variable for different vegetation types, depending on leaf surface moisture as well as biological exchange processes. Data are generally unavailable for the West but a very crude estimate of the deposition rate can be made indirectly from concentration data, yielding estimates of 7 kg ha^{-1} y^{-1} as SO_4^{2-}, and 5 kg ha^{-1} y^{-1} as NO_3^-.

Wet Deposition

The chemistry of precipitation in the West has been characterized recently, but unfortunately, virtually no observations have been taken in the central Great Basin. Therefore, we must rely on observations taken at locations that border the region. Concentrations of major anions and cations in precipitation are shown in Table 2.2, which also includes corresponding values for annual wet deposition of the same ions. Young et al. (1988) summarized the results of a much larger set of precipitation observations reported between 1981 and 1984 by the National Acid Deposition Program (NADP). Their results provide a broad picture of chemical precipitation in the western climate.

Variability in wet deposition can result from either changes in ionic species concentration or precipitation volume, or both. Analysis of deposition chemistry must consider variations in both of these parameters.

Table 2.2. Annual average chemical composition of precipitation and annual deposition of chemical species in regions bordering the Great Basin (Data from National Atmospheric Data Program 1983-84 and other sources, compiled by Young et al. 1988).

Site	Chemical species				
	pH/H^+	SO_4^{2-}	NO_3^-	NH_4^+	Ca^{2+}
Concentration (μM)					
Grand Canyon, AZ	5.35	7.3	11.0	5.0	6.5
Sand Springs, CO	5.15	10.4	14.7	9.4	7.3
Reynolds Creek, ID	5.65	6.8	9.2	7.8	5.0
Pendleton, OR	5.72	-	-	-	-
Pinedale, WY	5.10	12.6	15.7	8.3	10.5
Deposition (kg ha^{-1} yr^{-1})					
Grand Canyon, AZ	0.025	3.6	3.5	0.5	1.3
Sand Springs, CO	0.030	4.6	4.2	0.8	1.4
Reynold Creek, ID	0.010	2.0	2.0	0.4	0.6
Pendleton, OR	0.009	-	-	-	-
Pinedale, WY	0.026	4.0	4.0	0.5	1.3

Annual wet deposition data for H^+, SO_4^{2-}, NO_3^-, NH_4^+, and Ca^{2+} for 1983 are illustrated in Table 2.2. The annual deposition data do not support the spatial trends observed for concentration data. Instead, the deposition data are relatively uniform across the region. Deviations generally are associated with precipitation amount, urban influence, or other local factors, or are the result of averaging artifacts resulting from limited data.

The concentration of H^+ generally increases from west to east. In the Pacific coastal states, pH values generally range from 5.3 to 6.0 to between 5.4 and 5.0 in Colorado. Throughout the West, H^+ concentrations are much lower than the East; they are one fifth of the concentrations in the upper Ohio River Valley where annual average pH values are 4.2 or lower over large areas. Acidity of deposition in the West is more variable than that of other areas, ranging from 0.002 to 0.15 kg $ha^{-1}y^{-1}$. Very low precipitation acidity was found in Bishop, California in the eastern Sierra Nevada, because of low precipitation and this is believed to be typical of the Great Basin condition.

Concentrations of SO_4^{2-}, NO_3^-, NH_4^+ and Ca^{2+} also increase from west to east. Background SO_4^{2-} levels in the coastal states range from 2 to 4 μM. These levels can be compared with SO_4^{2-} concentrations in Colorado that range from 8 to 23 μM, depending on the location and year. Discounting urban sites, NO_3^- concentrations in the coastal states range from 3 to 6 μM as compared to concentrations in Colorado that range from 11 to 28 μM. Concentrations of SO_4^{2-} and NO_3^- in Colorado are roughly 20 to 50% of those measured in the East. Sulfate deposition is relatively uniform throughout the West, with most sites falling between 3 and 6 kg $ha^{-1}y^{-1}$. These are only 7 to 20% of values measured over large areas of the East that receive from 30 to 45 kg $ha^{-1}y^{-1}$. Similarly, NO_3^- deposition in the western coastal states ranges from 2 to 7 kg $ha^{-1}y^{-1}$, with values in the higher end of the range occurring sporadically. These too are only 10 to 30% of the NO_3^- deposition over large areas of the East.

Ammonium concentrations clearly follow the same trend, increasing from west to east. Ammonium background concentrations in the coastal states range from 1 to 7 μM, but are much higher in Colorado ranging from 5 to 40 μM. In Colorado the range is from 30 to 100% of measured concentrations in the eastern United States. Calcium concentrations are nominally 3 μM in the coastal states and 10 μl^{-1} in Colorado, roughly equal to those in the East. Ammonium deposition in California ranges from 1.3 to 2.7 kg $ha^{-1}y^{-1}$ and is elevated compared with the rest of the region, which ranges from 0.3 to 1.2 kg $ha^{-1}y^{-1}$. The elevated levels in California presumably arise from agricultural emissions and biological decay, and are similar to those measured in the East. Calcium deposition is generally less than 1.0 kg $ha^{-1}y^{-1}$ farther inland, especially in the Rocky Mountain region.

There is some seasonal variability among chemical species in the deposition data. Higher concentrations and lower precipitation amount in summer tend to balance lower species concentrations and high precipitation amounts in winter, producing relatively constant deposition from year to year. However, seasonal deposition data from individual sites can exhibit substantial variability from one year to the next.

There are very few data giving the chemical character of cloud or fog water in the West. The few measurements that have been taken suggest that ion concentrations are somewhat higher in cloud water than in precipitation. This is primarily the result of dilution during the accumulation of water on precipitation elements. Young et al. (1988)

Table 2.3. Estimates of potential sources of acidity produced from natural sources and from precipitation in the western US (Data of Page et al. 1986).

Source of acidity	Acidity $kg\ ha^{-1}\ yr^{-1}$	Reference
Precipitation, direct H^+	<0.01 - 0.02	NAPAP (1987)
from NH_4^+	0.02 - 0.4	USGS (1975)
Nitrification of symbiotically fixed N_2	2 - 24	Johnson et al. (1982) Nutman (1965) Nyatsanya and Pierre (1973)
Nutrient uptake	0.5	Bache and Rippon (1979)
Root and microbial respiration	0.2 - 1.0	Wiklander (1980)
Organic acids	0.2	Johnson et al. (1982)

argued that exposure of forests to cloud and fog droplets would become a serious air pollution consideration only in those forests in which pervasive in-fog conditions exist. This seems likely only along the crest of the Sierra Nevada and Cascades, the Transverse Ranges of southern California, and possibly in a few places along the crest of the Rockies. Comparable data for wet and dry deposition of acidity and of SO_4^{2-} and NO_3^- are not available. However, the calculations suggest that the two pathways of deposition may be similar in many parts of the West. In especially arid areas, such as the Great Basin, the balance may shift to dry deposition even when atmospheric concentrations are low, since precipitation is also low.

Chemical Input to Forest Systems

Once the deposition of chemicals takes place, they can react or migrate through a complex soil-biological system. Depending on the reactivity of this system, the flow rate or the contact time or transport medium (water percolation), the system may respond in different ways and alter the output of this system, surface water and groundwater. This perturbation may modify natural chemical processing. The degree of acceleration of natural chemical processing and the consequent biosystem response has yet to be determined. Indeed, this is the crux of the recent debate about the ecological consequences of acid deposition in the western U.S. (NAPAP 1987; Roth et al. 1985; Winner et al. 1985; Winner and Atkinson 1986). For example, natural processes in soils that lead to acidification include nitrogen fixation, nutrient uptake, microbial and soil respiration, and organic acid formation. This acidification is offset by natural buffering processes, including rock weathering, carbonate leaching, nitrate and sulfate reduction, cation exchange reactions and aluminum buffering. Acid deposition thus is superimposed on these natural soil processes, and some comparisons of acid deposition with estimates of natural processes are given in Table 2.3. The atmospheric component of hydrogen ion deposition, and atmospheric ammonium as an acidifier, is small relative to other soil processes. Similarly, the deposition of sulfur and nitrogen oxides is estimated to play a minor role in normal nutrient assimilation processes in western conditions (Page et al. 1986).

Only a minor fraction of western soils are considered susceptible to acidification. These include two classes having low cation exchange capacity and acid neutralization capacity, the inceptisols and ultisols, neither of which is found in the central Great Basin. Only along the western edge of the Sierra Nevada are they observed as a significant soil type, and current levels of deposition from the atmosphere represent a minor fraction ($<1\%$) of their indigenous soil acidity. Thus in the Great Basin, acid chemical climate should have little effect on even the most susceptible soil types.

Unmanaged western forests are often nitrogen deficient and thus there are potential beneficial effects of nutrient enhancement from N deposition. These potential beneficial effects of nitrogen may mitigate other potentially harmful effects of atmospheric pollution on vegetation, presumed to be mainly due to ozone exposure. Ozone is not an acidifier per se, but is a co-pollutant with sulfate and nitrate constituents.

Deposition rates of sulfate less than about 15 to 20 kg $ha^{-1}y^{-1}$ are not believed to have a negative effect on aquatic or terrestrial ecosystems, either directly or as a surrogate for acidity (Winner and Atkinson 1986). Western conditions for wet deposition are well below this level, and in the Great Basin, both wet and dry deposition should be less than this nominal threshold. From these considerations, it is believed that present levels of chemical deposition in the Great Basin represent a very mild stress factor compared with the severity of physical stress in the region. However, further studies are needed to assess specific local responses, and to study interactions between atmospheric and other stress factors.

References

Arno SF and Hammerly RP (1984) Timberline. Mountain and arctic forest frontiers. The Mountaineers, Seattle.

Bache BW and Rippon JE (1979) Ecological effects of acid precipitation. pp. D3.1-D3.5. Electric Power Research Institute Report SOA-77-403. Palo Alto, CA.

Barbour MG and Billings WD (1988) (editors) North American terrestrial vegetation. Cambridge Univ. Press, Cambridge.

Barry RG (1981) Mountain weather and climate. Methuen, NY.

Billings WD (1954) Temperature inversions in the pinyon-juniper zone of a Nevada mountain range. Butler Univ. Bot. Studies 11:112-118.

Brough RC, Jones DL and Stevens DJ (1987) Utah's comprehensive weather almanac. Publishers Press, Salt Lake City, UT.

Bryson RA and Hare FK (1974) (editors) World Survey of Climatology. Vol.II, Climates of North America. Elsevier, Amsterdam.

CGC (1988) Committee on Global Change: Toward an understanding of global change. National Academy Press, Washington, DC.

Clapp PF (1964) Global cloud cover for seasons using TIROS nephanalyses. Monthly Weather Review 92:495-507.

Hansen J, Johnson D, Lacis A, Lebedeff S, Lu D, Rink D and Russell G (1981) Climatic impact of increasing carbon dioxide. Science 213:957-966.

Heald W (1956) The proposed Great Basin National Park. Sierra Club Bulletin 41(10):46-53.

Hershfield DM (1962) A note on the variability of annual precipiation. J. Appl. Meteorol. 1:575-578.

Hidy GM (1984) Aerosols. An Industrial and Environmental Science. Academic Press, NY.

Houghton JG (1979) A model for orographic precipitation in the north-central Great Basin. Monthly Weather Review 107:1462-1475.

Houghton JG, Sakamoto CM and Gifford RO (1975) Nevada's Weather and Climate. Nevada Bureau of Mines and Geology, Univ. Nevada, Reno.

Houghton SG (1986) A Trace of Desert Waters. Howe Brothers, Salt Lake City, UT.

Johnson DW, Van Miegrott H, Cole DW and Richter DD (1982) Contribution of acid deposition and natural processes to cation leaching from forest soils: a review. J. Air Poll. Contr. Assn. 33:1036-1041.

Klieforth H (1987) Land-air interface. pp. 25-29. In Patten DT (editor) The Mono Basin Ecosystem. National Academy Press, Washington, DC.

Latimer D (1985) Modeling regional haze in the southwest: a preliminary assessment of source contributions. Report SYSAP:1895-1038 Systems Applications Inc., San Rafael CA.

McLane AR (1978) Silent Cordilleras. The Mountain Ranges of Nevada. Camp Nevada Monograph, 4, Reno.

Monson RK and Fall R (1989) Isoprene emission from aspen leaves. Plant Physiol. 90:267-274.

Mutel CF and Emerick JC (1984) From Grassland to Glacier. The Natural History of Colorado. Johnson Books, Denver.

NAPAP (1987) National acid precipitation assessment program: interim assessment: The causes and effects of acidic deposition. Interagency task force, Environmental Protection Agency, Washington, DC.

NOAA (1974) National Oceanic and Atmospheric Administration: Climatic Atlas of The United States. Department of Commerce, Washington, DC.

Nobel PS (1981) Wind as an ecological factor. pp. 475-500. In Lange OL, Nobel PS, Osmond CB, and Ziegler H (editors) Physiological Plant Ecology I: Response to the Physical Environment, Encyclopedia of Plant Physiology (New Series) Vol. 12A. Springer-Verlag, Heidelberg.

Nutman PS (1965) Symbiotic nitrogen fixation. pp. 363-383. In Bartholomew WV and Clark FE (editors) Soil Nitrogen. American Society of Agronomy, Madison, WI.

Nyatsanya R and Pierre WH (1973) Effect of nitrogen fixation by legumes on soil acidity. Agric. J. 65:936-940.

Page A, El Amamy MM, and Cole D (1986) Acid deposition and the west. Unpublished review. WEST Associates Environmental Research and Technology, Newbury Park, CA.

Powell D and Klieforth H (1989) The weather and climate of the White Mountains. In Hall C (editor) The Natural History of the White-Inyo Range. Univ. California Press Berkeley (in press).

Rasmussen RA and Went FW (1964) Volatile organic material of plant origin in the atmosphere. Proc. U.S. Natl. Acad. Sci. 53:215-220.

Roth P, Blanchard C, Harte J, Michaels H and El Ashry MT (1985) The American west's acid rain test. Research Report 1, World Resources Institute, Washington, DC.

Schulze E-D, Lange OL and Oren R (1989) (editors) Forest Decline and Air Pollution. A Study of Spruce (*Picea abies*) on Acid Soils. Ecological Studies, Vol. 77. Springer-Verlag, Heidelberg.

Trimble S (1989) The sagebrush ocean. Univ. Nevada Press, Reno.

U.S. Geological Survey (1975) The national atlas of the United States. U.S. Department of the Interior Geological Survey, Washington, DC.

Went FW, Slemmons DB and Mozingo HN (1967) The organic nature of atmospheric condensation nuclei. Proc. U.S. Natl. Acad. Sci. 58: 69-74.

Wiklander L (1980) The acidification of soil by acid precipitation. Grundforbattring 26:155-164.

Winner WE, Mooney HA, and Goldstein RA (1985) (editors) Sulphur Dioxide and Vegetation: Physiology, Ecology and Policy Issues. Stanford Univ. Press, Stanford, CA.

Winner WE and Atkinson CJ (1986) Absorption of air pollution by plants and consequences for growth. Tree 1:15-18.

Young JR, Ellis EC and Hidy GM (1988) Deposition of air-borne acidifiers in the western environment. J. Env. Qual. 17:1-26.

Chapter 3

The Mountain Forests of North America and Their Environments

W. D. Billings

Introduction

Prominent on a map of the North American continent are three great cordilleras trending north and south: the Appalachians, the Rocky Mountains, the Cascades-Sierra Nevada. All rise above lowland vegetation that, depending on latitude and position relative to the sea, ranges from wet forest through grasslands, to desert. All, in the coolness of their middle elevations, have forests distinctly different from anything below. These montane and subalpine forests consist mainly of evergreen conifers with the exception of those in the Appalachians. There, the montane or lower forests exhibit a rich diversity of deciduous trees while the subalpine forests are dominated by evergreen conifers of genera prominent in the subalpine forests of the Rocky Mountains and the Cascades-Sierra Nevada.

The stability of any vegetation-environment system is ephemeral in time and space. Damage to local populations, for any reason, may result in local or general extinction of a plant species within an ecosystem. Moreover, vegetational and climatic changes have become more rapid during the Pleistocene. This has been especially dramatic in the time elapsed since the Wisconsinan full glacial. For example, almost the entire vegetation of Canada south of the Arctic has developed by migration and evolution in the last 10-15,000 years as the continental ice melted. The forested slopes of the Northern Appalachians have been populated in the same way and time because they too were covered by the deep continental ice. At the other side of the continent, much of the Sierra Nevada was locally under alpine and montane glaciation. After melt, the lower glacial debris there has been populated by forest species, but high up on the glacially scoured granites and andesites, tree invasion has been very slow indeed. Any ecosystemic disturbance now in the Sierran forests could cause long-lasting damage.

The Environmental Gradients

Macrogradients

These gradients are occasioned by the configuration, height, and locations of the mountain systems on the continent. For simplicity, only three kinds of macrogradients will be mentioned: latitudinal, longitudinal, and elevational. All are complex, multifactorial gradients.

Latitudinal gradients are expressed primarily through solar radiation and temperature but many other factors are implicated also. The well-known major vegetational belts of the earth are largely determined by latitudinal gradients. However, such a gradient is by no means uniform along its length and the values of its principal components differ at any given latitude as shown in Figure 3.1 for the arctic-alpine zone above and beyond timberline.

Across North America from west to east, the longitudinal gradients are controlled by storm tracks in the westerly jet streams, by marine influences, and the rainshadow effects of the three big mountain systems particularly those of the Sierra Nevada-Cascades and the Rocky Mountains. The climates of the Great Basin valleys and mountains are influenced by the Sierra Nevada, and the climates of the prairies and plains are semi-arid because of the Rocky Mountains.

The elevational gradients are essentially telescoped and relatively smaller integrals of the latitudinal and longitudinal macrogradients. They are expressed by vegetational zonation on the sides of mountain ranges (Figure 3.2), as described by Merriam (1890) in one of the first studies of mountain vegetation in western North America. Such vegetational belts are unique to each mountain range, each mountain system, and since almost all North American mountain ranges trend north and south, to the west-facing and east-facing slopes of each range. The

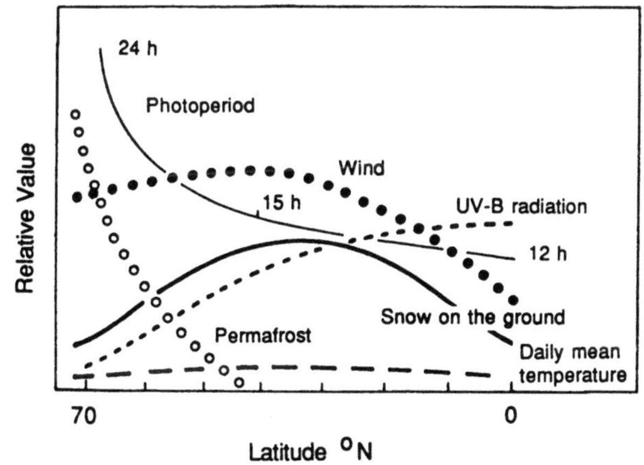

Figure 3.1. Relative values of six environmental factors in arctic-alpine ecosystems in North America along a latitudinal macrogradient from the Equator to the Arctic. All factor combinations differ considerably in total impact depending on latitude. Redrawn from Billings (1979).

principal exceptions to the latter point are the Brooks Range, the Alaska Range, the Uinta Mountains, the San Gabriel-San Bernardino Mountains, and isolated San Francisco Mountain itself, all of which trend east and west. On these, the zonation is tilted downward on the north slopes relative to the south slopes (Figure 3.2). This is particularly striking in the Brooks Range of northern Alaska where spruce forests rise to elevations of about 750 meters on the south-facing slopes but do not exist on the north slopes where they are replaced by tundra.

Mesogradients

These gradients of intermediate size are the results of local topography, as, for example, the temperature and radiation gradients between north- and south-facing slopes at different seasons of the year. Such topographic gradients are also particularly important in mountains from the standpoint of snow, rain, and the distribution and runoff of water in fairly open high country. Strong winter winds carry snow over ridges and leave it in long-lasting drifts on upper lee slopes. Such redistribution of precipitation plays an important role in the composition and character of vegetation in the upper parts of mountain ranges (Billings 1973).

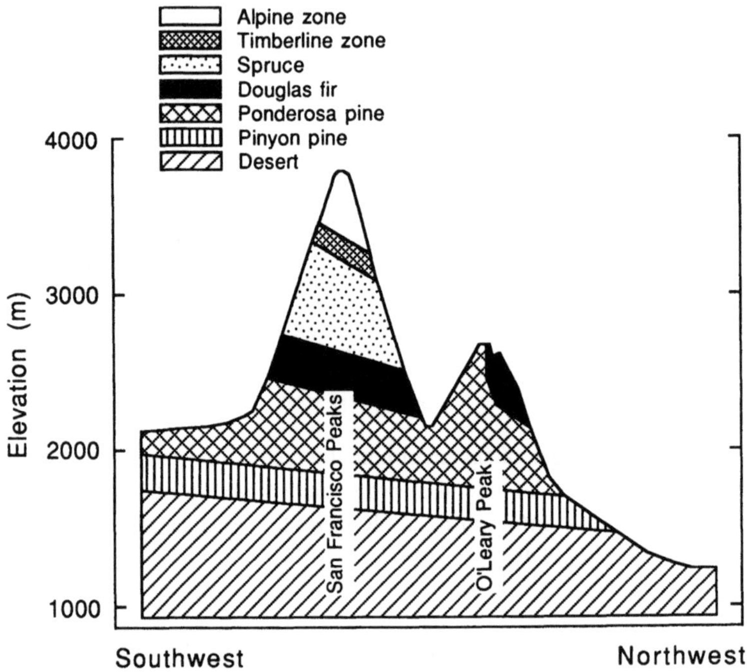

Figure 3.2. Vegetational zonation on San Francisco Mountain, Arizona, as viewed from the southeast. Redrawn from Merriam (1890).

Another topographic effect in mountains is expressed in a sharp drop in maximum daytime temperature as one climbs a ridge from the lee side. This is the result of cold winds at higher elevations. Such lower temperatures and wind stress play roles in tree growth near timberline. Reciprocally, cold air drainage at night keeps the ridges warmer than the hollows a few hundred meters lower where the cold air accumulates. Such meadows are frost-susceptible and the local environment there can be inimicable to tree growth and is usually occupied by hardy meadow vegetation.

Microgradients

These microtopographic gradients are important in regard to soil temperature and moisture distribution especially across furrows in croplands, and in grassland, tundra, and alpine ecosystems. They are of lesser importance in forests except near timberline where they play a role in ribbon forest (Billings 1969) and Krummholz patterning as influenced by snow catchment caused by the "snow fence effect" of the vegetation itself.

The Regional Mountain Forests

Appalachian Forests

The Appalachian Mountain system extends in a southwesterly direction from Newfoundland to northeastern Alabama. From the standpoint of geology, the Adirondack Mountains of New York are part of the Canadian Shield but their vegetation is Appalachian in character. So, for purposes of this paper, they will be considered Appalachian. The Appalachians are high in the north (up to 1,917 m on Mt. Washington in the White Mountains of New Hampshire), low in the middle (the mountains of Pennsylvania), and highest in the south (2,037 m on Mt. Mitchell in the Black Mountains of North Carolina.

These mountains are not high by western American standards. But, as Griggs (1946) has said: "The summit of Mount Washington, only 6,288 feet above the sea, stands several hundred feet lower than the valley floor at the foot of the Tetons. It is about at the same latitude--only a hundred miles farther north. Yet trees are forced to give way to alpine tundra 500 to 2,000 feet below the summit, in some places at elevations of less than a mile above sea level, whereas in the Rockies they reach more than twice as high. The climate of Mount Washington is, in fact, terrible. This little nubbin of a mountain . . . has taken more lives by far than any other American mountain. Dr. Charles Brooks has described its weather as the 'worst in the world.' Winds of hurricane violence occur in every month of the year, and in winter hurricanes often blow for many days at a time. At any season, summer as well as winter, the storms may be accompanied by snow or sleet. Whenever the temperature is below 32°F, the very prevalent cloud condenses on everything as frozen rime . . . known to build out to windward . . . as much as 7 feet in a single night. Such icing accompanied by the beating of high winds is extremely hard on trees."

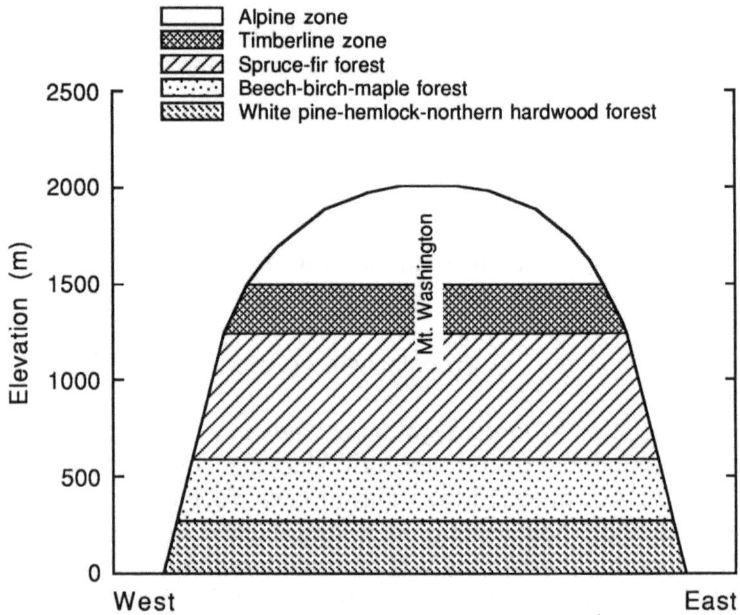

Figure 3.3. Vegetational zonation on Mt. Washington looking north, in the northern Appalachians as drawn from data of Griggs (1946), Oosting and Billings (1951), and Siccama et al. (1982).

The northern Appalachians rise across the broad ecotone between the transcontinental taiga in the north and the eastern American white pine-hemlock-northern hardwood forests to the south. The taiga here is characterized by white spruce (*Picea glauca*), black spruce (*P. mariana*), balsam fir (*Abies balsamea*), larch (*Larix laricina*), and white birch (*Betula papyrifera*) with numerous wet peat bogs. The white pine-hemlock-northern hardwood forest is a successional mixture of white pine (*Pinus strobus*), gray birch (*Betula populifolia*) and aspen (*Populus tremuloides*) followed in time by hemlock (*Tsuga canadensis*), sugar and red maples (*Acer saccharum*, *A. rubrum*), beech (*Fagus grandifolia*), northern red oak and other oaks (*Quercus rubra*, *Q.* spp.), ashes (*Fraxinus* spp.), elms (*Ulmus* spp.), and other deciduous taxa (Lull 1968).

Deciduousness also characterizes the forests of the lower slopes of the mountains (Figure 3.3). These forests are dominated by beech, birch (*Betula alleghaniensis*, *B. lenta*) and maples. It has a fairly diverse assemblage of other genera also.

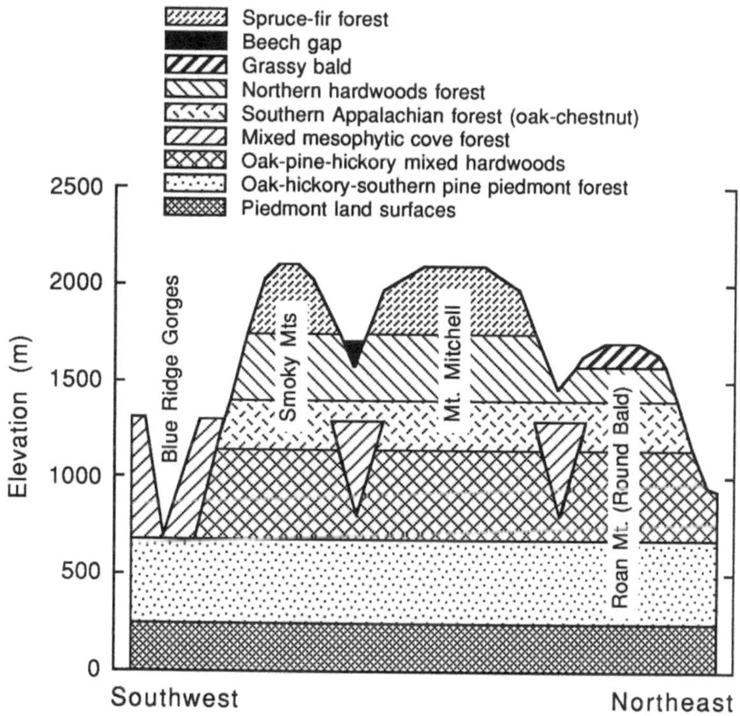

Figure 3.4. Vegetational zonation across a telescoped transect in the southern Appalachians across North Carolina from the Blue Ridge escarpment in the south to Roan Mountain in the north. Drawn from sketches by Paul Godfrey and Dwight Billings.

Above this hardwood forest is the true subalpine evergreen forest of spruce (*Picea rubens*), balsam fir (*Abies balsamea*), and birch (*B. papyrifera*) which extends to timberline. The higher elevations of the White Mountains of New Hampshire, the Green Mountains of Vermont, and the Adirondacks of New York are covered by this relatively simple forest in which the plant species are identical to or closely related to those of the lowland taiga to the north. These subalpine conifer forests have shown signs of rapid deterioration and death of trees in the last 20 years, particularly red spruce (Siccama et al. 1982).

The southern Appalachians of eastern West Virginia, southwestern Virginia, eastern Tennessee, and western North Carolina are the highest in the Appalachian cordillera (Figure 3.4). But, these mountains have no timberline at present; windswept spruce and fir forests cover even the highest peaks. About 20,000 years BP, the Wisconsinan continental glacier pushed up its terminal moraines across Indiana, Ohio, and Pennsylvania only 200-400 km to the north. During those times, lower regional temperatures, more snow, and strong winds off the continental ice must have resulted in forest retreat downslope in the southern Appalachians with a timberline a few hundred meters below the mountain crests. This would have been similar to the situation in the northern Appalachians today. However, the resulting alpine vegetation above timberline could have been richer in plant species than the present situation on Mt. Washington because of the very great diversity and richness of the total flora of the southern Appalachian region. In fact, in spite of the trees, a few alpine species still are to be found in the southern Appalachians on rocky cliffs and grassy balds where the forest has not shaded them out. Even though there is no palynological evidence in the mountains of this postulated timberline-alpine situation, there is corollary evidence for it from dated fossil pollen stratigraphy on the North Carolina coastal plain of an open boreal spruce-pine forest there during full glacial. This now-departed coastal plain forest was similar to that now present in northern Quebec.

The lower slopes and valleys ("coves") of these southern mountains up to about 1,500 m are covered with the earth's richest deciduous forest and one of the most colorful (Plate 5.1). There are more species of trees in this forest than there are in all of Europe, and the forest is more intact and preserved than are similar forests in eastern Asia. The reasons for its great floristic diversity as compared to that in Europe is the fact that the southern Appalachians were not glaciated during the Pleistocene. And, as compared to eastern Asian forests, the Appalachian forest has not been subjected to such intense "people pressure." Also, these moist southern mountains are not exposed to severe drought. The annual rainfall on the crests is over 255 cm. This is also true for the upper Blue Ridge gorges (Billings and Anderson 1966). Even during the severe American droughts of the 1930s, these mountains were wet (Thornthwaite 1941).

In general, below elevations of 1,065 m, the forest types are determined primarily by topography. Relatively dry slopes above 460 m are occupied by oak-pine-hickory forests with a number of species in each genus: *Quercus*, *Pinus*, and *Carya*. The deep, wet gorges in the Blue Ridge escarpment are occupied by a very rich mixed forest of hemlock, white pine, tulip tree, oaks, maples, birches, beech, magnolia, and rhododendron (*Tsuga*, *Pinus*, *Liriodendron*, *Quercus*, *Acer*, *Betula*, *Fagus*, *Magnolia*,

Rhododendron) with as many as 30 to 40 species of trees, mostly deciduous. Even richer are the very similar "cove" or mixed mesophytic forests. These are characteristic of the broader valleys and canyons throughout the southern Appalachians below 1,065 m (Figure 3.4). The cove forests are composed of all the tree species that are in the gorges and more. Some of these are *Aesculus octandra* (sweet buckeye), *Halesia carolina* (silverbell), *Magnolia fraseri*, *Magnolia acuminata* (cucumber tree), *Prunus serotina* (wild cherry), *Quercus rubra* (northern red oak), *Betula alleghaniensis* (yellow birch), and many others. In number of species, it is the richest forest in North America north of Mexico. The trees are tall and massive with at least some of the Liriodendrons being over 60 m tall and 3 m in diameter.

The slopes above the cove forests and extending to about 1,370 m are covered with the Appalachian oak-hickory-tulip tree forest. This used to be designated as the oak-chestnut forest; it was continuous northward along the crest of the Blue Ridge to Pennsylvania and southern New England. The chestnut, of course, is now virtually extinct due to the chestnut blight disease; its place in the forest has been taken by several species of oak, hickories, and by tulip tree (*Liriodendron*). However, the time since the death of the chestnut has not yet been long enough to say with certainty what the equilibrium floristic composition of this forest zone will be. At this forest's upper limits at the crest of the southern Blue Ridge Escarpment near 1,350 m elevation, the climate is foggy and wet (>200 cm mean annual precipitation). Here, *Tsuga* can be the dominant tree species (Oosting and Billings 1939).

Between about 1,370 m in elevation up to approximately 1,675 m is the southernmost extension of the northern hardwoods forest of New England and the northern Appalachians. This is the highest deciduous forest in these mountains. The principal dominant is the beech (*Fagus grandifolia*), but *Aesculus octandra*, *Acer saccharum*, *Betula alleghaniensis*, *Acer pensylvanicum*, *Tilia* spp., and *Halesia* spp;. are also common and important. Scattered individuals of red spruce (*Picea rubens*) occur at their lower elevational limits in this forest and are conspicuous in winter because of their dark evergreen crowns amid the matrix of leafless deciduous trees.

Along the ecotone between the northern hardwood forest and the evergreen Appalachian spruce-fir forest that lies above the deciduous forest on the mountains, are numerous grassy meadows. Since many of these meadows are atop mountains of intermediate elevation, 1,585 m to

1,800 m, they generally are called "grass balds" (Figure 3.4). Their origin can only be guessed at; Billings and Mark (1957) and Mark (1958) suggest that they may be the result of Holocene climatic changes resulting in differential retreats and advances by the beech-northern hardwoods forest vis-a-vis the spruce-fir subalpine forests above. Maintenance of the balds in the past has been due largely to grazing and browsing by wild and domestic herbivores. Now that grazing is reduced substantially, the balds are being invaded by spruce and fir (*Picea rubens* and *Abies fraseri*) where such a seed source is nearby. Isolated balds are maintaining themselves because of the lack of nearby seed sources of spruce and fir and because under the present climate the deciduous trees are relatively slow in advancing up-mountain into the balds.

Another ecotonal community, particularly in the Smoky Mountains, occurs in the passes or "gaps." Here, where clouds and fog drift through, are dwarf, open woodlands of small beech trees (*Fagus grandifolia*). The mountain people called these "beech orchards"; they have unique floras.

The spruce-fir subalpine forest of the southern Appalachian mountains is a disjunct extension of that forest in the northern Appalachians, and very similar to it (Oosting and Billings 1951). The principal difference is that the northern fir (*Abies balsamea*) is replaced in the south by the endemic *A. fraseri*. In both the south and the north, the fir is faster growing and shorter-lived than the spruce. Also, the white birch (*Betula papyrifera*) of the north is extremely rare in the south. Its role in the south is taken over by yellow birch (*B. alleghaniensis*.). There are more species of shrubs, herbs, and bryophytes in the southern spruce-fir ecosystem than in its northern counterpart. The soils under the southern spruce-fir forest are yellowish brown podzols as compared to gray, strongly leached podzols in such forests of the White Mountains in New Hampshire. Spruce-fir forest occupies a very wet, cloudy, foggy environment on the tops of the Black Mountains, the Great Smokies, the Balsams, the Unakas, and other high mountain ranges of western North Carolina and eastern Tennessee.

During the last 20 years, the spruce-fir forests of the southern Appalachians have been showing rapid deterioration (Plate 5.2). The causes of this are not known with any certainty even though some facts are known. First, the Fraser firs were attacked and damaged badly by the accidently introduced European woolly aphid. Since then, but not entirely the result of the woolly aphid, large areas of this forest, both spruce and fir, have died. This is particularly apparent on the highest peak in the southern Appalachians, Mt. Mitchell (2,037 m) in the Black Mountains of North Carolina. Much of the beautiful spruce-fir forest capping this mountain has died in the last 10-15 years with most of its associated species of shrubs, herbs, mosses. The opening up of these sites to sunlight has resulted in the spruce-fir forest herbaceous stratum being replaced by native Angiosperm species from subalpine meadows and open rocky places. However, there is some reproduction of the fir, less than 1 meter tall, dating back to the time of death of the overstory. Some of the dead overstory trees still stand but because of re-opening of the sites to mountain winds, the old trunks are now falling in windrows.

Rocky Mountain Forests

This great Cordillera extends diagonally from northwest to southeast across 55° of latitude from the arctic Brooks Range of Alaska at almost 70° N to Guatemala at 15° N in the tropics, a distance of ca. 7,200 km. Forests of related conifers in the genera *Pinus* (pines), *Picea* (spruces), and *Abies* (firs) extend from Alaska to Mexico, and even to Guatemala, along these high mountains. Through all of the climatic changes of the Cenozoic, the Rocky Mountains have provided a highway for the north and south migrations of plants and animals (Billings 1974, Axelrod and Raven 1985). This has been particularly apparent during the Pleistocene. No other montane migratory pathway connects the Arctic with the tropics; the only comparable system is that of the Andes, a much younger cordillera that does not extend as far into the antarctic polar regions.

As they exist at present, the forests of the Rocky Mountains can be sorted out latitudinally, rather arbitrarily, into four regional groups: Boreal, Central, Southern, and Mexican. The Boreal Rocky Mountain forests extend from the south slopes of the Brooks Range southeast along the mountains of east-central Alaska through the Mackenzie Mountains of Yukon Territory to the Peace River in northeastern British Columbia. The forests begin south of the Peace River at about latitude 55° N and continue southeastward through western Alberta, Montana, and into Wyoming including the Beartooth, Absaroka, Teton, Wind River, and Big Horn Mountains.

South of the high elevation sagebrush country of central Wyoming and the Snake River Plains of Idaho, the mountains and their forests are distinct enough from those farther north to be included in the Southern

Rocky Mountain Region. This extensive part of the Cordillera includes the Medicine Bow and Laramie Mountains of southeastern Wyoming, the Uintas and Wasatch Range of Utah, all the high mountains in the western half of Colorado, and the high mountains of eastern Arizona and New Mexico.

The deserts of southern Arizona and New Mexico separate the archetypical Rocky Mountains from the Sierra Madre Occidental, the farthest north range of the Mexican Region of the Rocky Mountains. The Sierra Madre leads southward into the great volcanic arc of central Mexico. South of these very high peaks, the Isthmus of Tehuantepec separates the mountains of Guatemala from the rest of the Mexican region. Yet, the vegetation of even the Guatemalan mountains shows a strong relationship with that of the Rocky Mountains farther north. Throughout the whole length of the Rocky Mountain system from Alaska to Mexico and Guatemala, the forests are dominated by evergreen conifers, mainly spruces, pines, and firs. Only in localized moist places and in the far north are deciduous trees conspicuous: groves of aspen (*Populus tremuloides*), white birch (*Betula papyrifera*), and larches (*Larix laricina, L. lyallii*). Of these broadleaved trees, only aspen occurs south as far as Utah, Colorado, Arizona, and New Mexico where it is abundant in the mountains.

Peet (1988) in his excellent description of the Rocky Mountain forests, subdivides the latitudinal regions in a slightly different way. He does not include those of the Brooks Range and the Mackenzie Mountains. His northern limit for Rocky Mountain forests is near the Liard River along the boundary between British Columbia and Yukon Territory. Actually, *Abies lasiocarpa* extends at least as far north as Keno Hill north of the Stewart River another 500 km to the northwest (Billings, field notes 1963). Also, *Pinus contorta* var. *latifolia* (lodgepole pine) is present almost 300 km north of Whitehorse, Yukon Territory.

Boreal Rocky Mountains

In Alaska and the Yukon, it is difficult to separate the Rocky Mountain forests from the extensive boreal forest (taiga). Throughout central and eastern Alaska (the northwestern limits of this forest) the winters are dark and bitterly cold with extreme temperatures down to lower than -50° C, and mean temperatures for the coldest month between -23° and -30° C. Summer temperatures are warm (25° to 35° C) with thunder

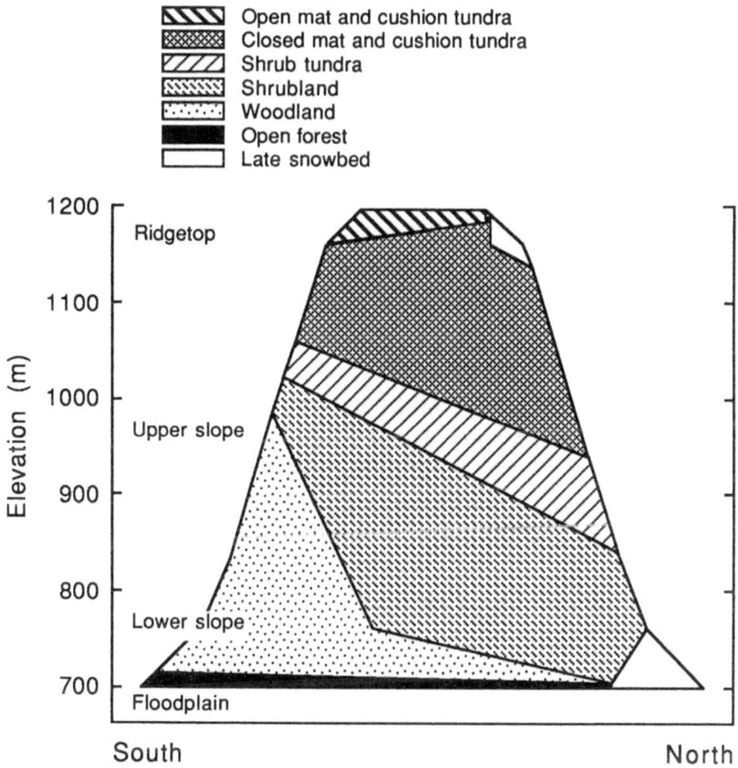

Figure 3.5. Vegetational zonation in the Dietrich River Valley, southern Brooks Range, at the northern limit of forest in the Boreal Rocky Mountains region. Redrawn from Densmore (1980).

showers and fire-causing lightning. On the south side of the Brooks Range in northern Alaska, at ca. 68° to 69° N latitude, the taiga forests are essentially subalpine in nature and rise on the slopes to elevations of between 700 to 800 m. A stunted woodland continues on up to about 915 m on south-facing slopes. These mountain forests are at their northern limits and consist of open aggregations of *Picea glauca* (white spruce), *P. mariana* (black spruce), *Betula glandulosa* (resin birch) and its hybrids with *B. papyrifera* (white birch), and *Populus balsamifera* (balsam poplar). Near its extreme limits, the forest is made up almost entirely of white spruce (Densmore 1980) with black spruce being left behind on the southern foothills. The subalpine forests in the Brooks Range are limited to streambeds in the mountains and to the lower south-facing slopes (Figure 3.5).

From the standpoint of tree floristics, there is a strong resemblance in the Tanana lowlands of central Alaska (never glaciated) and on the south side of the Brooks Range to the present-day forests of the northern Appalachians, a continent's width away. Some of the species held in common are *Picea glauca, P. mariana, Larix laricina* (but not in the Brooks), *Betula papyrifera*, and *Populus balsamifera*. Following deglaciation across Canada, there has been rapid migration of the spruces and birch toward the west and northwest from refugia south of the continental ice on the southeastern Coastal Plain of Virginia and the Carolinas and also in the southern Midwest (Jacobson, Webb, and Grimm 1987). The same migrational patterns could possibly hold true for *Larix* and *Populus*. Ager (1983) reports that *Picea* appeared in the middle Tanana Valley about 9,500 years BP and has been a constant feature of the central Alaskan landscape since about 6,500 to 6,000 years BP. The rate of migration from the eastern and midwestern North American refugia seems rapid indeed. Could there have been closer refugia to Alaska and the boreal Rocky Mountains?

In the White Mountains of east-central Alaska, between Fairbanks and Circle (on the Yukon River), the montane forest is represented by an upland spruce-hardwood forest, the most extensive forest type in the region. The principal tree species are *Picea glauca, P. mariana, Betula papyrifera, Populus tremuloides* (quaking aspen), and *P. balsamifera*. The birch and poplars are generally successional following fire. Timberline is near 850 m.

Eastward, in central Yukon Territory, the subalpine forest of the Rocky Mountains becomes somewhat more distinct from the lowland taiga. The Ogilvie Mountains, between the Yukon River on the west and the Mackenzie Mountains to the east, have a montane forest continuum rather typical of this boreal part of the Rocky Mountains. The lower slopes and river valleys are covered with a forest of white spruce, black spruce, larch (*Larix laricina*), lodgepole pine (*P. contorta* var. *latifolia*) which is here at its northern limits, aspen, balsam poplar, and paper birch. The winter temperatures in these valleys are frequently the coldest in North America with lows down to -45 to -50° C. As one goes higher in these mountains, the forest becomes subalpine and is characterized by small spire-shaped white spruces and subalpine fir (*Abies lasiocarpa*.) The latter, like lodgepole pine, is another typical Rocky Mountain species of Wyoming and Colorado that also reaches its northern geographic limits here near Keno Hill in central Yukon Territory (Billings, field notes 1963). Timberline of white spruce and subalpine fir here is at an elevation of 1,220 m with the fir forming a krummholz on windward slopes. A similar subalpine fir forest occurs around Macmillan Pass in the Mackenzie Mountains with firs as tall as 25 m and 45 cm in diameter (Porsild 1951).

This forest series, with the same tree species, continues southeastward along the mountains attaining greater elevational amplitude from timberline down to its lower reaches that occupy the river valleys such as that of the Liard. Frequent fires there allow the widespread coverage of lodgepole pine (*P. contorta*). Near Summit Lake, B. C., the subalpine forest ascends to about 1,465 m. This forest is similar to the subalpine forest of the Yukon. At timberline, it consists mostly of white spruce, some black spruce, and subalpine fir. It is not far enough south to have Engelmann spruce.

Central Rocky Mountains

South of the Peace River, the Canadian Rockies become much higher, decidedly more glaciated, and sharply rugged. These are the most northerly peaks and high ridges of what I have designated as the Central Rocky Mountains forests. This region extends south as far as central Wyoming. The Central Rocky Mountains correspond roughly to Peet's Northern Rocky Mountain region. When entering these mountains from the north, one almost immediately sees additional tree species that do not grow north of the Peace. Along the eastern slopes of the Rocky Mountains in east-central British Columbia and southwestern Alberta, the forest on the lower foothills between 915 to 1,220 m in elevation still resembles that farther to the northwest. The important tree species is lodgepole pine (*P. contorta* var. *latifolia*). Associated with it are quaking aspen and balsam poplar. All three, in general, represent fire successional forests. Older forests mainly are characterized by white spruce. Black spruce, larch, white birch, balsam fir (*Abies balsamea*), and subalpine fir are also present in smaller amounts (Rowe 1959). The upper foothills to 1,830 m are forested to their summits with conifers (Rowe 1959). The aspen, poplar, and birch of the lower hills are poorly represented in the upper foothills. The principal conifers are lodgepole pine and white spruce; there is some subalpine fir.

Above the foothill forests on the east sides of the Rockies, there is a true subalpine conifer forest from about 1,525 m to 2,075 m. It is distinguished by the presence of Engelmann spruce (*Picea engelmannii*) nearing its northern limits. Engelmann spruce hybridizes with white spruce here (Rowe 1959). Associated with the spruce in the subalpine forest are lodgepole pine and subalpine fir; this spruce-fir-pine forest is similar in composition to subalpine forests as far south as Wyoming and Colorado. Near treeline, whitebark pine (*P. albicaulis*) and limber pine (*P. flexilis*) characteristic of more southerly mountains, including the Sierra Nevada, reach their northern limits. Also present near timberline, and sometimes even above, is the endemic high elevation larch (*Larix lyallii*). The mountain forest tree species list increases dramatically compared to that north of the Peace River.

On the west slopes of the mountains in British Columbia, is a similar subalpine forest of Engelmann spruce, subalpine fir, and lodgepole pine which extends westward to the lee sides of the Pacific coastal mountains. It differs from the east-side forest of the Rocky Mountains by having some coastal tree species that are associated with the spruce; these coastal species are western hemlock (*Tsuga heterophylla*), western red cedar (*Thuja plicata*), and silver fir (*Abies amabilis*). Near timberline are alpine larch (*L. lyallii*), whitebark pine (*P. albicaulis*) and mountain hemlock (*Tsuga mertensiana*), the latter two extending southward as far as the southern Sierra Nevada.

Along the main continental divide in northwestern Montana (the Lewis Range of Glacier National Park), there is a sharp difference in forest composition between the west and east sides of the mountain range (Figure 3.6). The west-side forest is relatively luxuriant due to greater precipitation and contains a number of Pacific Coast species; the composition is similar to the forests of interior British Columbia as described in the above paragraph. The lower elevation forests have some large, "Pacific Coast-type" ponderosa pines. Above these in elevation is a mixed forest of grand fir (*Abies grandis*), western red cedar (*Thuja*), and western hemlock (*Tsuga heterophylla*) with quaking aspen and the western paper birch (*Betula papyrifera* var. *commutata*). In the higher subalpine zone, the forest consists of Engelmann spruce, subalpine fir, some mountain hemlock (*T. mertensiana*), with whitebark pine on the drier ridges of this very snowy western exposure. In spite of extremely cold winters at high elevations here, the moist Pacific air results in considerable packed snow and rime on the trees near timberline preventing much winter desiccation (Arno 1984).

East of the continental divide in the Glacier Park region, the climate is drier and more continental. The coastal tree species have dropped out

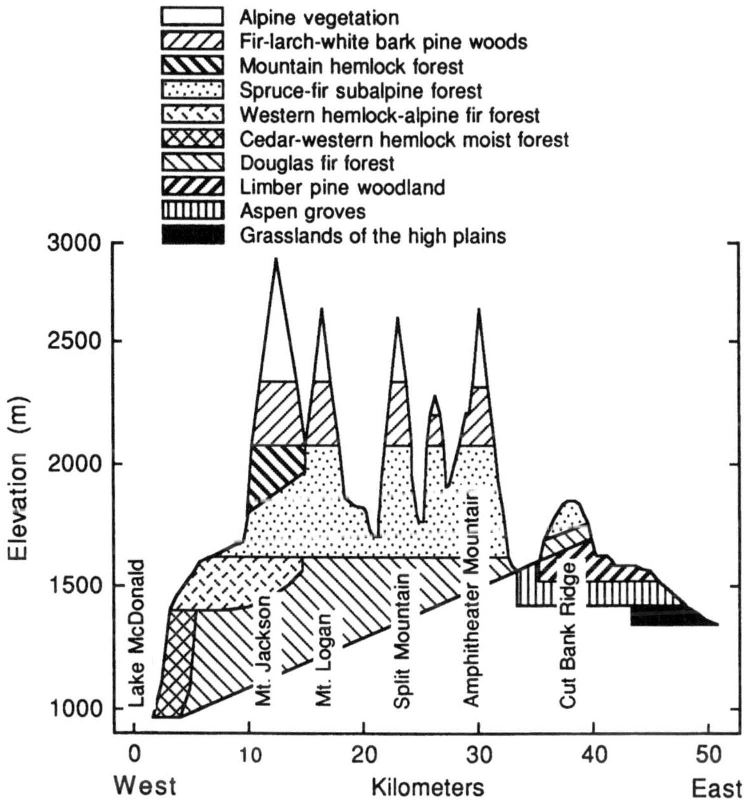

Figure 3.6. Vegetational zonation across the Lewis Range in Glacier National Park, Montana, in the Central Rocky Mountains region.

of the forests. The climate is very windy and the winters are bitterly cold even at the lowest elevations. Timberline is between 2,300 to 2,450 m and the forest consists of Engelmann spruce, subalpine fir, and whitebark pine as at the upper elevations on the west side. However, below that spruce-fir subalpine forest, Douglas-fir (Rocky Mt.-type) and lodgepole pine form extensive forests. Between these mountain forests and the grasslands of the High Plains, the lowest forest is scrubby and wind-battered; it consists of aspen, limber pine (*P. flexilis*) and some Douglas-fir (*Pseudotsuga menziesii*). The ecotone between this scrub forest and the grassland lies between 1,375-1,525 m. Ponderosa pine (Rocky Mt.-type) does not grow at this ecotone because of its susceptibility to damage by alternating warm and very cold winter temperatures in this "chinook" belt (Arno 1984). It does grow, however, along the lower slopes of the isolated mountain ranges, including the Big Snowy and Bull Mountains, farther east in Montana and thence southward into the Laramie Range of eastern Wyoming.

Following the continental divide southward, one comes to the high ranges along the Montana-Wyoming boundary: the Tetons, Absarokas, Beartooth, and the Wind Rivers. Across the deserts of the Bighorn Basin to the east are the Bighorn Mountains. These ranges, along with several smaller ones, are those farthest south in the Central Rocky Mountain Region.

The western ranges (Tetons, Beartooth, Absaroka, and Wind Rivers) near the continental divide have a rather narrow forested belt. Whitebark pine is associated with Engelmann spruce and subalpine fir in the subalpine forest particularly near timberline at about 3,050 m (Arno 1984; Reed 1976). At timberline along the northern edge of the Beartooth Plateau, whitebark pine reaches its largest size anywhere (12 to 15 m high, 0.5 to 1.5 m in diameter, with a broadly spreading crown). In this form, it is immediately adjacent to windswept alpine fellfields. Below the timberline and subalpine forests, most of the forests of these ranges along the continental divide are characterized by lodgepole pine with aspen groves, limber pine, and Douglas-fir here and there along the lower edge of the forested zone (Arno 1984; Reed 1952; Reed 1976). It is these forests which are so susceptible to fire and are successional after fires in the Yellowstone montane ecosystem. Here is where the great fires of late summer 1988 occurred (Christensen et al. 1989).

The large and isolated massif of the the Bighorn Mountains is separated from the Absarokas and the Wind Rivers by about 160 to 200 km of desert and semidesert. Its highest peaks are over 4,000 m in elevation. The bedrock of this range is mostly granite but other rock types are also present: shale, limestone, dolomite, sandstones. As in any relatively dry climate, geologic substratum plays a large role in determining vegetational types in the Bighorns (Despain 1973). Lodgepole pine forest covers almost all the granitic areas below timberline but is practically absent on the sedimentary rocks. The latter substrates have forests of Engelmann spruce and subalpine fir at the higher elevations but whitebark pine is absent (as contrasted to the situation in the Wind Rivers, Tetons, and all the ranges to the northwest). At the lower elevations on sedimentary rocks, the forests are of *Pseudotsuga menziesii* (Rocky Mt.-type Douglas-fir) and/or ponderosa pine. But, grassland is more common than forest at these lower elevations particularly on the fine-grained soils derived from sedimentary rocks (Despain 1973).

The Wind River Mountains and the Bighorn Mountains, in my opinion, represent the southernmost extension of the Central Rocky Mountain Region. Cold deserts and semideserts across central Wyoming separate them from the northernmost ranges of the Southern Rocky Mountains. Although there is considerable resemblance in the forest vegetation between these two ranges of northern Wyoming and that of the Medicine Bow Mountains, the Laramie Range, and the Uintas, there are some marked differences. Whitebark pine is lacking in the Southern Rocky Mountains except in the far northwest of the region in the Ruby Mountains of Elko County, Nevada, if one includes the Rubies as part of the Rocky Mountains botanically; I am so inclined. The bristlecone pines (*P. aristata* and *P. longaeva*) do not cross the Wyoming deserts in a northward direction, nor do the Colorado pinyon (*Pinus edulis*) and the one-seed juniper (*J. monosperma*). There is some overlap, of course, and this is most evident in three widespread (to the Pacific coast from the Plains) species of pine: limber (*P. flexilis*), lodgepole (*P. contorta*), and ponderosa (*P. ponderosa*). Also one could include the quaking aspen (*Populus tremuloides*) in this group since it, too, is very widespread. South of the Wyoming deserts, the Rocky Mountains are geologically, climatically, and botanically (herbaceous plants as well as trees) rather different than those mountains to the north.

All of the mountain ranges of the Central Rocky Mountain Region have very cold and snowy winters. The summers tend to be relatively dry with some thundershower activity. There is considerable "dry" lightning resulting in the forest fires mentioned earlier.

Southern Rocky Mountains

This region is nicely delineated by Figure 1 in Axelrod and Raven (1985). The northern Wasatch in Utah, the Medicine Bow Mountains, and the Laramie Range in southeastern Wyoming are those mountain ranges farthest north in this region. The latter two ranges rise from the high Wyoming tableland east of the continental divide. Their winters are cold, with heavy snow accumulations. This is particularly true of the Medicine Bow. Snow falls during every month of the year and the summers are marked by almost daily thundershowers and lightning because of the presence of airmasses from the Gulf of Mexico. Apart from the geology, the one aspect of the environment that is characteristic of the southern Rockies is the summer thundershower and lightning activity. Almost all forests burn within a fire cycle of about 200 to 800 years.

The Laramie Mountains, west of Cheyenne, extend from the Colorado line north for 160 km to Laramie Peak overlooking the North Platte River. These relatively low mountains are almost surrounded by the grasslands of the High Plains. The relatively young (mid-Tertiary) continental sedimentary rocks under the plains rise as a "gangplank" against the eastern side of this range and join the coarse pre-Cambrian granite (Sherman Formation) at about 2,135 m elevation. The Laramie Mountains are essentially a geologic extension northward of the Front Range of Colorado and its forests have some of the same characteristics as those of the latter range. The highest peak (Laramie Peak) is only 3,165 m in elevation, so there is no true upper timberline. The forests are mainly in the southern and northern ends of these mountains with the middle having forests only in the canyons, and grasslands, shrublands, and occasional aspen groves elsewhere. In the southern parts of the range, the forests are better developed on the granite, where as the sedimentary rocks are principally covered by grasslands. The overall effect is a patchwork mosaic. The principal trees are ponderosa pine, limber pine, Douglas-fir, and aspen, and on the higher north-facing slopes, a forest of lodgepole pine, Engelmann spruce, and subalpine fir. In the Laramie Peak area of the northern end of the range, the spruce-fir-lodgepole forest is more extensive. Lower timberline on the east side is around 2,150 m, but on the west side it is nearer to 2,450 m. Summer thundershowers are frequent over these mountains. Climatically and vegetationally, there is a strong resemblance to the situation in the Front Range rather than to the Medicine Bow only 50 km away to the west across the high Laramie Plains. Much of the resemblance to the Front Range seems to be due to the granitic core of the Laramie Mountains as compared to the hard quartzite core of the Medicine Bow. Sheep Mountain, only 8 km east of the base of the Medicine Bow also has a granitic core (again, the Sherman granite) and its forest vegetation is more like that of the Laramie Mountains rather than that of the nearby Medicine Bow. This illustrates again the importance of bedrock in governing forest composition in a semiarid climate.

The east face of the Medicine Bow rises abruptly from the high grasslands of the Laramie Plains, which are already at an elevation of 2,500 m (Figure 3.7). Within 15 km westward, the elevation is over 3,600 m at

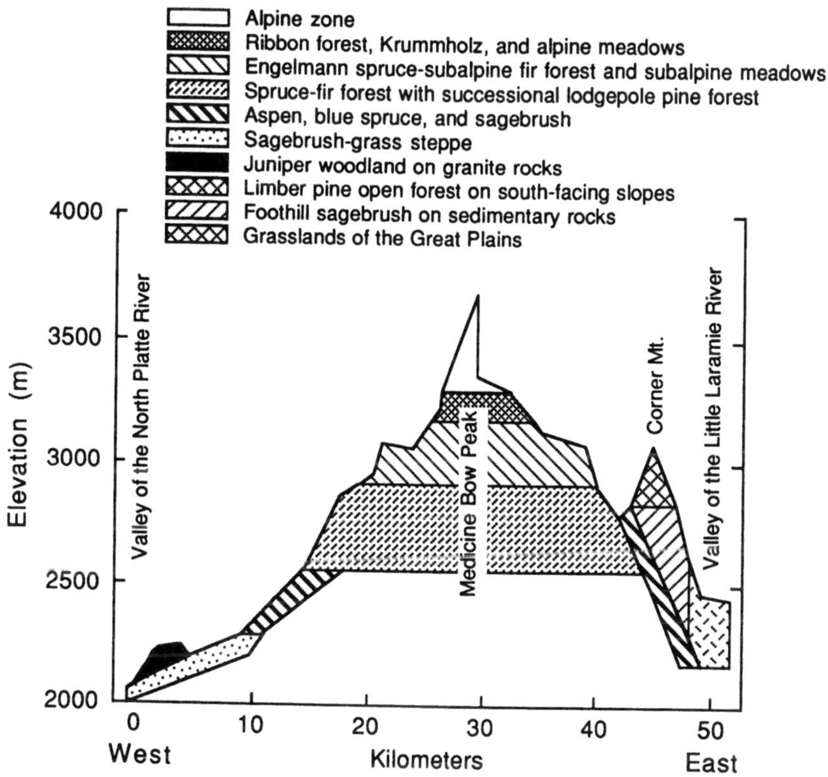

Figure 3.7. Vegetational zonation across the Snowy Range of the Medicine Bow Mountains, Wyoming, in the Southern Rocky Mountains region.

Medicine Bow Peak, part of the glaciated Pre-Cambrian quartzite core of the range, 375 m above timberline. The vegetation just above the foot of the range is a mosaic of sagebrush, aspen groves, and lodgepole pine forest. Ponderosa pine and Douglas-fir are essentially absent here where as Engelmann spruce and subalpine fir come down to the grasslands of the high plains in the moist canyons and along streams. One finds an occasional blue spruce (*Picea pungens*) in the stream-edge forest that is dominated by aspen and *Populus angustifolia* near the lower edge of the montane forest. But, this spruce is more common on the west side of the range. The greater part of the forest in the middle elevations is dominated by fire-successional lodgepole pine with some clonal groves of aspens that are also fire-successional. Near the upper edges of this forest, the drier windswept south-facing slopes have open stands of old limber pines. North-facing slopes at the same elevation (ca. 2,900 m) are covered with subalpine forest of Engelmann spruce and subalpine fir either in successional or climax condition. At 3,050 m, all slopes are occupied by the spruce-fir forest (Oosting and Reed 1952). This continues on up to timberline where both species exhibit ribbon forest and scattered Krummholz forms (Plate 6.1) across the alpine meadows at around 3,260 to 3,350 m (Billings 1969).

Descending the west, and wetter, side of the Medicine Bow, the same zonal patterns are repeated. However, the aspen and conifer groves descend about 150 m lower in elevation (to 2,400 m) than they do on the east side. Below this elevation, and down to the cottonwoods along the North Platte River at 2,135 m, the vegetation is sagebrush rather than the grasslands of the higher valleys on the east side of the range. Outcroppings of coarse pre-Cambrian granite are occupied by an open dry woodland of Rocky Mountain juniper (*J. scopulorum*) at elevations between 2,200 m and 2,300 m. This woodland is reminiscent of the open pinyon-juniper woodland of Nevada, but there is no pinyon, and the juniper is a different species.

Some 100 km or so southeastward from the Medicine Bow, the granites of the Laramie Range merge into those of the Colorado Front Range. This geologically complex and high mountain system is a great wall above the western end of the plains. Its base is at about 1,675 m and its high peaks well above 4,275 m in elevation. This range runs north and south through central Colorado; its escarpment "fronts" form the great array of high mountains that cover the western half of that state. Southward from this range, for another 400 km, the "fronting" is continued by the Sangre de Cristo Mountains through southern Colorado and into New Mexico above Santa Fe.

These high ranges are typical of the southern Rocky Mountains both climatically (with snow and lightning) and vegetationally. A transect across the Front Range from Boulder is useful because of the good set of environmental data gathered for the last 35 years by Dr. John Marr and his colleagues, particularly Barry (1973), at the Institute of Arctic and Alpine Research at the University of Colorado. The following descriptions of forest zonation along the Boulder transect (Figure 3.8) are adapted from Marr (1961).

The sharp topographic break between the plains and the foothills of the Front Range in the Boulder area is at an elevation of about 1,700 m. The rocky backbone of the foothills has scattered ponderosa pine woodlands with grasslands occupying the lowlands between the hills.

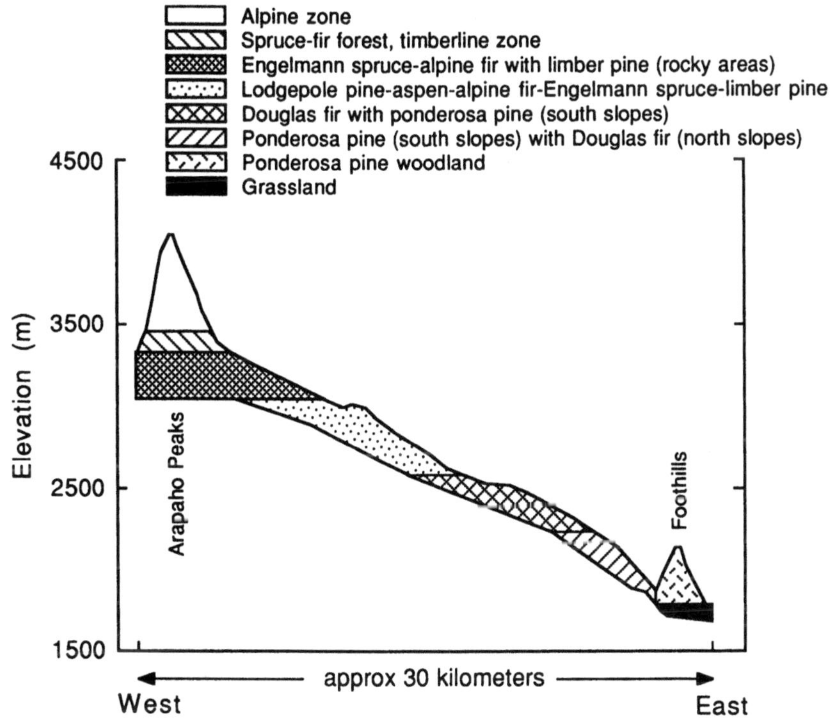

Figure 3.8. Vegetational zonation on the Front Range across Arapaho Peaks, Colorado. Drawn from data given by Marr (1961).

The pinyon-juniper woodlands so characteristic of the lowermost forested zone on the east slopes from Pike's Peak southward through the Sangre de Cristos are lacking in the Front Range except for a *Pinus edulis* population in Owl Canyon northwest of Ft. Collins. With increasing elevation in the Boulder foothills, the ponderosa pines become more abundant. As also seen farther north in Wyoming and Montana, these trees prefer the coarse geologic substrates such as granites while grasses occupy the finer soils. The pine stands are open and park-like on south-facing slopes while on north-facing slopes the forests are denser and usually contain an admixture of Douglas-fir.

Along this transect, forests cover the Front Range from just above the foothills at 1,830 m to tree-line at about 3,475 m. Between the upper edges of the foothills near 1,830 m and an elevation of about 2,350 m, the forest is mainly a mixture of ponderosa pine and Douglas-fir with the pines dominant on the south-facing slopes and Douglas-fir on the north-facing slopes. With increasing elevation, Douglas-fir becomes more common on the south-facing slopes. From 2,350 m to 2,750 m, the forest retains about the same floristic composition but becomes more dense.

Around 2,600 m, lodgepole pine and aspen appear on north-facing slopes in rather pure stands where there has been fire in past times. Such stands continue here and there to well over 3,050 m. But, the more common forest type above 3,050 m all the way to timberline is the climax subalpine forest of Engelmann spruce and subalpine fir (*Abies lasiocarpa*). These two species appear at about 2,750 m scattered in among the upper ponderosas, Douglas-fir, aspens, and lodgepole but they go much higher on the mountain than any of the others with the possible exception of limber pine (*Pinus flexilis*) which first appears in rocky places at about 2,440 m but goes all the way to timberline near 3,475 m. Limber pine has the greatest elevational range of any of the tree species in the Front Range. General environmental conditions along the transect are listed in Table 3.1 (adapted from Table 9 in Marr 1961). Minor changes in the data occur with the addition of observations over the last 25 years but there are no substantive differences. The principal exception is the increase of polluting smog in the valley from Denver northward through Boulder that strongly affects the foothill woodlands and lower montane forests, and quite possibly even the subalpine forest.

The high peaks of the southern Front Range and the Sangre de Cristo Mountains have essentially the same kind of forest zonation as in the mountains west of Boulder. However, there are additions both near upper timberline and lower timberline. Near upper timberline is the bristlecone pine (*Pinus aristata*), a characteristic component of the subalpine forest in exposed places in the southern Rocky Mountains from Mt. Evans to Truchas Peak in New Mexico. The lower timberline is extended downward by the addition of the pinyon-juniper woodland zone dominated by *Pinus edulis* and *Juniperus monosperma*; this woodland can extend downward to about 1,675 m elevation on the eastern foothills and bajadas. Lodgepole pine becomes scarce in the montane forests of southern Colorado and apparently is lacking in the mountains of northern New Mexico. However, there is an addition to these middle elevation forests in southern Colorado and New Mexico; it is the Rocky Mountain form of the white fir (*Abies concolor*).

Table 3.1. Annual means and totals of certain environmental conditions along the Front Range Transect shown in Figure 3.8 (adapted from Marr 1961) (Environmental factors measured in one vegetation site in each zone).

Zones:	Lower Montane	Upper Montane	Subalpine	Alpine
Vegetation type	Ponderosa Pine Forest	Douglas Fir-Ponderosa Pine Forest	Engelmann Spruce-Subalpine Fir Forest	Kobresia Meadow Tundra
Elevation (meters)	2,195	2,590	3,050	3,750
Air temperature, degrees C				
Maximum	36	32	24	17
Mean Max	15	12	7	0
Minimum	-19	-22	-26	-28
Mean minimum	2	-1	-4	-7
Mean	8	6	1	-4
Longest freeze-free period (days)	137+	100	87	45
Soil temperature, degrees C				
15 cm depth				
Maximum	27	22	22	12
Mean Max	12	8	8	1
Minimum	-4	-6	-8	-14
Mean Min	5	4	1	-3
Mean	8	6	4	-2
30 cm depth				
Mean	8	6	4	-2
Wind, total km	149,443	159,544	236,467	414,400
Velocity km hr^{-1}	18	18	26	47
Precipitation, cm	50.8	53.3	66.0	66.0
Maximum snow depth, cm (with min. of drifting)	30.5	27.9	152.4	all drifted
Soil moisture, % dry wt.				
13-18 cm depth, mean	7	7	10	32
28-33 cm depth, mean	8	7	11	16
Mean relative humidity	53	60	62	70

With the exception of lodgepole pine, the southern Rocky Mountain forest types extend into the San Juan Mountains of southwestern Colorado, the high plateaus of western Colorado, northern Arizona, southern Utah, and the Wasatch Mountains of central Utah. There are some species additions but the forest zonation and composition are essentially the same from the Wasatch southeastward to the mountains above Santa Fe, New Mexico. Amazingly, this also holds true for much of the Mogollon Rim in Arizona and even the strongly isolated San Francisco Mountain at the northwest end of the Rim. The vegetational zonation on this mountain was described as early as 1890 by Merriam. Accordingly, it is appropriate to use this diagram in Figure 3.2 as typical of the forest situation in the Southern Rocky Mountains Region.

The forests of this region extend to southern New Mexico along the Sacramento Mountains and the Black Range. In attenuated form, they enter Mexico along the Sierra Madre Occidental with some generic elements reaching the arc of the high Mexican volcanoes and even into Guatemala. But no attempt will be made here to describe the mountain forests of the Mexican Region.

Basin Range Forests

Between the southern Rocky Mountains and the Sierra Nevada there are many mountain ranges trending generally in a north-south direction. Most of these have some woodlands on them and a fair number are high enough to have extensive montane and subalpine forests. There is considerable overlap in floristic composition in the eastern Great Basin mountain forests with those of the Southern Rocky Mountain Region.

According to Axelrod and Raven (1985), certain Cordilleran taxa migrated across the Great Basin from the east to the western edge largely after 14 million years ago as the climate became drier and summer rains decreased. In the late Miocene (5-7 m. y. BP), some Cordilleran species became restricted to the eastern part of the Basin as summer rains became even more deficient. When the Southern Rocky Mountains were uplifted by 1,500-3,000 m following the later Miocene, the Great Basin flora differentiated from that of the Southern Rocky Mountains as a double rainshadow was produced by the rising Sierra Nevada on the west and the Wasatch Range - Colorado Plateaus on the east. This resulted in a marked floristic impoverishment among those surviving members of the Cordilleran flora still in the Great Basin. The

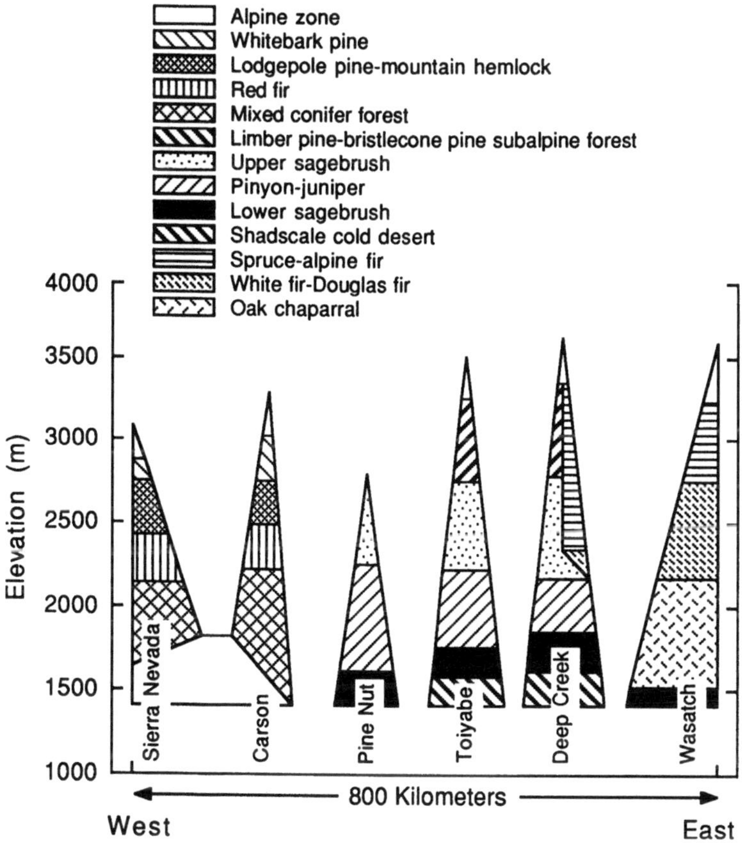

Figure 3.9. Vegetational zonation across the Basin and Range from the west face of the Wasatch Mountains to the east face of the Sierra Nevada and the Tahoe Basin. Note that this transect is telescoped with many mountain ranges omitted. Redrawn from Billings (1951).

remaining montane taxa continued to evolve in response to the present severity of the Basin's continental climate (Axelrod and Raven 1985).

On the eastern rim of the Great Basin, the Wasatch Range is vegetationally Southern Rocky Mountain in character with ponderosa pine and pinyon-juniper as the lower forest zones in the southern and central parts of the range, and oak-maple (*Quercus gambelii* and *Acer grandidentatum*) in the lower zones but becoming increasingly dominant toward the northern end of these mountains (Figure 3.9). Typical Rocky Mountain spruce-fir forest occupies the subalpine forest niche.

As one goes westward across the mountain ranges of the eastern Basin for about 360 km, there is a gradual loss of Rocky Mountain tree species (Billings 1951). The Deep Creek Mountains on the Utah-Nevada border (McMillan 1948) show Rocky Mountain forests mainly on the north-facing moister slopes where as the drier south-facing slopes are occupied by "mountain brush" and above that, typical Great Basin open subalpine forest of limber pine and bristlecone pine. Pinyon-juniper woodland now occupies the lower slopes but the species are different from those in the Wasatch and the rest of the southern Rocky Mountains region. They are the single-needle pinyon (*P. monophylla*) and the Utah juniper (*J. osteosperma*). These and their plant and animal compatriots occur around the base of every range rising above 1,900 m from western Utah to the Virginia Mountains near Reno, a distance of more than 500 km. It has been a very consistent native ecosystem, environmentally and biologically, that is now changing drastically due to fire especially in western Nevada (Billings 1989).

The larger mountain ranges of eastern Nevada (the Snake Range, the Ruby Mountains, the Jarbidge, the Egan Range, the Independence Range, and the Shell Creeks) have a modified Rocky Mountain forest zonation on their upper slopes but with Basin-type pinyon-juniper on their lower slopes. There is little or no ponderosa pine except in places in the Snake Range. The subalpine forests differ from range to range but most of the species are typical of the southern Rocky Mountains: Engelmann spruce, subalpine fir (*A. lasiocarpa*), limber pine, and bristlecone pine. But, with the latter, the species is the long-lived *Pinus longaeva* rather than *P. aristata* of the Rocky Mountains. In the Rubies, East Humboldts, and Jarbidge there is a prominent and abundant disjunct species in the subalpine forest from the central Rocky Mountains (Tetons, Wind Rivers, Beartooth) and the Sierra Nevada: whitebark pine (*P. albicaulis*) that is missing completely from the southern Rocky Mountains. Loope (1969) has described these mountain forests of northeastern Nevada in elegant fashion.

If one stands on top of the Ruby Mountains, Elko County, Nevada, and looks westward toward the Cortez Mountains and Shoshone Mountains, and southwestward toward the Toiyabe and Toquima Ranges rising above the hazy valleys 125-250 km away, one is looking from a modified Rocky Mountain forest zonational system into a truly Great Basin system. If I had to draw a line separating the two great biomes, I would draw it from the Humboldt River near Elko south between the Rubies and the Diamond Mountains and west of the White Pine Mountains

Plate 5.1. (upper) Mixed mesophytic forest near Whitewater Falls on the southern escarpment of the Blue Ridge mountains in western North Carolina at an elevation of about 725 - 925 m.

Plate 5.2. (lower) Damage to subalpine spruce-fir forest of *Picea rubens* and *Abies fraseri* on north ridge of Mt. Mitchell at about 2,025 m in the Black Range of western North Carolina due to the European woolly aphid and other causes of dieback.

Plate 6.1. (upper) Flagform Krummholz of *Picea engelmannii* near timberline at about 3,350 m in the Medicine Bow Mountains, Wyoming.

Plate 6.2. (lower) Subalpine woodland of whitebark pine, *Pinus albicaulis*, of the Central Rocky Mountains type, near the head of Lamoille Canyon (about 2,900 m) in the Ruby Mountains, Elko County, Nevada.

Plate 7.1. (upper) Great Basin woodland of single-needle pinyon (*Pinus monophylla*) in the Desatoya Mountains, central Nevada. Trees in the foreground are at about 2,250 m while those on the distant south-facing slope are at about 2,500 m.

Plate 7.2. (lower) Open subalpine forest of very old bristlecone pine (*Pinus longaeva*) on Boundary Peak, White Mountains, Nevada, at about 3,350 m.

Plate 8.1. (upper) Open forest of Sierra Nevada form of *Pinus albicaulis*, whitebark pine on north face of Round Top at about 2,800 m in the Sierra Nevada of Alpine County, California.

Plate 8.2. (lower) Krummholz of Sierra Nevada form of *Pinus albicaulis*, whitebark pine, near timberline at about 3,100 m on northwest side of Mt. Rose, Carson Range, in the Sierra Nevada of Washoe County, Nevada.

and the Grant Range. All mountain forests east of that line have a strong admixture of species and ecotypes from the southern Rocky Mountains. The ranges to the west have the typical sparse limber pine-bristlecone pine subalpine forest of the Great Basin (Plate 6.2). The rest of the Great Basin high mountain vegetation is "mountain brush" with aspen, sagebrush (*Artemisia* spp.), and mountain mahogany (*Cercocarpus ledifolius*); the ponderosa pine, white fir, Douglas-fir montane forest of the Rocky Mountains is absent completely. This situation prevails all the way to the westernmost of the Basin ranges: the Virginia Mountains, the Pine Nuts, the Whites (Figure 3.10). Only in the Sweetwater Mountains on the California-Nevada line is there an admixture of Sierran tree species from the west.

Probably the most typical series of mountain vegetational zones for the Basin ranges proper, and certainly the most complete, is that in the White Mountains on the Nevada-California border northeast of Bishop. Surprisingly, this typicality and completeness occurs at the western edge of the Great Basin just a few miles across Owens Valley from the distinctive and quite different forests of the Sierra Nevada (Figure 3.11).

The environments of the mountain forests of the Great Basin, in the rainshadow of the Sierra Nevada, are drier than comparable situations in the Sierra Nevada or the Rocky Mountains both winter and summer. There is considerably less snow than in the Sierra Nevada and even less than in the southern Rocky Mountains. The Ruby Mountains and the Jarbidge Mountains are quite snowy, however, and that accounts to some extent for the presence of Rocky Mountain tree species in their subalpine forests. The White Mountains at the western edge of the Basin are much drier since they lie in the immediate rainshadow of the High Sierra Nevada. There is a strong peak of precipitation during the summer months in the Rocky Mountains and just the reverse in the Sierra Nevada and Cascades where there usually is a great amount of winter snow followed by dry summers. The eastern high basin ranges from the Ruby Mountains east and southeastward do receive considerable snowfall but also get more summer rain than the White Mountains and the mountains around Reno. The Snake Range and Rubies receive 50-60% of their annual precipitation in the 6 months from 1 April to 30 September. This figure is about 40% for the Toiyabe Mountains, and 20-30% for the Sweetwaters and Whites.

The pinyon-juniper zone in the western Great Basin (Plate 7.1) is much drier than that of the bristlecone pine forests (Plate 7.2) and even the

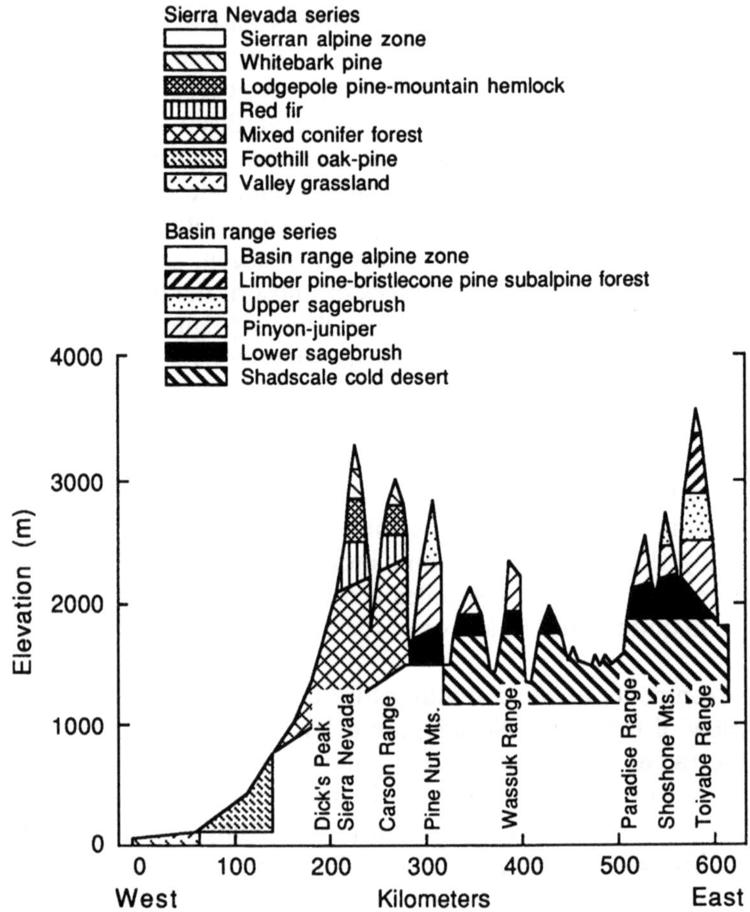

Figure 3.10. Vegetational zonation from the Sacramento Valley across the Sierra Nevada to the Toiyabe Range in the central Great Basin showing the principal differences between the Sierra Nevada and Great Basin series of mountain vegetation.

mountain brush or "upper sagebrush" zone above the pinyon-juniper. Also, the pinyons are indicative of a thermal belt around each range due to cold air drainage resulting in nocturnal temperature inversions during almost all months (Billings 1954). These inversions are particularly important to the pinyon-juniper ecosystem in regard to warmer temperatures during the winter. The valleys are much colder and so is the upper sagebrush or mountain brush zone above the pinyon-juniper. This inversion also results in smog and other atmospheric pollution lying in the valleys (winter and summer) but usually not in the pinyon-juniper. This

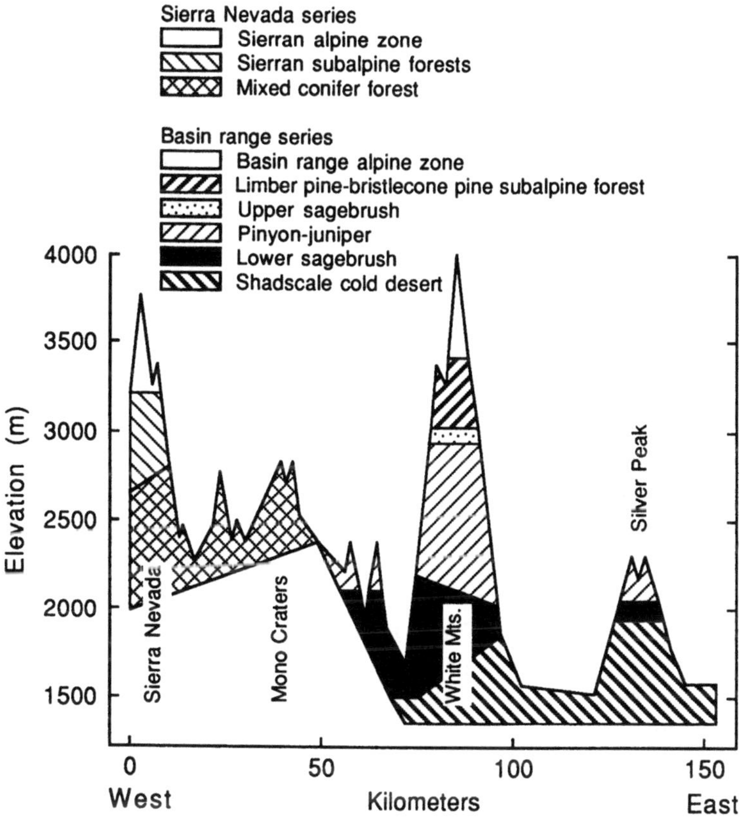

Figure 3.11. Vegetational zonation across the southern Sierra Nevada and the White Mountains to Silver Peak near Tonopah showing the sharp transition between the Sierra Nevada and Great Basin Range vegetational series.

is very apparent in the Reno-Carson City region. A similar but natural phenomenon is freezing fog (pogonip) in the valleys for many days during high pressure episodes in winter. The pinyon-juniper and higher zones are in sunshine during these freezing fog episodes (Billings 1954).

Forests of the Sierra Nevada

The Sierra Nevada is a young and biologically unique high mountain system. It extends about 565 km from the Feather River south to Tehachapi Pass. A number of the forest species extend into the San Gabriel, San Bernardino, and San Jacinto Mountains of southern

California. The Sierra Nevada is essentially a large granitic batholith tilted gently to the west and capped in places with the remnants of andesitic flows and metamorphic rocks. It is higher in the south (4,418 m at Mt. Whitney) and lowest in the north (3292 m at Mt. Rose). The western slopes receive heavy snowfall, but there is somewhat less snow on the steep east slope. The summers are almost rainless in the north but thundershowers become more frequent toward the south. Compared to the Rocky Mountains, however, there is relatively little summer rain. The forests are therefore dependent on meltwater from the winter snowpack which varies from year to year. Solar radiation is high in the summer through the clear, dry air.

Figures 3.10 and 3.11 show the forest zonation across the Sierra Nevada at two different latitudes. As can be seen, the zones are much broader on the relatively gentle western slopes and narrower on the steep, rocky eastern escarpment. There is also a sharp cutoff of Sierra Nevada tree species at the eastern base of the escarpment along the valleys from Reno south to Lone Pine. Almost the only crossing of this valley barrier by Sierran species into the Great Basin ranges occurs on the acidic hydrothermally altered rocks of the Virginia Mountains and Leviathan Ranges of western Nevada (Billings 1950; DeLucia, Schlesinger, and Billings 1988) and on other kinds of acidic rocks of the Sweetwater Mountains along the California-Nevada border. The phytogeographic boundary between the Sierra Nevada and the Great Basin is about as sharp as one finds anywhere between two large biological regions. This is due not only to a sharp difference in climates along this line but also to a difference in geology.

The Sierra Nevada has a large and distinct flora of mountain plants most of which are endemic to this mountain range or are of limited distribution northward into the Cascades. Very few of them enter the Great Basin; they are well-suited to the winter snow-summer bright environment of the Sierra Nevada. The forests are characterized by probably the most diverse and rich conifer flora anywhere on earth. And, the trees are big: the most massive on earth (*Sequoiadendron giganteum*), the largest soft pine on earth (*Pinus lambertiana*), the largest hard pine on earth (the Pacific form of *Pinus ponderosa*), the largest true fir (*Abies magnifica*), the most massive juniper (*Juniperus occidentalis*), and the largest incense-cedar (*Calocedrus decurrens*). The forests of the Sierra Nevada are unique; there is nothing like them anywhere, even in western North America. Barbour's (1988) excellent descriptions are definitive.

Referring to Figure 3.10, on the east side of the Sacramento Valley the lowest zone of oak-pine woodland appears on the foothills at an elevation of about 90 meters. The trees are scattered widely over grass and consist mainly of several species of oak (*Quercus*), both evergreen and deciduous. The principal ones are *Q. douglasii* (deciduous), *Q. kelloggii* (deciduous) and *Q. wislizenii* (evergreen). Also present are the western buckeye (*Aesculus californica*) and the characteristic gray and straggling digger pine (*P. sabiniana*). Quite a number of trees and shrubs of lesser importance are also present. This woodland extends the length of the Sierra Nevada on the west side up to elevations between 600 and 900 m.

By an elevation of 900 m or even lower, the ponderosa pines appear, and thus the beginning of the very rich and dense pine-fir-incense cedar forest. This is the most extensive and species-rich forest of the Sierra Nevada. It extends from the upper foothills to the Tahoe Basin, and in attenuated form down the east side of the range to about 1,500 m where it abuts on the sagebrush semi-desert. The principal tree species are *Pinus ponderosa*, *P. jeffreyi*, *P. lambertiana*, *P. contorta*, *Abies concolor*, *Pseudotsuga menziesii*, *Calocedrus decurrens*, and *Quercus chrysolepis*. In certain places, the bigtree (*Sequoiadendron giganteum*) forms extensive groves between 1,500 m-1,800 m. At higher elevations and on the eastern escarpment, Jeffrey pine replaces ponderosa pine. Lodgepole pine is not as common in this middle montane zone as it is in the Rocky Mountains and consists principally of large scattered trees in wet places.

Above this mixed conifer forest is an almost pure forest of red fir (*Abies magnifica*) from about 2,150 m up to almost 2,750 m particularly on north-facing slopes (Oosting and Billings 1943). There are occasional trees of other species mixed with the red firs as, for example, western white pine (*P. monticola*), lodgepole pine (*P. contorta*) in its large form, and the mountain hemlock (*Tsuga mertensiana*). This forest has deep and long lasting snow. The red fir forest extends in patches almost to the southern end of the Sierra Nevada but is best-developed in the north. Above the red fir, there is an open subalpine forest of *Pinus monticola*, *Pinus contorta*, *Tsuga mertensiana*, *Juniperus occidentalis*, and near its upper limits, *Pinus albicaulis* (whitebark pine), so familiar at timberline in the central Rocky Mountains from Wyoming to Alberta (Plate 8.1). The trees are surrounded by moist subalpine meadows that are filled with bright flowers in midsummer. The timberline is set by the uppermost whitebark pines sometimes as bushy trees of 5 to 15 m in height, and sometimes as dwarf Krummholz, as, for example, at an elevation of ca. 3,140 m on the west side of Mt. Rose in the Carson Range (Plate 8.2). Unlike the Rocky Mountains where Engelmann spruce is the usual timberline species from Alberta and British Columbia to New Mexico, there is no spruce in the Sierra Nevada.

Axelrod (1989) demonstrates with mid- and late Tertiary fossil floras from the western Great Basin and the Snake River drainage in Idaho that the present subalpine forest regions of the Sierra Nevada are geologically and ecologically young. This is due to mountain-building during Quaternary times. The high mountain climate became colder and thus, in space and environment, a major new forest zone, the subalpine came into existence with evolution and migration upward of taxa from the mixed montane forest.

Apart from the Sierra Nevada, other mountain ranges in California are forest-covered. These include the Klamath Mountains and the Trinity Alps in the northwestern part of the state and the Coast Ranges running the length of the coast. The Santa Lucia Mountains south of Monterey are precipitous but forest-covered (except where burned) and notable for their forests of Coulter pine (*P. coulteri*) and the narrow endemic Santa Lucia fir (*Abies bracteata*). For details, see Critchfield (1971), Griffin and Critchfield (1972), and Talley (1974). The first two references are useful for forests in much of California while Talley's thesis is excellent for the Santa Lucias. In southern California, the Cross Ranges (Santa Ynez, San Gabriel, San Bernardino, San Jacinto) have forests similar to, but not as rich, as the mixed conifer forests of the Sierra Nevada but with the addition of Coulter pine and the "bigcone spruce" (*Pseudotsuga macrocarpa*). These mountains of southern California receive considerable snow but not nearly so much as the west slope of the Sierra Nevada. They also are subject to man-made smog and other pollutants from the Los Angeles Basin.

Forests of the Cascade Mountains

The Cascade Range extends as a ridge punctuated by high and active volcanoes from northern California across Oregon and Washington to southwestern British Columbia. These mountains are wet and heavily forested on their western slopes and somewhat drier on the eastern slopes with open forests. The high volcanoes are covered on their upper slopes with permanent snow and alpine glaciers. Below the snow and ice there is only a narrow zone (usually patches) of alpine plants just

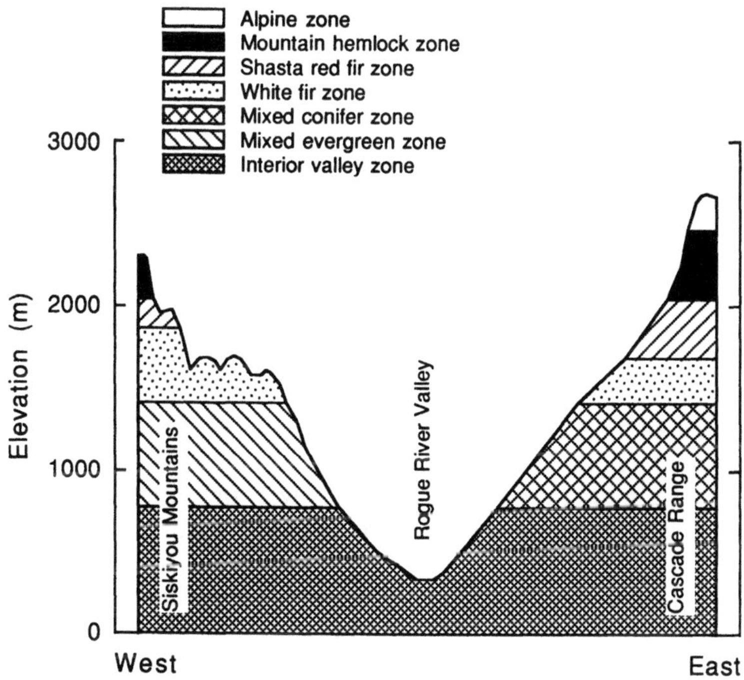

Figure 3.12. Vegetational zonation on the west face of the Cascade Mountains across the Rogue River Valley to the east face of the Siskiyou Mountains in Oregon. Redrawn from Franklin and Dyrness (1969), as seen from the south.

above timberline. Subalpine forests encircle each volcanic mountain below timberline. The two southernmost Cascadean volcanoes (Shasta and Lassen) are in northern California and have forest zones similar to those in the northern Sierra Nevada with forests of ponderosa pine, sugar pine, and Shasta red fir (*Abies magnifica* var. *shastensis*). Forests of these same species also occur in the Siskiyou Mountains of northwestern California and southwestern Oregon. The forest zonal relationships in southern Oregon are shown clearly in Figure 3.12 (from Franklin and Dyrness 1969). The vegetational resemblance of the west slopes of the southern Cascades to the Sierra Nevada is obvious, but the Siskiyous are rather different in forest cover. One unusual forest type on the lower slopes of the Siskiyous is an open Jeffrey pine woodland on xeric serpentine sites; this is quite a different situation for Jeffrey pine from that in the Tahoe Basin of the Sierra Nevada.

The forests of the Coast Ranges from western Oregon and the Cascades of central Oregon to northern Washington are unique and luxuriant wet forests dominated by long-lived coniferous trees. Only on the west coasts of New Zealand is there a comparable forest of conifers. But, in New Zealand, the trees are almost all members of the Podocarpaceae growing with Angiosperm trees in the genera *Nothofagus* (the southern beeches) and *Metrosideros* (the ratas). There are few floristic affinities between these two widely separated wet temperate conifer forests.

In a simplistic ordering of these forests, I shall list the forest zones that one would cross going from the Oregon coast across the Coast Range and the central part of the Cascade Range. Daubenmire (1969) has described, in a general way, the forest patterns of the Pacific Northwest. Franklin and Dyrness (1969) recognize four forest zones from the coast to the crest of the Cascades: (1) the coastal zone of *Picea sitchensis* (Sitka spruce), (2) the widespread *Tsuga heterophylla* (western hemlock) zone, (3) the montane *Abies amabilis* (silver fir) zone, and (4) the subalpine *Tsuga mertensiana* (mountain hemlock) zone.

The sealevel zone of Sitka spruce lies west of the coastal mountains and extends into them only in the river valleys. This long narrow zone reaches from northwestern California to southern Alaska. The spruce is the major species with considerable admixtures of western hemlock, Douglas-fir, and *Thuja plicata* (western red cedar). All are very large trees. The climate is wet, foggy, and cool.

The *Tsuga heterophylla* zone is the most extensive forest zone of western Oregon and western Washington and is extremely important in regard to timber production. In the Coast Range, its elevational limits are from just above sealevel to over 600 m. In the Cascades, these limits are from about 150 to 915 m. The zone extends northward into the mountains of British Columbia. The potential climax species is *Tsuga heterophylla* but extensive areas are dominated by the successional Pacific-coast form of Douglas-fir (*Pseudotsuga menziesii*). Such trees of the latter species are of very large size and usually reach 400 to 600 years of age. The climate is wet, mild, with some snow in its upper regions. The mean annual precipitation ranges from 125 to 305 cm per year. There is still a pronounced, but short, dry season in mid-summer.

The zone of *Abies amabilis* (silver fir) is the typical montane forest of the Pacific Northwest. It lies between the western hemlock zone and the subalpine mountain hemlock zone. Elevational limits are from about 915 to 1,375 m in Oregon and 610 to 1,160 m in northern Washington. The characteristic species are mainly firs: *Abies amabilis* and *A. procera* with some other associated trees such as Douglas-fir and western hemlock. The climate is cool with heavy precipitation and considerable accumulation of winter snow; the snowpack in winter can be as much as 3 to 4 m deep.

The *Tsuga mertensiana* zone is the highest forested zone along the western slopes and crest of the Cascades. In southern Oregon, it gets to over 1,950 m in elevation where as around Mt. Rainier, its upper limits are near 1,930 m. The principal species is the mountain hemlock, which is also one of the characteristic subalpine trees of the Sierra Nevada. Associated with it are *Chamaecyparis nootkatensis* (the Alaska yellow cedar) and *Abies lasiocarpa* (subalpine fir). Also present near timberline is whitebark pine. This zone is very snowy with level snowpacks up to 6 to 8 m deep and with great drifts. Dropping down the eastern slopes of the Cascades, the climate rapidly becomes drier in the rainshadow of the mountains. The result is a sharp change in forest composition from the hemlocks and firs to *Larix occidentalis* (western larch) and ponderosa pine, and finally to sagebrush desert. The Olympic Mountains of northwestern Washington have similar but slightly different forest zones to those on the west sides of the Cascades (Fonda and Bliss 1969).

Conclusions

One can conclude, very simply, that the mountain conifer forests of the Appalachians, the Rocky Mountains, and even those of the Cascades have strong relationships with the spruce and fir forests of the transcontinental taiga of the North, and with each other. The genera and even closely related species seem to bear this out. However, the boreal forests and those of the northern parts of the Cordilleras have only very weak relationships with the forests of the Basin Ranges which, with the exception of the subalpine forests of the Ruby Mountains, are mainly related to the forests of the southern Rocky Mountains and the Mexican mountains. Similarly, the forests of the Sierra Nevada show little relationship to those farther east or farther north. The Sierran forests consist mostly of truly endemic tree species in a unique climate. This needs to be remembered in any understanding of their ecological problems.

References

Ager TA (1983) Holocene vegetational history of Alaska. pp. 128-141. In Wright HE, Jr. (editor), Late Quaternary Environments of the United States. Vol. 2, The Holocene. Univ. Minnesota Press, Minneapolis.

Arno, SF (1979) Forest regions of Montana. USDA Forest Service Research Paper INT-218. Ogden, Utah.

Arno SF (1984) Timberline: Mountain and Arctic Frontiers. The Mountaineers, Seattle, Washington.

Axelrod DI (1989) Age and origin of subalpine forests. (submitted).

Axelrod DI and Raven PH (1985) Origins of the Cordilleran flora. J. Biogeog. 12:21-47.

Barbour MG (1988) Californian upland forests and woodlands. In Barbour MG and Billings WD (editors). North American Terrestrial Vegetation. Cambridge Univ. Press, NY.

Barry RG (1973) A climatological transect on the east slope of the Front Range, Colorado. Arctic Alpine Res. 5:89-110.

Billings WD (1950) Vegetation and plant growth as affected by chemically altered rocks in the western Great Basin. Ecology 31:62-74.

Billings WD (1951) Vegetational zonation in the Great Basin of western North America. pp. 101-122. In Compt. Rend. du Colloque sur les Bases Ecologique de la Regeneration de la Vegetation des Zones Arides. IUBS, Paris.

Billings WD (1954) Temperature inversions in the pinyon-juniper zone of a Nevada mountain range. Butler Univ. Bot. Studies 11:112-118.

Billings WD (1963) Field notes, Yukon Territory, and Diary. Duke Univ., Durham, NC.

Billings WD (1969) Vegetational pattern near alpine timberline as affected by fire--snowdrift interactions. Vegetatio 19:192-207.

Billings WD (1973) Arctic and alpine vegetations: Similarities, differences, and susceptibility to disturbance. BioScience 23:697-704.

Billings WD (1974) Adaptations and origins of alpine plants. Arctic Alpine Res. 6:129-142.

Billings WD (1979) Alpine ecosystems of western North America. pp. 6-21. In Johnson DA (editor) Special Management Needs of Alpine Ecosystems. Soc. for Range Management, Denver, CO.

Billings WD (1989) *Bromus tectorum*, a biotic cause of ecosystem impoverishment in the Great Basin. Chap. 12. In Woodwell, GM (editor), The Earth in Transition: Patterns and Processes of Biotic Impoverishment. Cambridge Univ. Press, NY (in press).

Billings WD and Anderson LE (1966) Some microclimatic characteristics of habitats of endemic and disjunct bryophytes in the Southern Blue Ridge. Bryologist 69:76-95.

Billings WD and Mark AF (1957) Factors involved in the persistence of montane treeless balds. Ecology 38:140-142.

Christensen, NL et al. (1989) Interpreting the Yellowstone fires. BioScience 39:678-685.

Critchfield WB (1971) Profiles of California vegetation. U.S. Forest Service Research Paper PSW-76.

Daubenmire R (1969) Ecologic plant geography of the Pacific Northwest. Madroño 20:111-128.

DeLucia EH, Schlesinger WH, and Billings WD (1988) Water relations and the maintenance of Sierran conifers on hydrothermally altered rock. Ecology 69:303-311.

Densmore D (1980) Vegetation and forest dynamics of the upper Dietrich River valley, Alaska. M.Sc. thesis, North Carolina State Univ., Raleigh, NC.

Despain DG (1973) Vegetation of the Big Horn Mountains, Wyoming, in relation to substrate and climate. Ecological Monogr. 43:329-355.

Fonda RW and Bliss LC (1969) Forest vegetation of the montane and subalpine zones, Olympic Mountains, Washington. Ecological Monogr. 39:271-301.

Franklin JF and Dyrness CT (1969) Vegetation of Oregon and Washington. U.S. Forest Service Research Paper PNW-80.

Griffin JR and Critchfield WB (1972) The distribution of forest trees in California. U.S. Forest Service Research Paper PSW-82.

Griggs RF (1946) The timberlines of northern America and their interpretation. Ecology 27:275-289.

Jacobson GL Jr., Webb T III, and Grimm EC (1987) Patterns and rates of vegetation change during the deglaciation of eastern North America. pp. 277-288. In Ruddiman WF and Wright HE Jr., (editors) The Geology of North America, Vol. K-3, Geological Society of America, Boulder, CO.

Loope LL (1969) Subalpine and alpine vegetation of northeastern Nevada. Ph.D. Dissertation, Duke Univ., Durham, NC.

Mark AF (1958) The ecology of the Southern Appalachian grass balds. Ecological Monogr. 28:293-336.

Marr JW (1961) Ecosystems of the East Slope of the Front Range in Colorado. Univ. Colorado Studies, Series in Biology, No. 8, Boulder, CO.

McMillan C (1948) A taxonomic and ecological study of the flora of the Deep Creek Mountains of central western Utah. M.Sc. thesis, Univ. Utah, Salt Lake City, UT.

Merriam CH (1890) Results of a Biological Survey of the San Francisco Mountain Region and the Desert of the Little Colorado, Arizona. North American Fauna, No. 3, USDA, Washington, DC.

Oosting HJ and Billings WD (1939) Edapho-vegetational relations in Ravenel's Woods, a virgin hemlock forest near Highlands, North Carolina. Amer. Midl. Nat. 22:333-350.

Oosting HJ and Billings WD (1943) The red fir forest of the Sierra Nevada. Ecological Monogr. 13:259-274.

Oosting HJ and Billings WD (1951) A comparison of virgin spruce-fir forest in the northern and southern Appalachian system. Ecology 32:84-103.

Oosting HJ and Reed JF (1952) Virgin spruce-fir forest in the Medicine Bow Mountains, Wyoming. Ecological Monogr. 22:69-91.

Peet RK (1988) Forests of the Rocky Mountains. pp. 63-101. In Barbour MG and Billings WD (editors), North American Terrestrial Vegetation. Cambridge Univ. Press, NY.

Porsild AE (1951) Botany of Southeastern Yukon Adjacent to the Canol Road. National Museum of Canada Bulletin No. 121.

Reed JF (1952) The vegetation of the Jackson Hole Wildlife Park, Wyoming. Amer. Midl. Nat. 48:700-729.

Reed RM (1976) Coniferous forest habitat types of the Wind River Mountains, Wyoming. Amer. Midl. Nat. 95:159-173.

Rowe JS (1959) Forest Regions of Canada. Department of Northern Affairs and National Resources, Forestry Branch, Bulletin 123, Ottawa.

Siccama TG, Bliss M, and Vogelmann HW (1982) Decline of red spruce in the Green Mountains of Vermont. Bull. Torrey Bot. Club 109:163-168.

Talley SN (1974) The ecology of Santa Lucia fir (*Abies bracteata*), a narrow endemic of California. Ph.D. dissertation, Duke Univ., Durham, NC.

Thornthwaite CW (1941) Atlas of climatic types in the United States 1900-1939. USDA Miscellaneous Publication 421. Washington, DC.

Chapter 4

Ecophysiology of High Elevation Forests

William K. Smith and Alan K. Knapp

Introduction

The high elevation forests surrounding the Great Basin of the western United States represent the most extreme cold-temperature environments of the area. Summer periods with non-freezing temperatures may be less than 50 days and, as described in Chapter 2, annual precipitation occurs primarily as winter snowfall. Because these forests are dominated by evergreen conifers, both summer drought, and winter drought caused by frozen soil and plant water, along with strong evaporative demand, may be especially important. In addition, potentially large increases in instantaneous solar irradiance at high elevations coupled with relatively tall tree height and open spacing, may lead to extreme variations in sunlight penetration within the forest community (Plate 9.1). Frequent midday and afternoon clouds, throughout the summer generate large variations in sunlight even in open habitats. This variability in sunlight contributes to similarly extreme variation in temperature, water and photosynthetic relations. Thus, the combination of cold temperatures, the potential for summer and winter drought, and extremes in sunlight provide one of the most challenging and variable environments within the Great Basin complex, or elsewhere. Moreover, the extreme upper boundary for forest tree growth occurs at alpine timberline in these high elevation forests. It is within extreme environments that ecophysiologists traditionally have searched for adaptations that enable plant survival. Many would argue that an understanding of adaptations to the extreme ends of environmental gradients provides the most informative perspective of evolutionary processes.

Autecological studies of species growing in the high elevation forests surrounding the Great Basin are skewed towards economically important species. For example, the majority of studies from the Cascade Range of the Pacific Northwest focus on species such as Douglas fir, *Pseudotsuga menziesii* (Lassoie et al. 1983). However, most Cascade forests are characterized by having particularly mild winters and higher annual precipitation relative to Sierra Nevada and, especially, Rocky Mountain forests. As much as 70% of the annual carbon gain of *P. menziesii* in the Cascades may come from non-summer growth compared to 0% for many Sierra Nevada and Rocky Mountain species (Grier et al. 1986; Emmingham and Waring 1977).

Little comprehensive autecological information is available for conifer species of the Rocky Mountains and even less is known about the relatively diverse species of the Sierra Nevada. The purpose of this chapter is to synthesize research concerning ecophysiological adaptations to the extreme conditions found in this high elevation environment. Both coniferous and nonconiferous species will be considered, including the forest understory community. An emphasis will be placed on identifying important limitations to photosynthetic carbon assimilation associated with the high elevation environment of the Rocky Mountains.

The Physical Environment

The wide range in elevation found within the Great Basin region, plus the primarily north-south orientation of the cordilleras, result in strong orographic effects on precipitation during winter and summer (Chapter 2). West-facing slopes receive considerably more winter snowfall and, thus total annual precipitation. Superimposed on these macroclimatic influences are the orientational effects on solar exposure due to a predominance of east or west exposures. During summer, morning sunlight will first heat eastern slopes when air temperatures are lowest and wind generally is reduced. Sunlight will strike western slopes more directly when air temperatures are higher in the afternoon, although windspeeds also are characteristically higher. Thus, in a general sense, plants in these forests will tend to have leaf temperatures further above air temperature in the morning on eastern slopes whereas western slope species will be warmer in the afternoon, but closer to air temperature. Thus, this north-south orientation may result in the relative magnitude of the differences in absolute leaf temperatures among trees being less than for an east-west-oriented cordillera. Evaporative demand could be enhanced on western slopes due to high insolation when air temperatures are warmer, although greater leaf-to-air temperature differences on eastern slopes in the morning also could lead to high evaporative demand. In most ranges soil moisture may be greater on the western slopes due to the dominant contribution of Pacific air masses to annual precipitation. This is especially true for the Cascade and Sierra Nevada ranges. A reverse situation often occurs in the southern Rocky Mountains due to the influence of southern fall and spring storms that bring moisture from the Gulf of California and Mexico.

Despite the possibility of potentially strong orientational and orographic effects on temperature and moisture, there are numerous elevational and microclimatic factors that can modify the actual conditions experienced by individual plants. These factors include influences due to microtopography and vegetation structure and several physical properties associated with lower air density and ambient pressure at higher elevations. Ultimately, microclimate potentially could overshadow macroclimatic and elevational effects at any elevation. Associating only macroscale climate phenomena with predictions regarding plant growth and distribution pattern can be quite misleading (Daubenmire 1978), especially when assumptions regarding limitations to physiological responses are often implicit.

As an example, one might suppose that atmospheric evaporative demand in deserts where summer air temperatures often exceed 40° C would be greater than in alpine zones where summer air temperatures rarely exceed 20° C due to the well-known decrease in temperature with elevation. In both cases, atmospheric relative humidity could be typically 20%. However, because of greater solar energy inputs at high elevation and the ease with which temperatures of leaves exposed to direct sunlight can exceed air temperature (Gates 1980), alpine plants can be typically much further above air temperature than desert plants and, thus, the leaf-to-air vapor difference, or evaporative demand, is surprisingly similar (Smith and Geller 1979). Transpiration is itself further enhanced at high elevation because mass diffusion rates in the gas phase increase substantially at elevations typical of subalpine/alpine zones of the Rocky Mountains. The observation that alpine plants are often cushion-like in form and strongly coupled to soil-heating effects further enhances this uncoupling from air temperature, a primary factor for evaluating evaporative demand. Although the greater air temperature lapse rate in drier mountain ranges may act to reduce diffusion rates somewhat, any increase in leaf temperatures would increase diffusion rates even further (Smith and Geller 1979).

Of course, the cushion-like growth form mentioned above could be important also for overcoming cool air temperatures that may limit carbon gain during the short summer growth season. A close proximity to the ground is also crucial for avoiding winter snow abrasion and for conserving tissue water via snow burial during potentially severe winter periods of desiccation (Hadley and Smith 1986; 1987). Even though snowfall and total annual precipitation may be much greater at higher elevations, the combination of extreme snow blowing and drifting and the possibility of both summer and winter periods of low moisture availability (due to poorly developed soils) produces a potentially, quite xeric environment over a microtopographic scale (Oberbauer and Billings 1986). The lesson of this rather cursory example is that macroscale differences in temperature and precipitation patterns may reveal little about the actual microclimatic conditions limiting plant growth and survival, especially at high elevation. Comparisons of the desiccating potential of alpine versus desert ecosystems represents only one example of the lack of resolution brought about by considering only macroscale climate differences.

Studies of biophysical and physiological effects of high elevation on plant growth are seldom comprehensive. The potential complexity of these effects is illustrated in Figure 4.1 using transpiration and photosynthesis as example physiological processes. Air temperatures are known to decline with elevation according to lapse conditions that are dependent primarily on the moisture content of the air, i.e., decreases in air temperature in drier, continental mountains can approach ideal, dry adiabatic conditions (0.01° C m^{-1}), while coastal mountains with more humid air frequently have much lower lapse rates (ca. 0.003° C m^{-1}). Substantial increases in maximum solar irradiance are also possible (> 40% increase above sea level values at about 3,500 m elevation; Smith and Geller 1979). The partial pressures of important gasses such as CO_2 and O_2 decline substantially with elevation in accord with decreases in atmospheric pressure (Figure 4.1). Similar decreases in H_2O vapor due to pressure are not expected because air temperature primarily influences the saturation kinetics of air humidity. Thus, the cooler air temperatures characteristic of higher elevations can be expected to result in lower ambient vapor pressures (at the same relative humidities) and higher transpiration rates only if all other factors influencing evapotranspiration are identical (Figure 4.2). Thus, the tendency

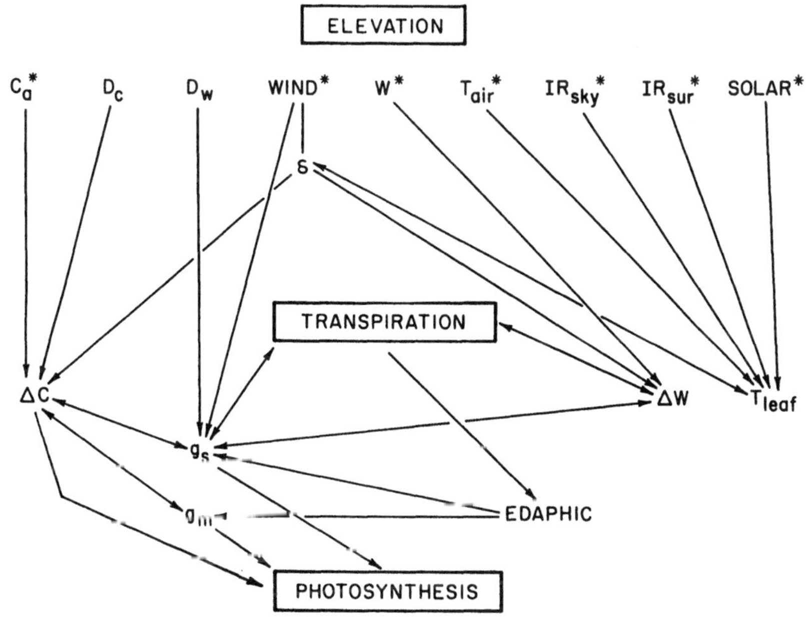

Figure 4.1. Schematic relationship of the physio-chemical environmental factors that can be influenced strongly by increases in elevation associated with high elevation forests. Variables with an asterisk are particularly susceptible to microclimatic influences that could overshadow changes due to elevation. C_a, CO_2 partial pressure of the air; D_c, diffusion coefficient for CO_2 in air; D_w, for H_2O vapor in air; WIND, wind speed; W, water vapor pressure of air; T_{air}, air temperature; IR_{sky}, infrared radiation from the sky; IR_{sur}, from the surroundings; SOLAR, solar irradiance; BL, leaf boundary layer thickness; ΔC, leaf-to-air difference in CO_2; ΔW, in water vapor partial pressure; T_{leaf}, leaf temperature; g_s, stomatal conductance; g_m, mesophyll conductance of CO_2; EDAPHIC, soil factors.

to higher leaf-to-air vapor pressure differences would be greater at higher elevations simply due to lower air temperatures. Consequently, condensation events like dewfall, ground fog, and cloud formation are also more likely at higher elevations. Ultimately, the sources of water vapor and the biophysical forces influencing mass mixing and exchange processes will most likely determine absolute humidity levels at any elevation.

Another feature of high elevation is the potential for leaf temperatures to decrease well below air temperature at night due to a high outward flux of longwave radiation (Figure 4.1). Reductions in leaf temperature

Table 4.1. Leaf surface water repellency (contact angle,°) stomatal frequency, and photosynthetic carbon uptake for alpine and subalpine species of the central Rocky Mountains. Species are grouped for comparison according to the surface water repellency (wettable <110 °, or nonwettable, >130 °) of the leaf side with the most stomata. Experimentally wetted plants in the field initially were measured dry and then following spray misting; naturally wet plants initially were measured with natural dew deposition and then after careful blotting with tissue paper until dry. Photosynthesis was measured only for surfaces with the higher stomatal frequency. The greatest standard errors within each group, 3 replicates from each of 3 leaves from a single plant, are shown. All other values are means of at least two measurements on each of 3 different leaves per species. Percentage differences in photosynthesis (Δ) between dry and wet leaves were significant for all categories at α = 0.05 (ANOVA, paired t-test) (Smith and McClean 1989).

		Contact Angle (°)	Stomatal Frequency (mm^2)	Photosynthesis ($\mu mol\ CO_2\ m^{-2}\ s^{-1}$)					
				Natural			Experimental		
	Surface and Species	Adax/Abax	Adax/Abax	Dry	Dew	Δ (%)	Dry	Wetted	Δ (%)
Nonwettable	*Arnica cordifolia*	74/140	7/56	4.1	5.0	22.0	7.3	9.8	34.2
	Arnica latifolia	70±12/144	12±4/69	4.8	4.6	-4.2	4.1	4.0	-2.4
	Berberis repens	66/150±24	0/283±32	4.8	5.3	10.4	5.7	5.5	-3.5
	Fragaria virginiana	55/133	0/210	4.6	5.6	21.7	4.9	6.1	24.5
	Acer glabrum	67/168	0/223	3.9	4.7	20.5 [1]	5.3	6.4	20.8 [2]
Wettable	*Pyrola secunda*	76/68	0/154±29	3.1	0.6	-64.5	—	—	—
	Caltha leptosepala	77/67	48±11/164	3.9	0.8	-79.5	3.6	0.8	-77.8
	Senecio crassulus	55/83	42/58	3.9	0.6	-84.6	3.2	0.3	-90.6
	Vaccinium scoparium	98/108±21	0/89	2.8	1.0	-64.3	—	—	—
	Rumex densiflorus	66/52	46/91	4.6	0.5	-89.1	5.2	0.6	-88.5
	Actaea rubra	101±14/103	0/39	4.7	1.4	-70.2	4.9	0.9	-81.6
	Ribes montigenum	30/30	0/144	5.4	0.6	-88.9 [3]	4.0	0.2	-90.5 [4]

[1-4] Mean differences 14.1, 14.2, -77.3, and -85.8, respectively.

below air temperature will be particularly pronounced close to the ground or in other wind sheltered situations such as within forest gaps. In combination with the potential for higher absolute humidity next to the ground, dew and frost formation may be especially common. Earlier frost formation on exposed plants appears to result in a significant shortening of the summer growth period (W.K. Smith, personal observation). Also, a recent study reported a high frequency of summer dewfall in the subalpine and several important physiological consequences for wet leaves (Smith and McClean 1989). Evidence for an evolutionary relationship between leaf water repellency, stomatal distribution pattern, and photosynthetic capabilities was presented (Table 4.1). Gas exchange was influenced strongly by leaf surface wetness caused by dewfall for a major portion of a day, especially the important early morning hours when plant water stress was lowest. As a result, leaf water repellency insures that a photosynthesis-inhibiting water film does not form over stomata (Smith and McClean 1989). Estimated influences on total daily carbon gain amounted to over 30% in some species with more wettable leaf surfaces. The influence of dew formation or other common leaf-wetting events has not been considered previously as potentially significant ecophysiological factors.

Figure 4.3 summarizes the current, although incomplete, view of possible interactions among plant and environmental variables that have been linked to elevational effects on photosynthesis. Ambient CO_2 levels decrease from near 340 μbar partial pressure at sea level to about 230 μbar at 4 km elevation. This decline in partial pressure could be expected to create substantial reductions in the capability for photosynthetic CO_2 uptake in high-elevation species. However, the diffusion coefficient for CO_2 in air increases by a similar proportion over this elevational increase, nearly compensating for lower partial pressures (Gale 1972, 1973). An effect on stomatal gas exchange could be expected due to interactive effects of the diffusion coefficient for CO_2, leaf temperature effects on this coefficient, and the assimilation capability of mesophyll cells (Figures 4.2, 4.3). In contrast to effects on transpiration, the absolute value of leaf temperature is of primary concern for CO_2 diffusion, not the difference between leaf temperature and air temperature.

Little is known about the specific components of microclimate generated by vegetation and topography in these forests. Yet, the potential alteration of macro- and mesoscale climate due to vegetation structure and relatively radical changes in microtopography that occur within mountain ecosystems is probably large relative to many other

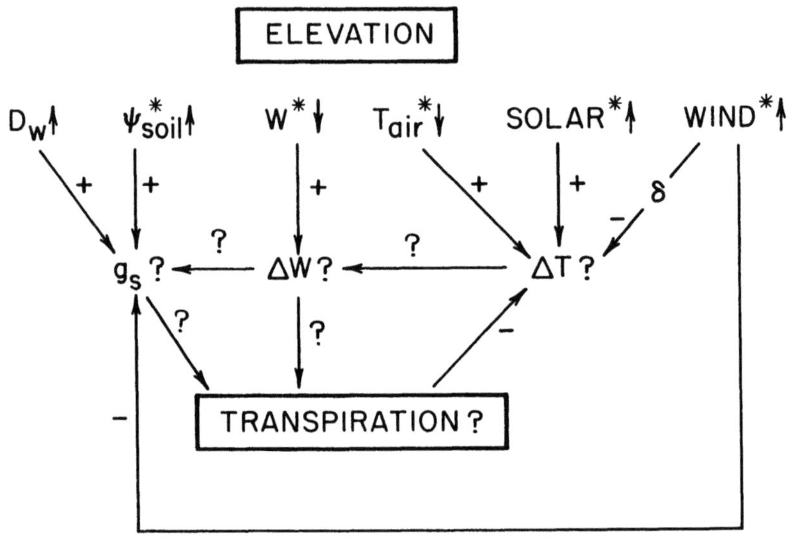

Figure 4.2. Proposed effects of elevation on primary factors influencing transpiration. ψ_{soil}, soil water potential, other parameters as defined in Figure 4.1. Vertical arrows indicate increasing effect due to greater elevation; plus and minus values indicate an increasing versus decreasing effect, respectively. Question marks indicate unknown effects of elevation.

ecosystems. The possibility exists that the elevational effects illustrated in Figures 4.1-4.3 could be entirely overshadowed by microclimatic effects due to vegetation structure and/or topography. A distinctive feature of these subalpine forests is the inhibition of wind penetration even though sunlight penetration may remain high due to the relatively tall stature of coniferous trees and their wide spacing (Plates 9.1, 9.2). This combination can result in a highly variable light microenvironment for adult trees as well as tree seedlings and other understory vegetation. A recent review of the ecophysiological significance of sunflecks in different forest types points out some of the unique aspects of sunlight penetration in coniferous forests compared to other forest types (Chazdon 1988). Young and Smith (1982) reported that longer-term, high irradiance sunlight periods ("sunpatches") dominated the understory solar regime in a subalpine spruce-fir forest (Plate 9.2) in contrast to the abundant, but short duration and low irradiance sunflecks of tropical forests. Also, cloudy days resulted in relatively high diffuse irradiance levels that were more uniform throughout microsites within the

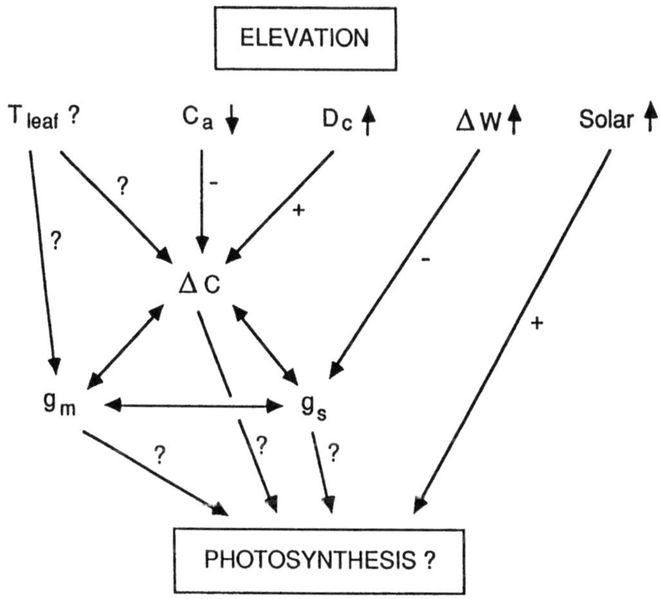

Figure 4.3. Proposed effects of elevation on primary atmospheric factors influencing photosynthesis. C_i, intercellular CO_2 partial pressure, other parameters as defined in Figure 4.1. Only D_c and C_a are known to increase and decrease, respectively, with increasing elevation. Contrasting changes in g_s and g_m with elevation have been reported, although a strong interaction between g_s, g_m, and C_i appears probable.

understory (Young and Smith 1983). Smith et al. (1989) also point out that the characteristics of sunlight penetration in these forests are based on the height and spacing of individual trees. Sunpatches are composed of both a full-sun center portion (numbra) and a shadow-edge (penumbra) of significantly reduced irradiance. Irradiance level and duration of individual sunpatches are determined by the height of the trees and the size of canopy openings. As the height of the trees increases or the canopy opening size decreases, the ratio of penumbra to numbra increases as does penumbral "spreading". The dominance of sunpatches and low windspeeds results in substantial periods of high insolation, extreme leaf temperatures and water stress alternating with prolonged periods of low irradiance with minimal water demand. Corresponding extremes in photosynthetic carbon gain and stomatal behavior are also possible (Young and Smith 1980, 1982). Additional information on microclimatic characteristics such as wind, humidity, and CO_2 levels is provided in recent reviews (Smith 1985a; Lassoie et al.

1983). Further discussion of the physiological influence of this type of variable sunlight regime is provided in the following sections.

These high elevation forests end at the uppermost boundary of conifer tree growth, the upper timberline. This alpine timberline (more aptly described as the timberline ecotone or "Kampfzone"; Tranquillini 1976) represents a vivid display of marginal growth and survival conditions for coniferous trees (Plate 10.1). The only forest trees commonly surviving under these extreme environmental constraints are those species capable of forming the distinct Krummholz ("twisted wood"; Tranquillini 1976) growth habit. This division between the alpine/subalpine zones provides a unique opportunity to study extreme limits of forest tree survival and the corresponding, minimum physiological requirements for growth and/or reproduction.

Recent studies of the physical environment of the alpine timberline ecotone revealed that blowing snow and snow accumulation are crucial features influencing shoot survival at these extreme upper elevations (Plate 10.2; Hadley and Smith 1987). Daily maximum winds can average over 15 m s^{-1} for most of the winter and even greater than 8 m s^{-1} for summer. These winds, plus air temperatures frequently below -40° C, generate extremely hard and abrasive ice crystals next to the snow surface (Plate 10.1, Hadley and Smith 1989a). The resulting damage to the leaf cuticle and ensuing leaf desiccation leads to severe leaf mortality of exposed leaves. The concomitant occurrence of frozen soils and vascular tissues in plants exacerbates desiccation. At the uppermost portion of the timberline ecotone, extremely cold temperatures may be responsible for needle mortality due to desiccation even without wind abrasion, possibly the result of a shortened summer growth period and inadequate cuticle development (Tranquillini 1976; Hadley and Smith 1986). Thus, both cold temperature and drought appear to be primary factors associated with plant growth and winter survival at high elevation.

Although the alpine timberline is an extreme microclimatic situation, the potential importance of winter stress on the growth and survival of forest trees may be a somewhat neglected ecological factor (Tranquillini 1982; Sakai and Larcher 1987). Recently, Grier (1988) measured large foliar losses (up to 42% of the total foliar biomass) during winter within 12 coniferous forests in the western U.S. He attributed this loss to wind, mechanical damage due to snow deposition, and leaf desiccation. Hadley and Smith (1989b) found that resistance to winter desiccation

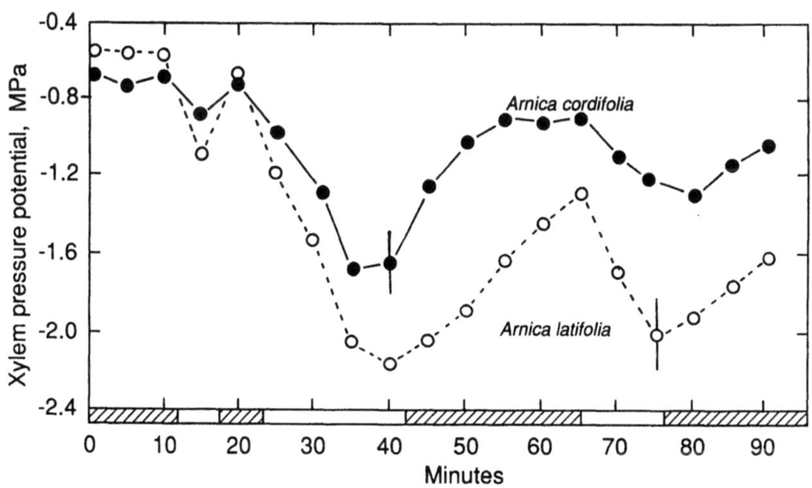

Figure 4.4. Changes in xylem pressure potential in two sympatric understory congeners, *Arnica cordifolia* and *A. latifolia*, during exposure to full sun and 15-20% of full sun (hatched bars) in a subalpine understory environment. Vertical bars represent the greatest 95% confidence intervals.

among six conifer species of the central Rocky Mountains corresponded to their relative drought tolerances ascribed to summer conditions. In general, a greater amount of epicuticular wax and lower cuticular conductance to water vapor, plus a lower ratio of leaf area to saturated leaf water content was associated with greater tolerance to winter desiccation. Thus, the similarity between winter desiccation tolerance and summer drought tolerance reflects a need for a more precise understanding of the relative ecological importance of winter versus summer water relations in these conifer species (Hadley and Smith 1989b).

Another dramatic illustration of the potential importance of temperature and water relations during summer within these forests is the now well-documented occurrence of water stress in understory shrub and herbaceous species (Young and Smith 1979, 1980, 1982; Smith 1981; Knapp and Smith 1989). These species are characterized by rapid decreases in xylem pressure potential of over 1.5 kPa s^{-1} and severe wilting when exposed to full sun (Plate 9.2; Figure 4.4). Periods of intermittent shade due to clouds or overstory shading appear crucial for maintaining high photosynthetic rates in full sun (Knapp et al. 1989). Experimentally extending natural sun periods by as much as 30% resulted in irreversible leaf necrosis (Young and Smith 1979). Leaf necrosis has been reported previously by Hadfield (1975) in *Camellia* and Rackham (1974) for understory species exposed to natural sunflecks in a deciduous forest.

One might consider the timberline ecotone as representing one extreme end of a microhabitat spectrum within the subalpine zone while the forest understory represents the other. At timberline, snow accumulation is critical for avoiding severe cold, abrasion, and desiccation and insuring survival during winter (Plate 10.1). In the understory, wind-sheltering results in a deep and seasonally prolonged snowpack due to an abundance of shade. At timberline, too little snow accumulation can result in severe damage and, possibly, greater soil drought in summer. In the understory, too much snow can shorten the growth season, although soil water should be more abundant (Billings and Bliss 1959). In several ways a timberline Krummholz mat reflects a microscale version of these comparisons. On the windward side of the mat very little snow accumulates and severe needle abrasion and desiccation occurs. On the leeward side, where the mat is expanding, less needle mortality occurs, but a limitation to growth results due to snow mold (e.g., *Herpatrichia nigra*) that infects branches that remain snow-covered well into early summer in deeper leeward snowbanks. These microclimatic effects due to whole-plant and/or stand structure have strong ecophysiological implications that easily could overshadow elevational effects even at the highest subalpine boundary.

In summary, the forests surrounding the Great Basin provide an opportunity to study the dominating effects of an extreme physical environment associated with high elevation, continental mountains of the temperate zone. Sharp distributional boundaries often occur at the uppermost limits of these forests, as well as within, providing a natural laboratory for studying physiological/morphological adaptations. Possible elevational effects such as increased sunlight, leaf temperatures, the leaf-to-air temperature and vapor differences, and increases in the rate of water vapor diffusion, plus potentially strong microclimatic effects due to topography and vegetation structure, collectively provide a stressful, yet highly diverse physical environment. Increased cloudiness and lower partial pressures of CO_2 also may impose limitations that could be particularly important for understanding photosynthetic effects at high elevations. A more definitive understanding of these variables, coupled with evaluations of the physiological capabilities of different species, is needed before a more comprehensive understanding of adaptive mechanisms will be possible. Conclusions regarding elevational zonation patterns, geographic distribution, and succession will be possible only after differences in physiological capabilities are better understood.

The following sections attempt to synthesize our current understanding of the important physiological adaptations of these high elevation species. A common theme throughout this chapter, and elsewhere in the field of ecophysiology, is the need for more integrative studies which consider interactive effects of more than a single variable (Mooney et al. 1987). However, our approach will be to identify environmental factors that appear to be, quantitatively, the greatest limitations to carbon gain at a certain time of year. We feel this approach is worthwhile in that predictions can be made concerning the most advantageous evolutionary outcomes of the past and future. A somewhat similar approach has been used to compare C_3 and C_4 species (see Table 7.1 in Schulze and Hall 1982) and forest species (Landsberg 1986).

Environmental Limitations

Identification of patterns in physiological responses to the environment provides evidence for major adaptive avenues. Specific patterns of physiological response often can be associated with specific environmental variables and/or different plant types. Moreover, physiological differences among plant types might be linked to phylogenetic relationships or, possibly, functional differences in plant form, e.g., trees versus shrubs versus herbs. It is these latter relationships between form and function that may reflect evolutionary convergence in adaptive morphology or physiology to some fundamental limitation to fitness. With regard to the high-elevation forests considered here, temperature and water appear to be sometimes independent but, more often, they are tightly coupled factors that appear as dominant limitations to growth and reproduction (Smith 1985b; Lassoie et al. 1983; Luxmore et al. 1986). However, sunlight availability can strongly influence this coupling of temperature and water relations, as can numerous morphological/physiological factors. Thus, of the four basic environmental factors influencing physiology (temperature, water, sunlight, and nutrients), only nutrients appear to be of lesser importance in these forests. Admittedly, this conclusion result of the limited amount of information available concerning nutrient limitations in these species.

Table 4.2. Hypothetical interaction between plant and environmental water parameters within the spruce-fir zone during the growth season. Double asterisks indicate a primary limiting effect, either directly on stomatal opening (S, stomatal) or indirectly through limitations on water uptake from soil and/or movement to the leaves (R, recharge). Single asterisks denote important, but less significant effects (from Smith 1985b).

	Late Spring		Early Summer		Midsummer		Late Summer		Fall	
	S	R	S	R	S	R	S	R	S	R
Air temperature	**	*		*					**	*
Soil temperature[1]		*		**		*				*
Xylem pressure potential										
morning	*		*				*			
afternoon					*		**			
Soil water potential[1]								**	*	
LAVD [2]			*		*		*			

[1] In root zone.
[2] Leaf-to-air water vapor deficit.

Investigators comparing a number of conifer forest ecosystems have recognized for some time that high-elevation, continental forests probably experience the most extreme environmental limitations to conifer tree growth (Emmingham 1982; Waring and Schlesinger 1985). Winter photosynthesis approaches zero in these forests and low winter snow accumulation coupled with dry summer conditions can result in decreases in potential summer carbon gain of over 50%. As a result, annual production is characteristically less than 0.8 kg m^{-2} yr^{-1}, aboveground biomass is below 20 kg m^{-2}, and trees are smaller than those in coastal ranges, even within mature climax forests (Grier et al. 1986). This low potential for productivity leads to the relative openness of the canopies and wide spacing between trees which are unique characteristics of these high-elevation, subalpine forests surrounding the Great Basin (Smith 1985a).

Recent studies have focused on qualifying and quantifying specific environmental factors that may limit growth in these coniferous forests. One study proposed that the summer growth period can vary considerably from year-to-year due to the interaction of specific environmental factors that may become primary limitations at different times during

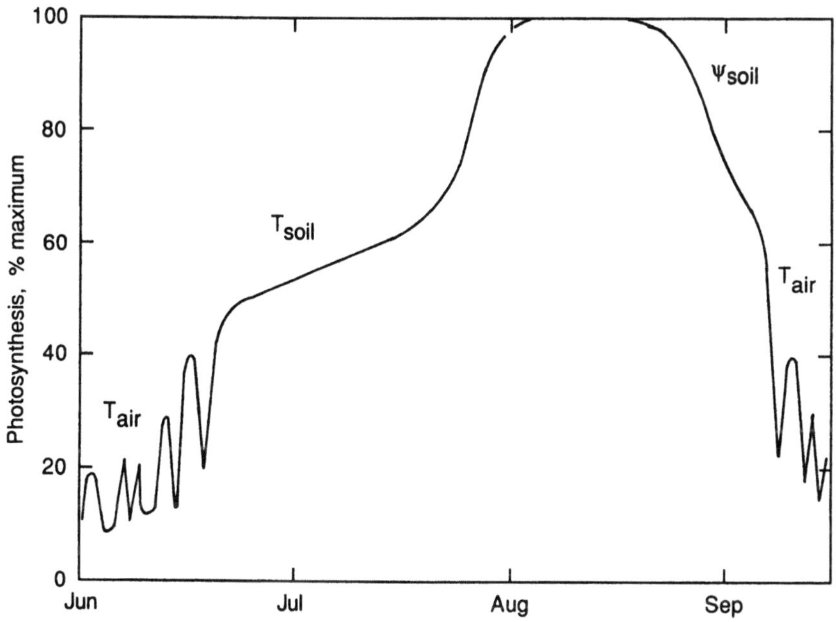

Figure 4.5. Idealized changes in photosynthetic capacity in a high-elevation forest based on field measurements on coniferous species. Freezing air temperatures (T_{air}) initially create a highly variable photosynthetic period in early summer followed by a persistent cold soil temperature (T_{soil}) limitation. A period of maximum capacity can be terminated by soil drought (ψ_{soil}), with the return of freezing air temperatures ending the summer period of carbon gain. Depending on elevation and year-to-year variation, the duration and amplitude of these periods may vary.

summer (Smith 1985b; Table 4.2). Although this study focused on summer variations in leaf conductance to water vapor (i.e., the degree of stomatal opening), similar interpretations could be made concerning photosynthetic CO_2 uptake, given the usual linear relationship between photosynthetic CO_2 assimilation and stomatal conductance (Schulze and Hall 1982). It was proposed that stomatal opening in early summer in the subalpine forests of southeastern Wyoming initially was limited by below-freezing air temperatures at night (Figure 4.5). These cold nighttime temperatures may persist throughout much of early summer, even after daytime air temperature becomes adequate for stomatal opening and photosynthesis (Figure 4.5). A significant period of growth limitation due to soil temperatures below about 8° C has also been documented, both during the early period of air temperature limitation as

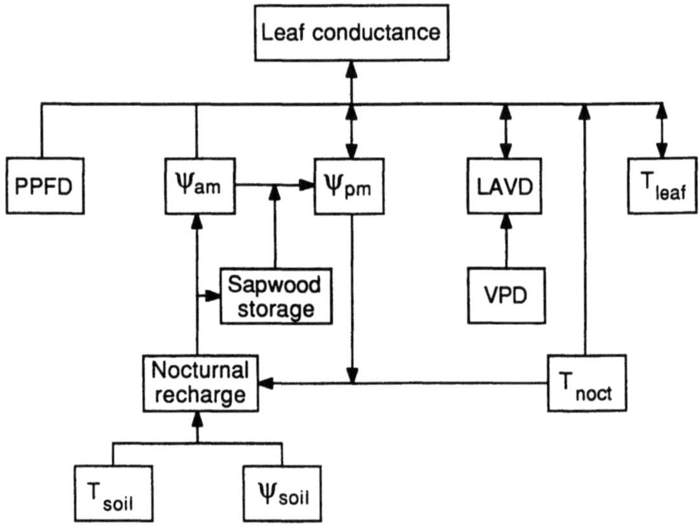

Figure 4.6. Schematic of the relationship between the potential interaction of cold temperature limitations with water-relation limitations (from Smith et al. 1984). Symbols are as follows: PPFD, photosynthetic photon flux density; ψ_{am}, morning xylem pressure potential; ψ_{pm}, afternoon xylem pressure potential; LAVD, leaf-to-air water vapor difference; T_{leaf}, leaf temperature; Sapwood storage, water storage capacity of plant; VPD, air vapor pressure deficit; Nocturnal recharge, uptake of water at night; T_{noct}, minimum nighttime air temperature; T_{soil}, soil temperature; ψ_{soil}, soil water potential.

well as later in the summer (Smith 1985b; DeLucia 1986; DeLucia and Smith 1987). As midsummer approaches, maximum stomatal opening occurs and some plateau of maximum stomatal opening, without cold temperature limitation, becomes possible (Figure 4.5). This maximum plateau in photosynthetic capacity is probably constrained by plant nutrient concentration, particularly N (Schulze and Hall 1982; Field and Mooney 1986). Short-term reductions in CO_2 uptake on a daily basis may result from stomatal responses to the leaf-to-air water vapor difference (LAVD). However, the water conservation that results from these daily savings in transpirational water loss could extend photosynthetic carbon gain into drier periods of summer. Limitations to CO_2 uptake in late summer and early autumn could consist of either soil water drought, or the return of freezing nights. At the highest elevations near timberline, freezing nights can return early in August before soil water stress develops, where as at lower elevations soil drought can persist for several weeks before subfreezing air temperatures return. Thus, the primary deterrents to stomatal opening and photosynthetic carbon gain may include a successive progression of low air and soil temperatures, and low soil water, plus a more persistent limitation due to LAVD (Figure 4.6 and Table 4.2). These variables may interact to varying degrees depending on elevation and time of year. A more mechanistic explanation of these effects has been provided which describes autumn stomatal closure in subalpine conifers based on interactions between cold temperature effects on stomatal behavior and plant water status (Figure 4.6).

Recent work on conifers, as well as other species in the subalpine, also revealed a potentially strong limitation to summer carbon gain due to the common occurrence of shade generated by such factors as shoot, canopy and stand architecture, plus cloudcover. Reviews by both Lassoie et al. (1983) and Waring and Schlesinger (1985), dealing with forest ecology and ecophysiology point out the importance of shading to leaf development and photosynthesis, as well as the important influence of canopy and stand structure on limiting annual net production, especially in mature forests. During most afternoon periods, predictable cloud patterns may generate major decreases in incident sunlight in the subalpine (Knapp and Smith 1987). Moreover, the combination of rapid full sun/shade cycles and relatively slow stomatal response patterns of these species may result in substantial periods of non-steady-state physiological responses (Knapp and Smith 1987, 1988). These non-steady-state periods are especially evident for stomatal behavior because stomatal apertures may change more slowly than natural variations in sunlight. Thus, not only are decreases in total sunlight important, but the nature of these sunlight fluctuations may involve a totally different physiological response pattern than traditionally interpreted for steady-state responses (Knapp and Smith 1989). Knapp et al. (1989) point out the importance of intermittent shade to the ecophysiology of certain subalpine species. Optimal photosynthetic performance in full sun was highly dependent on the periodic occurrence of shade intervals. Thus, a dynamic sunlight regime can be included with air and soil temperatures, air humidity and soil water as major environmental factors influencing annual carbon gain in these high elevation forest species. Although these environmental variables may interact, their specific effects on physiological performance can often be traced to a single, critical factor. For example, wind is of crucial importance at timberline where conifers may suffer high leaf mortality when leaves are not buried beneath snow in winter (Larcher and Siegwolf 1985; Hadley and Smith

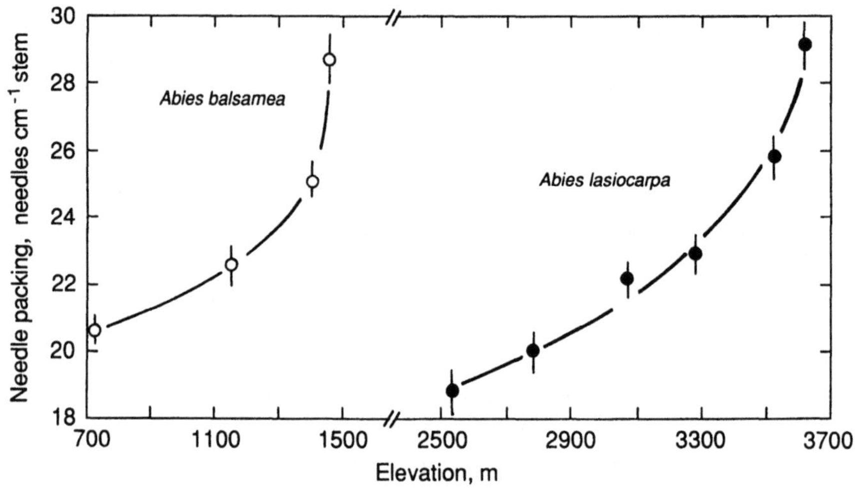

Figure 4.7. Changes in the number of leaves (needles) per stem length with elevation for Balsam fir (*Abies balsamea*) from the White Mountains of the northeastern U.S. and subalpine fir (*A. lasiocarpa*) from the central Rocky Mountains of the western U.S. All shoots sampled were sun-exposed and collected from within the subalpine forest (except for the highest elevations; 1,455 m; 3,540 m; 3,615 m), which were from timberline Krummholz trees. Data for Balsam fir taken from DeLucia and Berlyn (1984). Vertical bars are ± 2 SEM.

1986, 1987). The importance of prolonged snowpack (Billings 1969), or even the occurrence of natural leaf-wetting events such as dewfall (Smith and McClean 1989), are other factors that may dominate CO_2 exchange characteristics for substantial periods.

Elevational Effects on Morphology and Growth Form

Numerous structural changes have been observed at the leaf, shoot, whole-tree, and stand level for high elevation species. These appear to have potentially strong influences on physiological capabilities related to photosynthetic carbon gain. Increases in specific leaf area and greater palisade and leaf thickness with elevation have been documented for numerous species (Woodward 1986; Körner and Diemer 1987), although a decline in specific leaf area was also reported by Woodward (1983). Corresponding increases in leaf nitrogen have also occurred with increases in specific leaf area and elevation (Woodward 1986). As mentioned previously, these morphological changes of individual leaves have important implications for increases in the capacity of mesophyll cells for CO_2 assimilation (Nobel and Walker 1985). Most of these morphological changes have been linked previously to specific environmental factors related to sun/shade leaf development such as the increase in mesophyll cell area to leaf surface area with increasing sunlight during leaf development (Nobel et al. 1975; Nobel and Walker 1985). Other factors such as temperature and moisture appear to have a strong interactive influence on both internal and external leaf morphology (Smith and Nobel 1978a, b). Differences in leaf morphology for coniferous tree species has become more evident (Leverenz and Jarvis 1980). More work is needed before a synthetic understanding of the influence of elevation on leaf structure and the functional importance to carbon gain in high elevation species will be possible. Surprisingly, little information exists for the conifer tree species of the subalpine zone.

At the shoot level, there are data (Figure 4.7) that show rather dramatic increases in the number of leaves per unit stem length in conifers with elevation (Zelawski and Niwinski 1966; DeLucia and Berlyn 1984; Smith and Carter 1988). The functional significance of increased needle clustering could result from increased needle warming and, thus, carbon gain in a cold-temperature-limited system (Smith and Carter 1988). Dramatic changes in individual leaf (needle) orientation on shoots can occur, resulting in horizontal, nonoverlapping needles on shade shoots and more upright, clustered leaves on sun shoots (Smith and Carter 1988). It appears as though the degree of sun/shade exposure at any elevation plays a dominant role in determining shoot structure. However, the most extreme clustering of leaves on individual shoots occurs in the Krummholz growth forms at timberline (Hadley and Smith 1987; Figure 4.7). Here, the adaptive role of structure and morphology at three levels (leaf, shoot, whole-tree) is crucial to the survival of these forest trees in such a marginal environment. A similar structural change at high elevation is observed in the many alpine tundra species with cushion-like growth forms (Bliss 1956; Billings and Mooney 1968). At extreme high elevations, a low stature and/or effective capture of snow (to insure burial) is essential for surviving winter snow abrasion and desiccation (Hadley and Smith 1987). Tight leaf packing also enhances survival of individual leaves on the leeward sides of shoots, contributing to the familiar "flagged" appearance of timberline trees (Tranquillini 1979).

Stand-level changes are also noticeable at higher elevations within subalpine forests. For example, the well-known ribbon-like appearance of

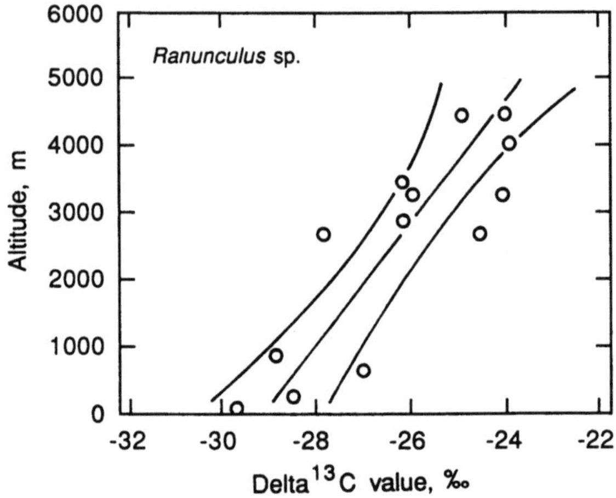

Figure 4.8. Variation in carbon isotope discrimination with elevation in the genus *Ranunculus* from six mountain regions from both hemispheres. A greater value of delta ^{13}C indicates a lower nonstomatal limitation to photosynthetic CO_2 assimilation (from Körner et al. 1988).

the spruce-fir forest just below the alpine zone coincides with the initial occurrence of flagged trees and, thus, the beginning of the timberline ecotone (Billings 1969). Here, snow deposition by strong winds results in ribbon-like stands of trees generated from massive snow accumulation just beyond the windward edges of forest-meadow boundaries. Major changes in shoot structure and branching patterns of these ribbon forest trees should result in major influences on whole-tree carbon-gain capacity. The importance of these types of structural effects on summer photosynthesis has been reported for the more extreme Krummholz form (Hadley and Smith 1987). The prostrate, compressed growth forms of nonconiferous, alpine, and meadow species at high elevations recently has been associated with adaptive differences in leaf extension rates (Woodward et al. 1986; Körner and Woodward 1987). The importance of other commonly observed differences in stand structure, such as tree spacing, tree heights during different successional stages, and other physiographic features, has not been evaluated as to effects on carbon gain.

In addition to above-ground changes in plant structure with elevation, below-ground changes also may be vital. With regard to water relations at high elevations, a very large increase in fine roots may occur with elevation (Körner and Renhardt 1987; Körner and Woodward 1987). Again, this increase could be related to differences in developmental processes and carbon allocation strategies unrelated to metabolic performance. Additional information is needed to clarify the qualitative versus quantitative importance of these types of structural changes in plant form which can provide major avenues of adaptation (Tilman 1988). In all habitats such an avenue appears to be based on morphological plasticity and not physiological adjustments in metabolic response patterns. Attempts at quantifying these morphological influences on carbon-gain capacity can only sharpen our perspective of their relative importance compared to adaptations leading to physiological adjustments in gas exchange metabolism.

Elevational Effects on CO_2 Exchange

A continuing interest in the influence of elevation on physiological adaptation has resulted in numerous comparisons among species growing at high versus low elevation. This more empirical approach attempts to correlate similarities in physiological response with elevational occurrence among species. For example, Körner et al. (1988) used an analysis of carbon isotope discrimination to conclude that higher elevation plants tended to have a lower CO_2 concentration inside the leaf relative to low elevation plants, indicating a lower limitation to photosynthesis via CO_2 carboxylation (Figure 4.8). The authors concluded that this increase in CO_2 uptake capabilities more than offsets the decline in the gas-phase conductance of the stomata that occurs due to the lower partial pressure of CO_2 as elevation increases. Körner and Diemer (1987) came to similar conclusions for 12 pairs of herbaceous congeners after comparing intercellular CO_2 levels (C_i) and the slope of the photosynthetic CO_2 assimilation (A) versus C_i curves. Higher elevation congeners had lower C_i but steeper A versus C_i curves, indicating a lower limitation to CO_2 uptake at the mesophyll cell level. Greater leaf thickness and leaf nitrogen content also indicated a potentially important influence of leaf anatomy (Björkman 1981; Nobel and Walker 1985). Despite this documented increase in photosynthetic efficiency, previous measurements of A do not appear to be consistently greater in all high elevation species, for example forest trees (Reneke et al. 1981). Körner and Diemer's (1987) altitudinal comparisons of congeneric herbs did show a consistently higher assimilation at saturating sunlight and natural ambient CO_2 levels for the high elevation congeners. Moreover, the gas exchange measurements were made in the field with an effort to insure that maximum photosynthetic capacities were being compared equally for each congener of the high and low elevation pair. Differences in microclimate, such as canopy air temperature and solar irradiance, were significant between the high and low elevation sites, although specific information on incident sunlight and leaf temperatures was not provided. Also, a wide variation in A at light saturation occurred that was independent of altitude. Temperature optima for A also showed variability according to microsite preferences.

The results of the studies described above are difficult to interpret in terms of the specific environmental factors associated with elevation that lead to the observed changes in physiological performance. The lower C_i and ratio C_i/C_a with elevation and greater mesophyll uptake efficiencies, as indicated by the steeper A versus C_i response curve, points to more efficient CO_2 uptake at high elevations. The question arises as to why C_i and C_i/C_a are lower at high elevation. Not only could C_i and C_i/C_a be reduced due to greater mesophyll efficiency for CO_2 uptake, but greater stomatal limitations could also reduce C_i. In general, high elevation has been shown to create a potentially much greater leaf-to-air vapor pressure deficit as well as diffusion rate for water vapor (Smith and Geller 1979). This increase in diffusion rate with elevation may be especially great when leaf temperatures are high. Thus, for identical stomata and the same degree of stomatal opening considerably more water could be transpired at high elevations resulting in greater water stress.

Although higher stomatal conductances have been reported for several species at high elevation (Körner et al. 1983; Körner and Diemer 1987), reduced stomatal opening would be needed to maintain the same A/E (E = transpiration) in high elevation plants relative to a lower elevation, and a lower C_i would result. The effect of generally lower stomatal opening in these plants to offset excessive transpiration (Smith and Geller 1979) could very likely select for enhanced mesophyll uptake efficiency at lower C_i. It is interesting that several congeneric pairs with wide elevational ranges did not show differences in the slope of the A vs. C_i curve (Körner and Diemer 1987). Moreover, the initial slopes of A/C_i were relatively small, indicating a lower mesophyll uptake capacity than that more typical of lower elevation species.

In summary, there is currently a very active concern for the influence of elevation on plant gas exchange (Figures 4.2, 4.3). A crucial feature of understanding these effects is a correct interpretation of environmental changes as well as corresponding physiological responses. To date, it appears as though changes in leaf conductance to water vapor with elevation are still enigmatic and can vary oppositely depending on species, microsite, and geographic location (see also Körner et al. 1979; Körner et al. 1986; Woodward 1986). Other investigators also have pointed recently to the somewhat smaller contribution of stomata to CO_2 exchange relative to the importance of the mesophyll cells for a variety of plant species, including conifers (Teskey et al. 1987). Yet, differences in stomatal control of water loss, especially at high elevations, could be eliciting physiological adjustments by the mesophyll cells. A potentially close coupling of transpirational water loss and CO_2 exchange is now a well-recognized phenomena (Wong et al. 1979).

The large increase in fine root length per unit leaf area recently reported for the same herbaceous species discussed above provides evidence for increased water stress at higher elevation (Körner and Renhardt 1987). Decreases in abaxial stomatal density and increases in the smaller number of adaxial stomata (typical of many hypostomatous plants) has also been reported as elevation increases (Körner et al. 1979; Woodward 1986). However, it is the abaxial stomata that generally dominate transpirational effects, especially during periods of high transpiration (Foster and Smith 1986). More research is needed before further conclusions regarding specific elevational effects on gas exchange can be clarified (Figures 4.3, 4.5).

Temperature and Moisture Limitations to Carbon Gain in Forest Trees

There are recent publications that review the community structure, elevational zonation, and geographic boundaries of the forests of the Sierra Nevada (Barbour 1988), the Cascades (Franklin 1988), and the Rocky Mountains (Peet 1988). Each of these allude to the primary importance of temperature and/or moisture as causal factors explaining differences in plant productivity and distributional pattern. This is particularly true for the Sierra Nevada/Cascade species, but the data for Rocky Mountain species (Smith 1985a) and the more economically important Pacific-Northwest species are far from comprehensive (Lassoie et al. 1983). Other recent publications have summarized the autecology of some Rocky Mountain (Smith 1985a) and Cascade forest tree species (Lassoie et al. 1983), as well as other species found within understory and meadow communities (Smith 1985a).

Table 4.3. Importance of elevated needle temperatures for conifer shoot photosynthesis in *P. engelmanni* (S), *A. lasiocarpa* (F), and *P. contorta* (P) during the summers of 1982-1986. Table shows temperature ranges during which photosynthesis was >90% of the maximum rate, and the average mean maximum ambient air temperatures that occurred during the periods of photosynthetic measurement. It also shows the percentage of the day (08:00-18:00h) that *A. lasiocarpa* shoots would have been within these temperature ranges for maximum photosynthesis if needles were at air temperature (T_a), as deduced from the temperature-photosynthesis response curves measured that day. Plus and minus values are 95% confidence intervals (from Smith and Carter 1988).

Period	Species	Number of Shoots	Temperature Range for Maximum Photosynthesis	Mean Maximum Daily Air Temperature (°C)[1]	Maximum photosynthesis (% of Day)		Percent Increase in Daily CO_2 Uptake[2]
					Actual	T_a	
9-23 June	S,F,P	2,2,2	13-22,15-24,13-25	13.8 ± 2.3	64	32	26
1-16 June	S,F,P	3,3,2	14-24,14-26	15.6 ± 1.9	59	30	29
16-24 July	F	2	14-26	16.7 ± 1.6	56	29	33
1-14 July	S,F,P	2,2,1	15-27,14-25,13-24	17.4 ± 1.8	65	32	35
16-28 July	S,F,P	2,2,2	16-26,15-26	16.9 ± 1.6	49	26	20
17-24 Aug	S,F,P	4,4,4	15-26,14-24,16-23	14.6 ± 1.9	53	31	29
		$\bar{x}$ =	14.1-25.1	15.8 ± 1.2	58	30	29

[1] 12-year means (1962-1974, Wyoming Solar Observatory; see text).
[2] Each value is computed for the indicated number of fir sun shoots only.

According to the limitations idealized in Figure 4.5, both air and soil temperatures probably interact to create the boundary limitations that determine the length of the summer period of net CO_2 assimilation (DeLucia and Smith 1987). The ecophysiological importance of low temperatures to annual production is paramount in these high elevation forests, limiting the growth period to a very short time in summer. During three different winters (1980 - 81 - 83) and up to three sampling dates per winter, stomatal opening was not detected during relatively warm periods when conifer needle temperatures ranged from 5-22° C (W.K. Smith, unpublished data). This is in agreement with the measurements of Tranquillini (1959), Bauer et al. (1969), and Larcher and Bauer (1981), wherein CO_2 exchange of timberline conifers did not rise above zero during most of the winter. Differences in autumn stomatal closure and photosynthetic decline has been measured recently for a variety of evergreen species, including conifers, herbs, and shrubs in the subalpine (M. Cui and W.K. Smith, unpublished data).

Although differences in responses to freezing nights appeared evident among the species, virtually no recovery in photosynthesis or stomatal opening occurred for any species following a relatively lengthy, warm period that occurred within about one week after photosynthesis initially declined to zero. Although certain species of evergreen angiosperms have been reported to respond to favorable periods in winter (Tranquillini 1979; Larcher 1985; Larcher and Siegwolf 1985), winter carbon gain appears unlikely for the evergreen species of these high elevation forests. This conclusion emphasizes the importance of adaptations which could accelerate the early-summer emergence from winter dormancy or delay entry into dormancy during autumn.

In early summer, additional limitations due to high LAVD and shading due to clouds, or other sources of shade, may also occur. Daytime air temperatures can also limit the maximum potential for carbon gain depending on how often actual leaf temperatures fall outside the temperature range for maximum photosynthesis. During both day and night, leaf temperatures outside the respiratory optimum could also inhibit efficient carbon metabolism and growth, leading to a reduction in daytime photosynthetic potential. Smith and Carter (1988) recently reported that subalpine conifers had temperature optima for photosynthesis considerably above daytime air temperature maximums (Table 4.3). However, leaf clustering on conifer shoots led to warmer leaf temperatures that were within the optimal photosynthetic range more often during the day. This finding is in contrast to the common assumption that conifer needle temperatures depart only slightly from air temperatures (e.g., Waring and Schlesinger 1985).

Additional studies on temperature limitation have shown that conifer photosynthetic capacity declines immediately in response to below-freezing air temperatures at night (Salo 1974; Lassoie et al. 1983; Fahey 1979; Kaufmann 1982a,b; Smith 1985b; DeLucia and Smith 1987). Between daytime air temperatures of about -3° to -8° C, conifer leaves freeze (Doehlert and Walker 1981; Larcher 1969) and net photosynthesis approaches zero (Larcher 1969, Ludlow and Jarvis 1971, Neilsen et al. 1972). The effects of these near or below-freezing air temperatures may persist for many days after the return of above-freezing temperatures. A host of indirect effects of low temperatures, such as freezing of soil or stem water, are probably strongly involved in limiting photosynthetic capability via negative effects on water status (Tranquillini 1979, 1982; Smith 1985b; Larcher 1985; Richards and Bliss 1986; Sakai and Larcher 1987). The schematic diagram in Figure 4.6 illustrates possible interactions of freezing temperatures with water relations and stomatal behavior, reflecting the potentially strong coupling of cold temperature and water relations at the beginning and end of the summer growth period. We know very little about the recovery capabilities of individual species after exposure to intermittent episodes of freezing nighttime temperatures that may occur during a major portion of the early, and possibly late, summer growth period. At high elevations where solar irradiance may be especially high and air temperatures low, non-stomatal limitations to CO_2 assimilation could be substantial (Öquist 1983). Interspecific variation in this limitation could contribute to differences among *Pinus* and *Picea* species in the recovery of their photosynthetic biochemistry in spring (e.g. Leverenz and Öquist 1987). At higher elevations, particularly high insolation and cold air temperatures could occur throughout much of the entire summer growth period. Data on the frequency, amplitude, duration, and recovery of the photosynthetic decline (both stomatal and non-stomatal) are needed to evaluate the quantitative importance of freezing air temperatures to summer carbon gain at high elevations.

Concomitant with the limitations to CO_2 assimilation in summer due to freezing air temperatures are potential limitations due to cold soil temperatures (Smith 1985b; DeLucia 1987; DeLucia and Smith 1987; Day et al. 1989). Seasonal changes in soil temperature may lag behind changes in air temperature by several months in areas with heavy snow accumulation. Thus, cold soil temperatures may persist into early

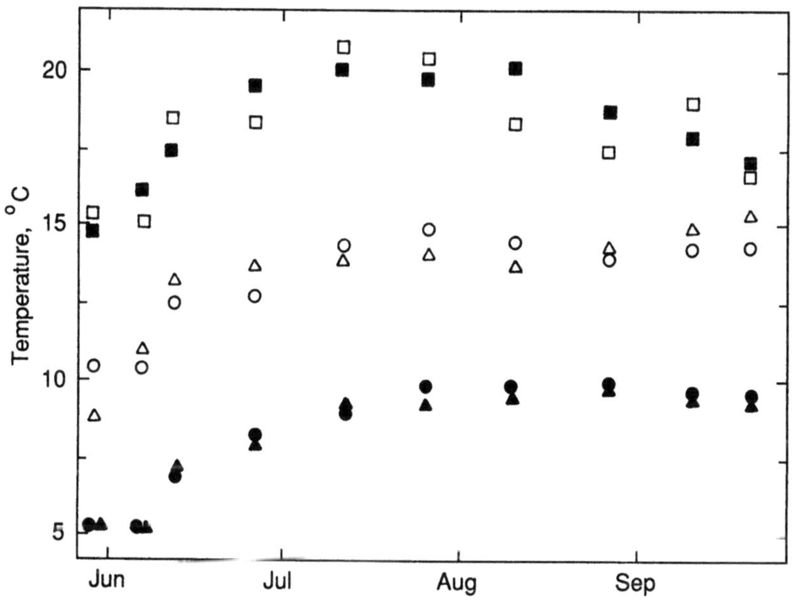

Figure 4.9. Air (10 cm height, □) and soil temperature (10cm depth, o, Δ) at 2,630 m and 3,664 m elevation during summer 1983. Soil temperatures were measured in the open (open symbols) or forest understory (closed symbols) at the high (Δ) versus low (o) elevation. High elevation (□) and low elevation (■) air temperatures are from the open sites. Microclimatic effects appear to overshadow any elevational effects for this situation over this elevation range.

summer despite warming air temperatures (Figure 4.5). As alluded to previously, elevation per se may have much less influence on soil temperature profiles than do microclimatic effects. Soil temperatures with depth appear very similar at any elevation depending on exposure to direct sunlight (Figure 4.9). More open locations had significant soil warming which quickly surpassed air temperature maximums at depths shallower than 15 cm. However, shallow subsurface temperatures (e.g., 5 to 15 cm) did not rise above 12° C the entire summer on forested sites, regardless of elevation. Also, air temperatures measured at plant heights did not show the typically high lapse rates expected for these mountains (Figure 4.9).

In contrast to air temperature effects, a number of investigators have reported severe restrictions to photosynthesis that occur at soil temperatures well above freezing (Running and Reid 1980; Kaufmann 1979a; Teskey et al. 1983, 1984; DeLucia 1986; DeLucia and Smith 1987). At soil temperatures of somewhat less than 10° C initial inhibition of photosynthesis has been measured in high elevation conifers. Within forests at higher elevations this could result in carbon gain limitations over a much greater portion of the growth season than depicted in Figure 4.5. More work is needed before a quantitative estimate of the importance of these limitations to summer carbon gain will be possible. The strong interactions with air temperature, and LAVD at this time of year, provide a challenging research scenario. In fact, recent evidence suggests that some compensation for cold soil temperatures beneath snowpack may come from increases in photosynthesis due to increased reflected solar irradiance (Day et al. 1989).

During and after the dimunition in carbon gain due to cold air and soil temperatures, there may be a period of limitation caused by LAVD effects on stomatal opening (Figure 4.5). Much recent attention has been focused on the capability of species to respond to variations in LAVD (Schulze and Hall 1982; Schulze 1986; Schulze et al. 1987) which would optimize water use per amount of CO_2 assimilated. Cowan (1982) developed a model of an interactive coupling between stomatal behavior and photosynthetic capabilities in plants that would act to optimize water use. A coupling between mesophyll cells and stomatal guard cells via the CO_2 concentration inside the leaf has been suggested. However, recent studies have suggested that leaf conductance and CO_2 assimilation may not be coupled solely to the internal CO_2 concentrations (Morison 1987). In the high elevation forests around the Great Basin, variation in LAVD is probably more dependent on changes in leaf temperature and not ambient changes in absolute ambient humidity (as discussed in the previous section). Exposure to direct sunlight will act to raise leaf temperature above air temperature with corresponding exponential increases in LAVD. Here again, temperature and water relations are strongly coupled, although the initiating variable is probably leaf temperature. The loss in carbon gain due to stomatal responses to LAVD may be offset and/or secondary to the beneficial effects of enhanced plant water status resulting from partial stomatal closure, particularly later in the summer when soil water may become limiting (Figure 4.5). The relative importance of the LAVD response to summer carbon gain has been determined for only a few plant species (e.g., Schulze 1982).

Numerous investigations have documented stomatal responses to increasing LAVD among high elevation conifers (Grace et al. 1975; Running 1976, 1980; Rutter 1977; Jarvis 1980; Knapp and Smith 1981;

Leverenz 1981a; Kaufmann 1982a, b; Sanford and Jarvis 1986). This LAVD response may depend strongly on overall plant water status determined by soil water availability. Thus, a strong coupling between daily leaf temperature regimes and LAVD may be a predominant factor in early summer, where as the important coupling of the stomatal response to LAVD with plant/soil water status may become more important later in summer (Schulze 1986). The specific limitation of the LAVD response to carbon gain may be considerably less in early summer when low temperature effects dominate and before significant water status limitations occur.

The tight coupling between elevated leaf temperatures and LAVD requires a rather precise knowledge of actual leaf temperature patterns under natural conditions before an understanding of LAVD effects on transpiration is possible. Current researchers working on coniferous trees continue to assume that due to the small aerodynamic dimension, needle-shaped leaves remain closely coupled to air temperature (Kaufmann 1982a, b; Waring and Schlesinger 1985; Lassoie et al. 1983). However, recent studies indicate that elevated leaf temperatures may be relatively common, especially for sun shoots with characteristically high needle densities per stem length (Smith and Carter 1988). This needle clustering on individual shoots resulted in reduced convective cooling and leaf temperatures that were as much as 14° C above air temperature with mean values of over 8° C (Figure 4.10). Similar observations of high leaf temperatures in conifers have been reported in several studies (Smith and Carter 1988). Although these temperatures appeared to enhance photosynthesis, substantial increases in LAVD and potential stomatal closure were not considered. This interaction of shoot structure, leaf temperature, and LAVD, along with potentially large influences on photosynthesis and transpiration, indicates the need for a greater understanding of the interaction of several micro-environmental and physiological variables. It is conceivable that increased carbon gain due to elevated leaf temperatures could be more important than water conservation at certain times of the year when soil water is abundant. Later in summer when air temperatures are greater and are closer to leaf temperatures, a greater sensitivity in the stomatal response to LAVD could conserve water at some short-term cost to carbon gain. This capability for precise seasonal adjustments in the stomatal control of water-use efficiency has already been proposed as a possible general mechanism among terrestrial plants (Cowan and Farquhar 1977; Cowan 1982; Morison 1987; Schulze et al. 1987).

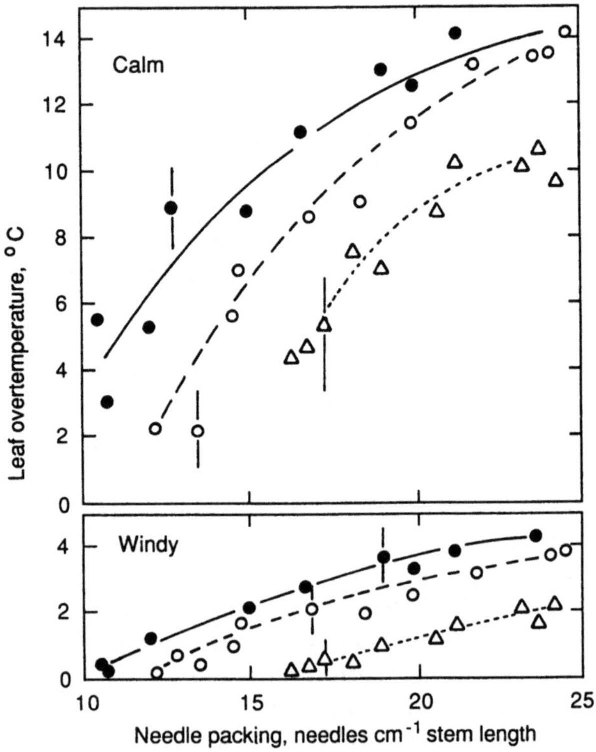

Figure 4.10. Leaf overtemperatures (T_{leaf} - T_{air}) vs. the degree of needle packing (number of needles per cm stem length) under simulated full-sun field conditions in the laboratory. Calm refers to an early morning windflow of ca. 0.3 m s^{-1}; windy refers to an experimental wind of ca. 3.0 m s^{-1}, typical of afternoon periods. Ten different shoot tips were monitored for each species with ten needles per shoot measured for temperature changes. Vertical bars are greatest 95% confidence intervals for each data set. Open circles, *Picea engelmannii*; closed circles, *Abies lasiocarpa*; triangles, *Pinus contorta* (from Smith and Carter 1988).

Seasonal variations in stomatal conductance in response to natural variations in plant water status, as indicated by measurements of xylem pressure potentials and estimated effects on carbon gain, are well documented in the literature (Schulze 1982; Schulze et al. 1987). However, results from these studies are often based on greenhouse-grown plants with artificially induced water stress. There are fewer data regarding the natural occurrence of xylem pressure potential variations and quantitative influences on carbon assimilation. There does appear to be a threshold response to xylem pressure potential in tree species, and others, that corresponds to an initial decline in assimilation potential (Lassoie et al. 1983). This threshold ranges from about -1.0 to -2.0 MPa depending on species and individual. The rate of decline in stomatal conductance to decreasing xylem pressure potential may also depend strongly on LAVD (Losch and Tenhunen 1981). As LAVD increases, conductance declines more rapidly with declining xylem pressure potential, i.e. stomatal sensitivity to xylem pressure potential increases as LAVD increases. It is interesting that stomatal conductance probably dictates the minimum daily value of xylem pressure potential when soil water availability is high, but xylem pressure potential becomes progressively more of a determinant of maximum daily conductance as soil water becomes limiting. A further complication to understanding stomatal behavior and xylem pressure potential relationships is the observation that predawn values may often be lower than midday values in sub alpine trees during periods when subfreezing nights occur (Smith et al. 1984). Apparently, restrictions to nocturnal water recharge, which are normally expected to give minimum values of xylem pressure potential just before stomatal opening in the morning, may have been caused by either frozen leaves, stems, roots, or some combination. As a result, predawn values could be lower than afternoon values, especially if stomatal opening was restricted the following day (Smith et al. 1984).

Some investigators have ranked individual tree species according to drought tolerance (ability to maintain photosynthesis at lower xylem pressure potential and a lack of sensitivity of stomatal conductance to xylem pressure potential or LAVD) versus drought avoidance (ability to close stomata earlier at higher xylem pressure potential and lower photosynthesis at a relatively high xylem pressure potential). Lassoie et al. (1983) classified *Pinus* spp. as more drought tolerant than *Picea* and *Abies*, and *Abies* spp. as better drought avoiders than *Pinus* spp. or *Picea* spp. Similar findings were summarized in Smith (1985a) when comparing the lower elevation *Pinus contorta* with the subalpine *Abies lasiocarpa* and *Picea engelmannii*. Other shade-intolerant species, such as *Pinus ponderosa* and *Pseudotsuga menziesii*, also close stomata at higher xylem pressure potential than do more shade-tolerant species and tend to have lower osmotic potentials to maintain turgor (Lassoie et al. 1983). However, studies that compare the specific differences in gains or losses in carbon assimilation due to responses of stomata to variations in xylem pressure potential over a season are lacking. Thus, the importance of xylem pressure potential limitations to carbon gain either through stomatal effects or more direct effects on photosynthesis, cannot be quantified relative to other limiting factors. Recent evidence for a direct influence of root water relations on stomatal behavior without effects on xylem pressure potential presents another interesting, although complex interaction between plant water relations and stomatal behavior (Schulze 1986).

The termination of the active growth period in these high elevation conifers appears to be due to the return of subfreezing air temperatures at night (Figure 4.5). Smith et al. (1984) observed a rather gradual, two month decline in conductance during autumn for six Rocky Mountain conifer species at a relatively low elevation that corresponded to an increased frequency of below-freezing air temperatures at night. By the time nighttime minimums were consistently below 0° C, stomatal opening did not occur. At this same time, early morning xylem pressure potential was often less than that in the afternoon, indicating possible effects of xylem pressure potential as well as temperature on stomatal behavior. Thus, as shown in Figure 4.5, the onset of seasonal stomatal closure appears due to the return of sub-freezing nighttime minimums in air temperature, but the prior limitations due to xylem pressure potential could also play a role. A drier year could result in a greater sensitivity to water stress induced by nocturnal freezing temperatures. Again, the interactive effects of the return of freezing air temperatures and water stress on carbon assimilation capacity which already may have developed to varying degrees due to soil water limitations, remains virtually unstudied. A similar increase in non-stomatal limitations to photosynthesis has been found for 4 of the 6 species mentioned above (author's unpublished data). These two factors may be the primary influences that interact to effectively terminate the annual carbon-gain period. A similar scenario for timberline species has been proposed by Schulze (1982). The possibility of direct biochemical effects on photosynthetic capacity due to the return of subfreezing temperatures also exists (Öquist 1983).

Sunlight Limitations to Carbon Gain

The amount and quality of incident sunlight can influence carbon gain through direct stomatal effects on CO_2 exchange as well as effects on the cellular biochemistry involved in the photosynthetic light reactions. In addition, the amount of incident sunlight can have direct influences on leaf temperature and, thus, affect stomatal (guard cell) and mesophyll cell biochemistry. Temperature may also influence stomata indirectly via effects on LAVD. Qualitative changes in sunlight can influence growth, development, and phenology such that major effects on physiological performance result (Björkman 1981; Morgan and Smith 1981). Often, measured differences in carbon gain and water relations among species can be due in large part to the physiological effects of differences in morphology. The following discussion will focus on stomatal, nonstomatal, and morphological adaptations that may strongly influence photosynthetic carbon gain in these high elevation species. We will attempt to evaluate limitations in carbon gain due to sunlight within the context of the scheme of temperature and moisture limitations proposed in Figure 4.5.

Generally, it is observed that stomata open very early in the morning at relatively low sunlight levels such that differences in daily carbon gain are not influenced by the time of opening (Squire and Black 1981; Willmer 1983; Tenhunen et al. 1987). Yet, there is a remarkable paucity of information regarding the dynamics of initial stomatal opening in the morning for different species within a particular community or for different leaves of the same plant (Hinckley et al. 1975). These data are of particular importance because CO_2 uptake in the morning (when water status is optimal and LAVD low) may account for a large portion of the daily carbon gain, especially in more xeric habitats. As already mentioned, leaf wetness due to dewfall may also play an important role during morning hours (Smith and McClean 1989). A similar lack of information exists for stomatal closure in the evening although stomatal opening at night has been reported in some coniferous trees (e.g., Leverenz 1981a). Disregarding the interesting ecophysiological features of nighttime stomatal opening, stomatal opening in the morning usually begins when sunlight levels are quite low (<20 μmol m^{-2} s^{-1} photosynthetic photon flux density, PPFD) and continues until maximum opening occurs at a PPFD value considerably less than full sun (200-600 μmol m^{-2} s^{-1}). These boundary values, as well as the rate of stomatal opening or closing, can vary dramatically depending on a variety of plant and environmental parameters such as temperature, water status, and age (Sharkey and Ogawa 1987). Stomatal opening and closing could be strongly influenced by vegetation structure in forests composed of very tall trees where within-canopy and understory sunlight is highly variable and often below 30% of full sun. Diffuse PPFD values in shade in the understory often range from 150-250 μmol m^{-2} s^{-1} on clear days versus 200-400 μmol m^{-2} s^{-1} on cloudy days (Young and Smith 1983). However, only recently researchers have begun to address the dynamics of stomatal opening or closing according to natural microhabitat differences in sunlight exposure, physiological status, or leaf position in the canopy. A discussion of the importance of non-steady-state stomatal behavior in a highly variable environment such as the subalpine is presented in the following section.

Most studies of sunlight effects on stomatal opening and carbon gain involve measurements made under saturating sunlight levels. Differences in stomatal conductance among species are assumed to be due to other stress factors such as water availability, leaf age, and developmental history (i.e., sun versus shade leaves), that can influence these daily maximums in conductance (Jarvis and Mansfield 1981; Willmer 1983; Zieger et al. 1987). As always, acclimational effects are reported which can strongly influence maximum conductance. Despite these complexities, maximum conductance appears generally greater for herbaceous plants from sunny habitats (ca. 4 to 17 mm s^{-1}) than for conifers (ca. 1.5 to 5 mm s^{-1}) (Körner et al. 1979; Larcher 1980). However, there appears to be considerable overlap among these two groups as well as other types of plants.

One particular problem with measuring and computing stomatal conductance values for conifers, or other species with complex shoot structure, is appropriate leaf area (Figure 4.11; Carter and Smith 1985). Conductance values reported in the literature are often based on the total or projected leaf areas of leaves detached from the measurement shoot. In the case of complex shoots, considerable mutual shading by individual leaves may result in unequal conductance among leaves on the same shoot, e.g., lower values for shaded leaves. Thus, an underestimate of maximum conductance can easily result from measurements on complex shoots. Recent work has recognized this problem for conifers as well as other species where complex shoots are placed in gas exchange cuvettes or diffusion porometers to determine stomatal conductance (Carter and Smith 1985; Smith and Hollinger 1990). These same factors can have identical effects on the measurement and computation of photosynthetic CO_2 flux density (Szawiawski and Wierzbicki 1978; Leverenz and Jarvis 1979, 1980; Carter and Smith 1985). The problems associated with shoot measurements and corresponding computations of gas exchange characteristics are quite similar to problems associated with extrapolating branch measurements to the whole-plant or stand level. More work is needed at all structural levels to elucidate the influences of stomatal behavior and sunlight limitations to daily carbon gain. This information will enable much more accurate comparisons among different species or plant types than are possible currently.

Incident sunlight can also influence water vapor exchange, independent of stomatal effects. Elevated leaf temperatures will have an exponential effect on increasing LAVD and, thus, transpiration. This increase can be potentially large due to substantially increased solar irradiance at higher elevations and the relative ease with which leaf temperatures can rise above cooler air temperatures (Smith and Geller 1979). As discussed earlier, even conifer leaves (needles) with small aerodynamic dimensions may be frequently elevated above air temperature in full sun due to leaf-clustering effects on shoots (Smith and Carter 1988). Thus, shading by other leaves on the same shoot, other shoots, or adjacent trees could have substantial influences on transpirational water loss not related directly to stomatal function.

Very few studies have measured the relative influences of shaded versus sunlit leaves within plant canopies on whole-plant transpiration (Squire and Black 1981) or similar effects due to differences in leaf age (Field 1987). This attempt at relating leaf or shoot measurements to whole-plant processes has been recognized as important to understanding the ecophysiology of most species, especially those with more complex structure (Mooney and Chiariello 1984). A recent example of an extrapolation from shoot to whole plant effects in a desert shrub is given in Neufeld et al. (1988). Ultimately, this approach may bridge the gap between the now abundant studies on shoot-level physiology and canopy-exchange processes (Monteith 1976; Hutchinson and Hicks 1983; Sharma 1984), providing a more complete integration between shoot, whole plant, and stand influences.

Given that considerable variation in irradiance exists within high elevation forests, what is the potential influence of sunlight limitation relative to other possible limitations to annual carbon gain? At present, there is insufficient information available to answer this question for conifer tree species, especially since most data are based on individual shoot measurements. These measurements include estimates of light saturation and compensation points for photosynthetic CO_2 uptake, differences between sun and shade leaf photosynthesis, and effects of leaf age. The majority of these data come from the Pacific Northwest species (Lassoie et al. 1983), followed by the Rocky Mountain species (Smith 1985a). In general, light levels for photosynthetic compensation and saturation in conifers appear similar to many other C_3 plant types except for the highest values recorded for certain herbaceous heliophytes (Larcher 1980). Only a few studies have pointed to distinct differences in the light saturation levels for high elevation species, despite

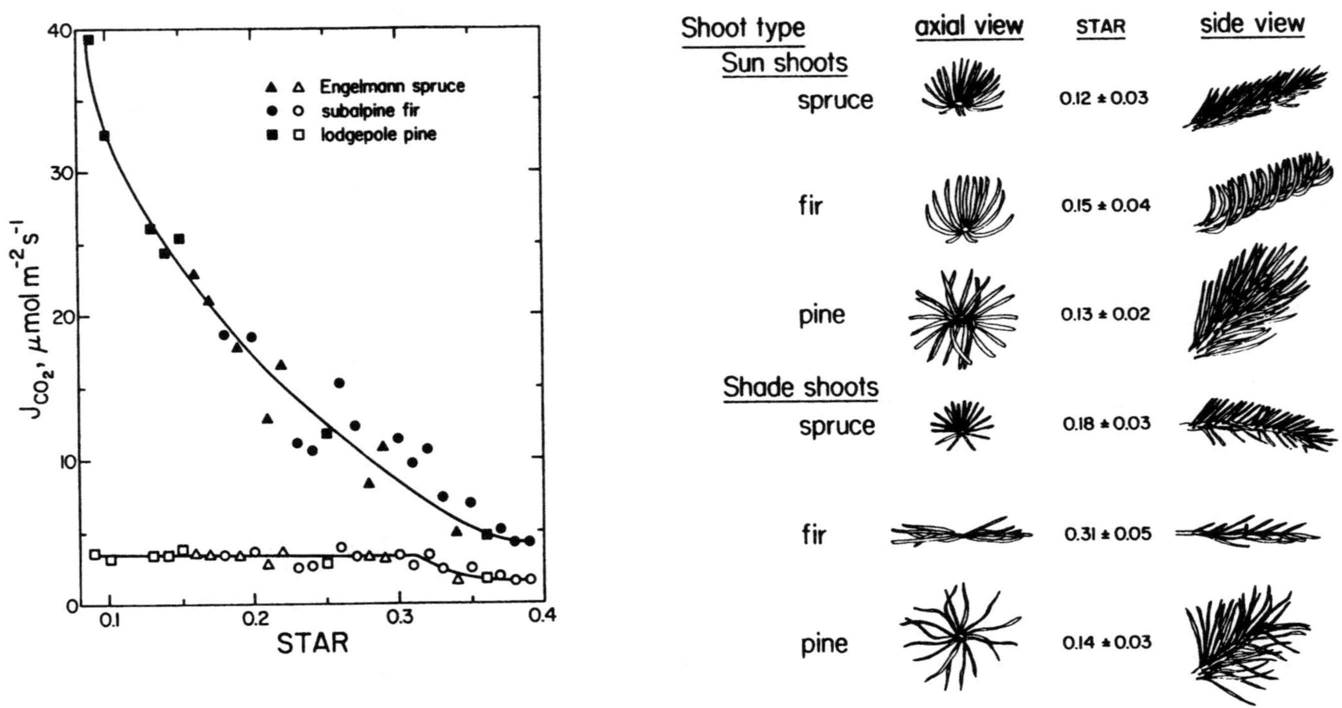

Figure 4.11. Representative photosynthetic flux (J_{CO_2}) for sun and shade shoots of *Picea engelmannii* (Engelmann spruce), *Abies lasiocarpa* (subalpine fir), and *Pinus contorta* (lodgepole pine). Dark symbols are for J_{CO_2} expressed per unit of silhouette leaf area (projected area of needles attached to the shoot in its natural orientation to the sun); open symbols are for J_{CO_2} expressed per total leaf area. STAR is the silhouette-to-total leaf area ratio (from Carter and Smith, 1985).

potentially large increases in solar irradiance with elevation (Smith and Geller 1979). Smith and Carter (1988) compared photosynthetic characteristics between *P. contorta*, *A. lasiocarpa* and *P. engelmannii* according to microhabitat and found small differences among the species in their photosynthetic response to sunlight level. Greater water-use efficiency in *P. contorta* was calculated and appeared to be due primarily to reductions in stomatal opening at similar photosynthetic capacities. Similarly, high water-use efficiencies were reported for *P. contorta* when expressed as bole volume versus volume of water transpired (Kaufmann 1985). Growth efficiency (annual volume growth/potential radiation absorbed) was also higher in *P. contorta*. However, potential radiation absorbed was estimated only according to an arbitrary canopy value representing crown position.

The potentially important effects of mutual shading become especially important when determining canopy leaf areas to be used for estimates of growth efficiency. The importance of the silhouette-to-total leaf area ratio (STAR) for computing accurate CO_2 or water vapor flux densities is addressed in Carter and Smith (1985) for conifer shoots (Figure 4.11). Silhouette leaf areas (shadow cast by leaves on intact shoots) may be much more meaningful than estimates of projected leaf areas that are customarily determined for detached leaves lying flat on conveyor belts of leaf area meters. Data are needed on actual STAR values for shoots and entire plants so that the more common and easier physiological measurements at the shoot level can be extrapolated to whole-tree and even stand-level effects. Because irradiance levels during development can have major influences on leaf and shoot structure and physiological capabilities, important whole-canopy and stand effects on sunlight absorption can directly influence physiological performance, as well as the developmental constraints on morphology and physiology.

Another aspect of the solar radiation regime that is often considered a characteristic of high elevations involves qualitative changes in sunlight. Potential increases in ultra-violet wavelengths with elevation, especially the UV-B wavelengths, may be particularly damaging to nucleic acids (Caldwell 1981). Because the atmospheric ozone layer is located primarily in the stratosphere the higher elevations of mountains do not alter the pathlength through this layer. However, such factors as cloud-cover and the occurrence of a high surface albedo could enhance UV-B wavelengths, possibly doubling the downward sky flux. Caldwell (1981) concluded that UV-B sensitivity varied considerably among different plant species and resulted in rather subtle influences on plant growth.

High altitude plants at low latitudes, where UV-B is highest, are expected to be most influenced by UV-B effects.

In summary, evidence suggests that sunlight effects on carbon gain may originate from physiographic, stand, whole-tree, and shoot differences among the same or different species. A comprehensive synthesis of the relative ecological importance of these different structural levels to sunlight absorption and physiological performance is currently not possible in high elevation forests, or elsewhere. Recent attempts at understanding the interaction of these different structural levels have included such factors as tree leaf area and physiographic conditions (e.g. Kaufmann and Ryan 1986). The relative contribution of shoot versus whole-tree or stand effects to growth efficiency may vary among species. A recent workshop addressing the role of branch studies to whole-tree physiology pointed out the potential importance of branch-tree interrelationships (Winner and Phelps 1988). Future studies that focus on the integration of all of these factors are needed to provide a comprehensive evaluation of adaptive mechanisms on both a physiological and structural level.

Nutrient Effects

As stated earlier, soil nutrients have been considered less of a limitation to growth compared to the strong physical limitations of temperature, water, or sunlight in these high elevation forests. This appears true even for the relatively moist forests of the Pacific Northwest (Franklin and Waring 1979). However, virtually all of the possible physiological/morphological characteristics listed in the previous sections could change significantly in many species according to nutrient status. Most studies in high elevation forests have dealt with the economically important *P. menziesii* and the influence of nitrogen fertilization (Lassoie et al. 1983). A host of branch and tree growth parameters were shown to increase with fertilization (Brix and Ebell 1969; Brix 1981a). Leaf chlorophyll and nitrogen levels also increased with nitrogen fertilization (Brix 1971), as did photosynthetic capacity (Brix 1981b). Improvements in tree water status were also measured in response to nitrogen fertilizer (Brix 1972), possibly reflecting improved root growth (Hermann 1977).

The ecophysiological effects of nutrients such as nitrogen do appear to be particularly sensitive to the availability of other factors such as sunlight, temperature, and water (Luxmore et al. 1987). Even when temperature and water are not limiting, the absence of adequate sunlight (e.g. frequent cloudcover) could still relegate nutrient requirements to a subsidiary role. However, particularly good years for growth, in terms of temperature and especially water, might be expected to entail a greater limitation due to nutrients. Moreover, these unusual years of more optimal growth conditions could be as important as unusual suboptimal years. The excessively high transpiration rates measured in some species with low water-use efficiency (e.g. fast growing herbs) could ultimately be adaptive for increasing nutrient uptake and storage needed to support such high photosynthetic rates. Much less information on nutrient relations is available for the high elevation forests of the continental Rocky Mountains than other forest systems. Fahey and Knight (1986) concluded that nitrogen may limit the growth of *P. contorta* forests in addition to temperature and water. They reported considerable variability in nitrogen availability within these forests which was due, in part, to differences in soil types and corresponding decomposition activity. However, nitrogen fertilization studies were not attempted. Recent reviews by Waring and Schlesinger (1985) and Lassoie et al., (1983) cite numerous examples of increasing site production after fertilization. Moreover, these increases may be complicated by changes in allocation patterns of carbohydrates to different plant parts. In particular below-ground allocation to fine roots has often been observed after fertilization (Axelsson and Axelsson 1986). Actual increases in wood production may lag considerably behind increases in leaf nutrients and photosynthetic carbon uptake. Despite the studies cited above and many additional studies on nutrient cycling in forest systems, the relative importance of nutrients versus other environmental factors that may limit carbon fixation has not been determined for high elevation forests, or any other ecosystem. Interactions between temperature and water relations with nutrient effects on carbon gain may be particularly complex due to the potentially important influence of transpiration and soil water on nutrient uptake and translocation.

Non-Steady-State Responses to Variable Sunlight

Relatively few studies of high elevation forests have focused on the dynamics of variable sunlight regimes in these high elevation forests. As mentioned previously, the subalpine forest understory may be unique in that sunlight penetration is substantial and, yet, wind is diminished considerably because of the combination of tall trees with porous leaf canopies, large crown size, and relatively wide tree spacing. As a result, prolonged snowpack may occur in shaded microsites in a patchy, mosaic pattern that contributes to variability in the total summer growth period. Extreme microsite differences in soil temperature and water relations also can occur during the summer. In mid-summer, the understory ground surface temperatures may vary by over 40° C (20-70° C) according to sun exposure, litter layer thickness, and other surface properties (W.K. Smith, unpublished data). Even in open areas, the extreme variation in incident sunlight throughout summer due to predictable midday and afternoon clouds also contributes to an extremely variable sunlight regime (Young and Smith 1983; Knapp and Smith 1988).

Other factors, such as the interception of summer rainfall, dew formation, and the seasonal onset of freezing leaf temperatures at night are potentially important features that may interact with the degree of openness of a particular microsite. Energy balance relationships at high elevation dictate that an organism will often be considerably warmer than air temperatures if it is fully sunlit and wind-sheltered (Smith and Geller 1979; Gauslaa 1984). Thus, organism temperatures in these environments are coupled strongly to radiation exposure, where as increased wind exposure can reduce large temperature differences relative to air temperature (Figure 4.10). High transpiration due to leaf temperatures 20° C above air temperature may occur, and the more rapid diffusion of water vapor at high elevation can also result in large depressions in leaf temperature below air temperature in shade. Longwave radiation exchange at night with cold skies (ca. -60° C) can also lead to nocturnal reductions in leaf temperatures, enhancing dew and frost formation on the leaf surface.

All of these effects combined can produce a highly variable environment characterized by frequent and relatively large changes in incident sunlight, leaf temperature and, thus, LAVD. These parameters can have immediate and major effects on transpiration and photosynthesis. Furthermore, the possibility exists that non-steady-state responses, i.e. responses that do not approach a steady-state before another environmental change and illicited response occurs, may comprise a substantial portion of the day (Knapp and Smith 1987; 1988). If this is true, the importance of non-steady-state response patterns may be an unappreciated ecological factor. These high elevation forests provide a natural laboratory for testing the importance of non-steady-state physiological responses to adaptive biology.

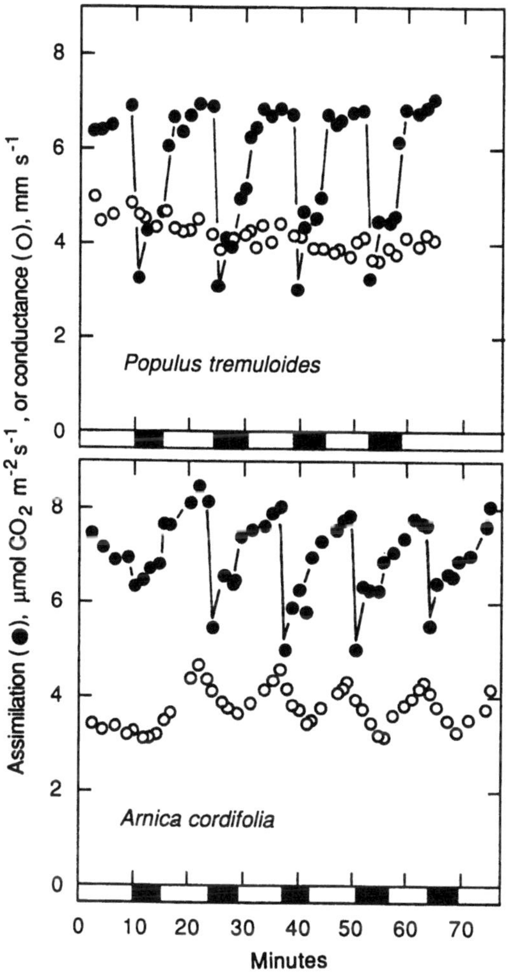

Figure 4.12. Responses in net photosynthesis (A, as CO_2 uptake; μmol m^{-2} s^{-1}) and stomatal conductance to water vapor (mm s^{-1}) to alternating periods of full sunlight and shade (ca. 20% of full sunlight) in a subalpine tree, *Populus tremuloides*, and an herb, *Arnica cordifolia*. The durations (8 ± 1 min full sun, 5 ± 1 min shade) and light levels during sun/shade periods simulated a typical period of intermittent cloud-cover in the subalpine zone of southeast Wyoming, USA. Dark bars along bottom axis denote shade periods (from Knapp and Smith 1989).

Recent research on the non steady-state responses of subalpine species has identified major differences in stomatal responses to variable sunlight among different plant types (Figure 4.12; Knapp and Smith 1987,

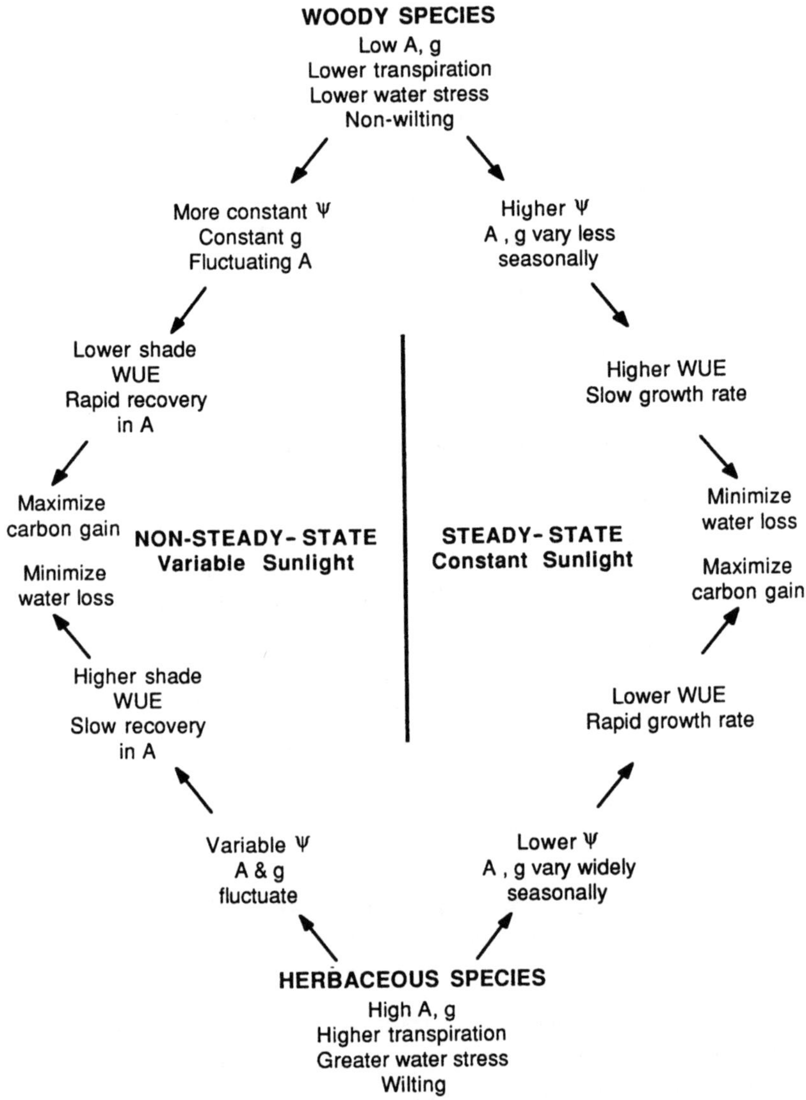

Figure 4.13. Flow diagram of representative response patterns and growth strategies in woody and herbaceous subalpine plants under steady and non-steady-state conditions . Steady-state responses are derived from literature generalizations and data on subalpine plant responses (A.K. Knapp and W.K. Smith, unpubl. data). Non-steady-state responses are based on data in Knapp and Smith (1987, 1988, 1989). A, net photosynthesis; g, stomatal conductance; ψ, xylem pressure potential; WUE, water-use efficiency; other symbols as defined in figure 4.1 (from Knapp and Smith 1989).

1988, 1989). Frequent cycles of full sun-to-shade and back to full sun are common in the subalpine due to clouds, overstory shading, or shading among leaves of the same leaf canopy. These variations occur in cycles of from about 3 to 10 min in the case of clouds, most commonly with 3 to 5 min of shade alternating with 8 to 10 min of full sun. In certain species such as faster growing herbs with high photosynthetic and transpirational flux, stomatal responses are fairly rapid and able to "track" changes in CO_2 assimilation and sunlight, i.e. stomata close in response to shade and open again when full sun returns (Knapp and Smith 1988). This results in increases in WUE during shade, although reductions in carbon gain can result when full sun returns and stomata are still partially closed. Other species such as trees and shrubs, have stomatal responses that are slower and do not track changes in sunlight and photosynthetic potential. In this case water is not conserved, but the potential for carbon assimilation is maximized.

This example is contrary to traditional dogma concerning steady-state responses and the water-conserving growth strategies exhibited by the slower growing tree and shrub species compared with the maximal carbon gain and profligate water-use strategy of a herbaceous species (Figure 4.13). In fact, herbaceous species may become more efficient trackers as seasonal water stress increases (Knapp and Smith 1989). A recent study has also pointed out the crucial importance of intermittent shade to steady-state responses in full sun (Knapp et al. 1989). Periodic shade periods were necessary for avoiding severe water stress (wilting) and corresponding reductions in steady-state stomatal opening and photosynthetic CO_2 assimilation in two herbaceous congeners. Thus, the non-steady-state response patterns and the interaction of steady-state responses in full sun with intermittent shade periods indicates a rather complex array of potentially important adaptive patterns in these species. Currently, little is known regarding non-steady-state responses to the highly variable sunlight environments of high elevation, subalpine forests. Most studies on short-term sunlight variation have involved tropical understory species where low intensity sunflecks do not pose serious water relations problems (Chazdon 1988). All of the potentially limiting factors mentioned previously need to be evaluated in terms of their interactive effects on non-steady-state as well as steady-state response patterns before a comprehensive evaluation of adaptive responses will be possible.

Concluding Remarks

In recent years, advances in more mechanistic research have shown promise for interpreting the interactive effects of multiple environmental variables and the integration of physiological mechanisms. The importance of this type of research is in its ultimate ability to transcend plant or habitat types and provide a more generalized predictive capability. Today, most ecophysiological models for predicting response patterns in plants, especially forest trees, involve only empirical relationships that may be highly species and habitat specific. They also tend to lack the resolution required to accurately simulate the more variable and rapidly changing environments of high elevation forests. As an example, ecosystem models of stomatal response patterns are often based on steady-state values of daily sunlight level, ambient humidity regime, and soil and plant water status, without any interactive influences based on actual microclimatic conditions or non-steady-state influences during a day. Not only are many of the daily changes in sunlight availability due to various sources of shading not considered, but corresponding influences on leaf temperatures and LAVD are also neglected.

An example of a more mechanistic approach that involves stomatal function is that presented in reviews by Schulze (1986; 1987). This approach attempts to integrate stomatal responses according to variations in LAVD, plant and soil water status, and CO_2 uptake capacity. These results suggest such provocative ideas as the lack of influence of bulk leaf water potential on stomatal behavior and, instead, a stomatal sensitivity to the maintenance of water flux in the epidermis and roots. In some species, the CO_2 uptake capability by mesophyll cells appears to be coupled to stomatal function such that the intercellular CO_2 level is conservative (Morison 1987). The internal CO_2 level may also regulate the coupling of mesophyll to guard cell function. However, other links between guard cell and mesophyll physiology appear important. The longer-term influences of plant hormones on stomatal function during drought also appear to be important. This type of interactive analysis which incorporates many physiological variables ultimately will provide a basis for understanding the key variables for predicting physiological response patterns. Coupling more mechanistic physiological approaches with similar approaches for evaluating whole plant and stand effects on key environmental variables will, potentially, lead to more realistic and accurate predictions of ecophysiological processes over a broad spectrum of plant types.

It appears as though a host of environmental factors and plant properties may undergo characteristic changes with elevation, although qualitative and quantitative differences may still occur depending on a specific geographic location, topography, and microclimate. In fact, microclimatic differences may override many effects due to elevation *per se*, although a precise comparison of microclimatic factors between two identical microhabitats, but at different elevations, has not been attempted. In some sense this may not be entirely possible. Regardless, this experiment is necessary before the specific environmental changes associated with elevational change can be identified and quantified. Without this information we cannot predict which physiological adaptations might be necessary to insure ecological fitness at high elevation. The possibility exists that the breadth of microclimate variation at a given elevation is much greater than are changes occurring between sea level and the highest alpine habitats.

So far, only a few patterns have begun to emerge in the physiological responses of high elevation species, and some of these seem inconsistent, if not contradictory. This could be due to the fact that microclimate can overshadow elevational effects. In contrast, certain structural/morphological responses appear characteristic of high elevation, although many of these appear associated with the extreme environments of the timberline ecotone and alpine zone. Changes in structure are often viewed as avoidance mechanisms compared to the increased tolerance resulting from physiological adaptations in metabolism. The functional significance of these structural features in terms of ecophysiological effects is just beginning to be elucidated. Despite this lack of appropriate data and a generally high degree of complexity due to many interacting variables, there are certain responses which may differ in their potential adaptive significance. These include any changes in physiology or structure that would minimize reductions in annual carbon gain due to:

1. Low temperatures of both air and soil;
2. Atmospheric and/or soil water deficits during winter and summer;
3. Limitations in absorbed sunlight;
4. Limitations in nutrients during high growth years.

Numerous adaptive avenues available primarily through structural changes could outweigh any physiological differences influencing carbon gain. As an example, these might include the necessity for snow burial to avoid winter desiccation and mortality mentioned earlier. There is an abundance of observations on the growth and phenology of high elevation species that may also be unrelated to carbon gain, but that could have primary influences on reproductive biology and, thus, ecological fitness. Although carbon gain certainly can influence reproductive fitness, smaller amounts of carbon gained, but more effectively allocated, may confer greater fitness in one species than another. Such factors as a predominance of asexual reproductive biology in subalpine understory species and alpine species versus the sexual reproduction of the forest trees (except *Populus tremuloides*) indicates a potentially strong evolutionary relationship that is polyphyletic.

Finally, virtually no ecophysiological information concerning natural seedling establishment is available for these high elevation forests. Most studies are for mature individuals or for greenhouse grown seedlings, especially among the tree species. Yet, recent work indicates that mortality is extremely high (>80%) for current-year conifer seedlings under field conditions (W.K. Smith and G.A. Carter, unpublished data). Thus, most interpretations of adaptive patterns in forest trees are based on measurements of adults where selective pressures may be minimal compared to seedlings. Recently, dramatic differences in water stress among very young, natural seedlings (current and first-year) have been measured that appear to be correlated with microsite conditions (Cui and Smith 1990). Such high mortality during this initial life stage might reflect the underlying ecophysiological limitations leading to most of the observed successional and distributional patterns of mature trees. Additional research on these very young seedlings may provide a more lucid interpretation of the specific environmental limitations to these ecological patterns. This important coupling of population biology with ecophysiology appears greatly neglected, especially for plant species that are long-lived perennials with episodic seedling recruitment (Osmond et al. 1980; Schulze 1982).

References

Axelsson E and Axelsson B (1986) Changes in carbon allocation patterns in spruce and pine trees following irrigation fertilization. Tree Physiol. 2:189-204.

Barbour MG (1988) Californian upland forests and woodlands. pp. 131-164. In Barbour MG and Billings DW (editors) North American Terrestrial Vegetation. Cambridge Univ. Press, Cambridge.

Bauer H, Huter M and Larcher W (1969) Der Einfluss und die Nachwirkung von Hitzeund Kaltestress auf den CO_2-Gaswechsel von Tanne und Ahorn. Ber. Dtsch. Bot. Ges. 82:65-70.

Benecke U, Schulze E-D, Matyssek R, and Havranek WM (1981) Environmental control of CO_2 assimilation and leaf conductance in *Larix decidua* Mill. I. A comparison of contrasting natural environments. Oecologia 50:54-61.

Billings WD and Bliss LC (1959) An alpine snowbank environment and its effect on vegetation, plant development, and productivity. Ecology 40:388-397.

Billings WD and Mooney HA (1968) The ecology of arctic and alpine plants. Biol. Rev. 43:481-529.

Billings WD (1969) Vegetational pattern near alpine timberline as affected by fire-snowdrift interactions. Vegetatio 19:192-207.

Björkman O (1981) Responses to quantum flux densities. pp. 57-108. In Lange OL, Nobel PS, Osmond CB, and Ziegler H (editors) Physiological Plant Ecology I: Response to the Physical Environment, Encyclopedia of Plant Physiology (New Series) Vol. 12A. Springer-Verlag, Heidelberg.

Bliss LC (1956) A comparison of plant development in microenvironments of arctic and alpine tundras. Ecol. Monogr. 26:303-337.

Brix H (1971) Effects of nitrogen fertilization on photosynthesis and respiration in Douglas-fir. Forest Sci. 17:407-414.

Brix H (1972) Nitrogen fertilization and water effects on photosynthesis and early-wood-latewood production in Douglas-fir. Can. J. For. Res. 2:467-478.

Brix H (1981a) Effects of thinning and nitrogen fertilization on branch and foliage production in Douglas-fir. Can. J. For. Res. 11:502-511.

Brix H (1981b) Effects of nitrogen fertilizer source and application rates on foliar nitrogen concentration, photosynthesis, and growth of Douglas-fir. Can. J. For. Res. 11:775-780.

Brix H and Ebell LE (1969) Effects of nitrogen fertilizer on growth, leaf area, and photosynthesis rate in Douglas-fir. For. Sci. 15:189-195.

Caldwell MM (1981) Plant response to solar ultraviolet radiation. pp. 169-198. In Lange OL, Nobel PS, Osmond CB, and Ziegler H (editors) Physiological Plant Ecology I: Response to the Physical Environment, Encyclopedia of Plant Physiology (New Series) Vol. 12A. Springer-Verlag, Heidelberg.

Carter GA and Smith WK (1985) Influence of shoot structure on light interception and photosynthesis in conifers. Plant Physiol. 79:1038-1043.

Carter GA and Smith WK (1987) Influence of microhabitat on water and photosynthetic relations of three codominant conifers of the central Rocky Mountains. Can. J. Bot. 66:963-999.

Chazdon RL (1988) Sunflecks and their importance to understory plants. Adv. Ecol. Res. 18:1-63.

Cowan IR and Farquhar GD (1977) Stomatal function in relation to leaf metabolism and environment. pp. 471-505. In Jennings DA (editor) Integration of Activity in the Higher Plant. Univ. Press, Cambridge.

Cowan IR (1982) Regulation of water use in relation to carbon gain in higher plants. pp. 589-614. In Lange OL, Nobel PS, Osmond CB, and Ziegler H (editors) Physiological Plant Ecology II: Water Relations and Carbon Assimilation. Encyclopedia of Plant Physiology (New Series), Vol. 12B. Springer-Verlag, Heidelberg.

Cui M and Smith WK (1990) Photosynthesis and water relationships of young seedlings of *Abies lasiocarpa* with high natural mortality. Can. J. For. Res. (in press).

Daubenmire R (1978) Plant Geography. Academic Press, NY.

Day TA, DeLucia EH, and Smith WK (1989) Influence of cold soil and snow cover on photosynthesis and conductance in two Rocky Mountain conifers. Oecologia 80:546-552.

DeLucia EH (1986) Effect of low root temperature on net photosynthesis, stomatal conductance and carbohydrate concentration in Engelmann spruce (*Picea engelmannii* Parry ex Engelm.) seedlings. Tree Physiol. 2:143-154.

DeLucia EH and Berlyn GP (1984) Effect of climate on the photosynthesis of *Picea mariana* at the subarctic treeline 1. Field measurements. Can. J. Bot. 62:2423-2428.

DeLucia EH and Smith WK (1987) Air and soil temperature limitations on photosynthesis in Engelmann spruce during summer. Can. J. Bot. 17:527-533.

Doehlert DC and Walker RB (1981) Photosynthesis and photorespiration in Douglas-fir as influenced by irradiance, CO_2 concentration, and temperature. For. Sci. 27:641-650.

Emmingham WH (1982) Ecological indices as a means of evaluating climate, species distribution and primary production. pp. 45-67. In Edmonds RL (editor) Analysis of Coniferous Forest Ecosystems in the Western US. US/IBP Series Vol 14, Dowden, Hutchinson, and Ross, Stroudsburg, PA.

Emmingham WH and Waring RH (1977) An index for photosynthesis for comparing forest sites in Oregon. Can. J. For. Res. 7:165-174.

Fahey TJ (1979) The effect of night frost on the transpiration of *Pinus contorta* ssp. *latifolia*. Oecol. Plant 19:483-490.

Fahey TJ and Knight DH (1986) Lodgepole pine ecosystems. Bioscience 36:610-617.

Field CB (1987) Leaf-age effects on stomatal conductance. pp. 367-384. In Zieger E, Farquhar GD, and Cowan IR (editors) Stomatal Function. Stanford Univ. Press, Stanford, CA.

Field C and Mooney HA (1986) The photosynthesis-nitrogen relationship in wild plants. pp. 25-55. In Givinish TJ (editor) On the Economy of Plant Form and Function. Cambridge Univ. Press, Cambridge.

Foster JR and Smith WK (1986) Influence of stomatal distribution on transpiration in low-wind environments. Plant Cell Env. 9:751-759.

Franklin JF (1988) Pacific northwest forests. pp. 103-130. In Barbour MG and Billings DW (editors) North American Terrestrial Vegetation. Cambridge Univ. Press, Cambridge.

Franklin JF and Waring RH (1979) Distinctive features of the northwestern coniferous forests: Development, structure, and function. pp. 83-105. In Waring RH (editor) Forests: Fresh Perspectives from Ecosystem Analysis. Oregon State Univ. Press, Corvallis, OR.

Gale J (1972) Availability of carbon dioxide for photosynthesis at high altitudes: theoretical considerations. Ecology 53:494-497.

Gale J (1973) Experimental evidence for the effect of barometric pressure on photosynthesis and transpiration. pp. 289-294. In Plant Response to Climate Factors, UNESCO, Uppsala Symposium on Ecology and Conservation 5.

Gates DM (1980) Biophysical Ecology. Springer-Verlag, Heidelberg.

Gauslaa Y (1984) Heat resistance and energy budget in different Scandinavian plants. Holarct. Ecol. 7:1-78.

Grace J, Malcolm DC, and Bradbury IK (1975) The effect of wind and humidity on leaf diffusive conductance in Sitka spruce seedlings. J. Appl. Ecol. 12:931-940.

Grier CC (1988) Foliage loss due to snow, wind, and winter drying damage: its effect on leaf biomass of some western forests. Can. J. For. Res. 18:1097-1102.

Grier CC, Lee KM, Nadkarni NM, and Klock GO (1986) Productivity of forests of the United States, its relation to site factors and management practices: a review. USDA, Pacific Northwest Forest and Range Experiment Station, Portland, OR.

Hadfield W (1975) The effect of high temperatures on some aspects of the physiology and cultivation of the tea bush, *Camellia sinensis*, in North East India. pp. 477-496. In Evans EC, Bainbridge R, and Rackham O (editors) Light as an Ecological Factor II. Blackwell Scientific Publications, Oxford.

Hadley JL and Smith WK (1983) Influence of wind exposure on needle desiccation and mortality for timberline species in Wyoming. Arctic Alpine Res. 15:127-135.

Hadley JL and Smith WK (1986) Wind effects on needles of timberline conifers: seasonal influence on mortality. Ecology 67:12-19.

Hadley JL and Smith WK (1987) Influence of krummholz mat structure on microclimate and needle physiology. Oecologia 73:82-90.

Hadley JL and Smith WK (1989a) Influence of leaf surface wax and the leaf area to water content ratio on cuticular transpiration in western conifers, USA. Can. J. For. Res. (submitted).

Hadley JL and Smith WK (1989b) Wind erosion of leaf surface wax in alpine timberline conifers. Arctic Alpine Res. (In press).

Hermann HK (1977) Growth and production of tree roots. pp. 44-57. In Marshall JK (editor) The below-ground ecosystem: a synthesis of plant-associated processes. Range Science Department Series 26, Colorado State Univ., Fort Collins, CO.

Hinckley TM, Schroeder MO, Roberts JE, and Bruckerhoff DN (1975) Effects of several environmental variables and xylem pressure potential on leaf surface resistance in white oak. For. Sci. 22:201-211.

Hutchinson BA and Hicks BB (1983) The Forest-Atmosphere Interaction. D. Reidal, Boston, MA.

Jarvis PG (1980) Stomatal responses to water stress in conifers. pp. 105-122. In Turner NC and Kramer PJ (editors) Adaptations of Plants to Water and High Temperature Stress. Wiley, NY.

Jarvis PG and Mansfield TA (editors) (1981) Stomatal Physiology. Cambridge Univ. Press, Cambridge.

Kaufmann MR (1979) Stomatal control and the development of water deficit in Engelmann spruce seedlings during drought. Can. J. For. Res. 9:297-304.

Kaufmann MR (1982a) Leaf conductance as a function of photosynthetic photon flux density and absolute humidity difference from leaf to air. Plant Physiol. 69:1018-102.

Kaufmann MR (1982b) Evaluation of season, temperature, and water stress effects on stomata using a leaf conductance model. Plant Physiol. 69:1023-1026.

Kaufmann MR (1985) Annual transpiration in subalpine forests: large differences among four tree species. For. Ecol. Manag. 13:234-246.

Kaufmann MR and Ryan MG (1986) Physiographic, stand and environmental effects on individual tree growth and growth efficiency in subalpine forests. Tree Physiol. 2:47-60.

Knapp AK and Smith WK (1981) Water relations and succession in subalpine conifers in southeastern Wyoming. Bot. Gaz. 142:502-511..

Knapp AK and Smith WK (1987) Coupling of stomatal and photosynthetic responses to sun/shade transition in subalpine plants. Oecologia 74:62-67.

Knapp AK and Smith WK (1988) Influence of water stress on stomatal and photosynthetic responses to cloud pattern. Amer. J. Bot. 75:851-888.

Knapp AK and Smith WK (1989) Influence of growth form on the ecophysiology of subalpine plants during variable sunlight. Ecology 70:1069-1082.

Knapp AK, Smith WK, and Young DR (1989) Importance of intermittent shade to the ecophysiology of subalpine herbs. Funct. Ecol. 3:753-758.

Körner C and Diemer M (1987) *In situ* photosynthetic responses to light, temperature and carbon dioxide in herbaceous plants from low and high altitude. Funct. Ecol. 1:179-194.

Körner C and Renhardt U (1987) Dry matter partitioning and root length/leaf area ratios in herbaceous perennial plants with diverse altitudinal distribution. Oecologia 74:411-418.

Körner C and Woodward FI (1987) The dynamics of leaf extension in plants of diverse altitudinal ranges II. Field studies in *Poa* species between 600 and 3200 m altitude. Oecologia 72:279-283.

Körner C, Allison A, and Hilscher H (1983) Altitudinal variation of leaf diffusive conductance and leaf anatomy in heliophytes of montane New Guinea and their interrelation with microclimate. Flora 174:91-135.

Körner C, Bannister P, and Mark AF (1986) Altitudinal variation in stomatal conductance, nitrogen content and leaf anatomy in different life forms in New Zealand. Oecologia 69:577-588.

Körner C, Farquhar GD, and Roksandic Z (1988) A global survey of carbon isotope discrimination in plants from high altitude. Oecologia 74:623-632.

Körner C, Scheel JA, and Bauer H (1979) Maximum leaf diffusive conductance in vascular plants. Photosynthetica 13:45-82.

Landsberg JJ (1986) Physiological Ecology of Forest Production. Academic Press, NY.

Larcher W (1969) The effect of environmental and physiological variables on the carbon dioxide gas exchange of trees. Photosynthetica 3:167-198.

Larcher W (1980) Physiological Plant Ecology. Springer-Verlag, Heidelberg.

Larcher W (1985) Winter stress in high mountains. pp. 11-17. In Turner H and Tranquillini W (editors) Establishment and Tending of Subalpine Forest. Eidg. Anst. forstl. Versuchsw. Ber. 270.

Larcher W and Bauer H (1981) Ecological significance of resistance to low temperature. pp. 403-437. In Lange OL, Nobel PS, Osmond CB, and Ziegler H (editors) Physiological Plant Ecology I: Response to the Physical Environment, Encyclopedia of Plant Physiology (New Series) Vol. 12A. Springer-Verlag, Heidelberg.

Larcher W and Siegwolf R (1985) Development of acute frost drought in *Rhododendron ferrugineum* at the alpine timberline. Oecologia 67:298-300.

Lassoie JP, Dougherty PM, Reich PB, Hinckley TM, Metcalf CM and Dinn SD (1983) Ecophysiological investigations of understory eastern red cedar in central Missouri. Ecology 64:1355-1366.

Leverenz JW (1981a) Photosynthesis and transpiration in large forest-grown Douglas-fir: diurnal variation. Can. J. Bot. 59:349-356.

Leverenz JW (1981b) Photosynthesis and transpiration in large forest-grown Douglas-fir: interactions with apical control. Can. J. Bot. 59:2568-2576.

Leverenz JW and Jarvis PG (1979) Photosynthesis in Sitka spruce (*Picea sitchensis* [Bong.] Carr.) VII: the effects of light flux density and direction on the rate of photosynthesis and stomatal conductance of needles. J. Appl. Ecol. 16:919-932.

Leverenz JW and Jarvis PG (1980) Photosynthesis in Sitka spruce (*Picea sitchensis* [Bong.] Carr.) IX: the relative contribution made by needles at various points on the shoot. J. Appl. Ecol. 17:59-68.

Leverenz JW and Öquist G (1987) Quantum yields of photosynthesis at temperatures between -2° C and 35° C in a cold-tolerant C_3 plant (*Pinus sylvestris*) during the course of one year. Plant Cell Env. 10:287-295.

Losch R and Tenhunen J (1981) Stomatal responses to humidity/phenomenon and mechanism. pp. 137-161. In Jarvis PG and Mansfield TA (editors) Stomatal Physiology. Cambridge Univ. Press, Cambridge.

Ludlow MM and Jarvis PG (1971) Photosynthesis in Sitka spruce (*Picea sitchensis* [Bong.] Carr.). J. Appl. Ecol. 8:925-953.

Luxmore RJ, Landsberg JJ and Kaufmann MR (editors) (1986) Coupling of carbon, water, and nutrient interactions in woody plant soil systems. Tree Physiol, vol 2. Heron Publishing, Victoria, Canada.

Monteith JL (editor) (1976) Vegetation and the atmosphere, vol 2: case studies. Academic Press, NY.

Mooney HA and Chiariello NR (1984) The study of plant function - the plant as a balanced system. pp. 305-323. In Dirzo C and Sarukhan T (editors) Perspectives on plant population ecology. Sinauer, Sunderland, Massachusetts.

Mooney HA, Pearcy RW, and Ehleringer J (1987) Plant physiological ecology today. BioScience 37:18-20.

Morgan DC and Smith H (1981) Non-photosynthetic responses to light quality. pp. 109-134. In Lange OL, Nobel PS, Osmond CB, and Ziegler H (editors) Physiological Plant Ecology I: Response to the Physical Environment, Encyclopedia of Plant Physiology (New Series) Vol. 12A. Springer-Verlag, Heidelberg.

Morison JL (1987) Intercellular CO_2 concentration and stomatal response to CO_2. pp. 229-252. In Zieger E, Farquhar GD, and Cowan IR (editors) Stomatal function. Stanford Univ. Press, Stanford, CA.

Neilson RE, Ludlow MM, and Jarvis PG (1972) Photosynthesis in Sitka spruce (*Picea sitchinsis* [Bong.] Carr.) II. Responses to temperature. J. Appl. Ecol. 9:721-745.

Neufeld HS, Meizner FC, Wisdom CS, Sharifi MR, Rundel PW, Neufeld MS, Goldring Y, and Cunningham GL (1988) Canopy architecture of *Larrea tridentata* (DC.) Cov., a desert shrub: foliage orientation and direct beam radiation interception. Oecologia 75:54-60.

Nobel PS and Walker DB (1985) Structure of leaf photosynthetic tissue. pp. 501-536. In Barker J and Baker NR (editors) Photosynthetic mechanisms and the environment. Elsevier, Amsterdam.

Nobel PS, Zaragoza LJ, and Smith WK (1975) Relation between mesophyll cell surface area, photosynthetic rate, and illumination during development for leaves of *Plectranthus parviflorus* Henckel. Plant Physiol. 55:176-183.

Oberbauer SF and Billings WD (1981) Drought tolerance and water use along an alpine topographic gradient. Oecologia 50:325-331.

Öquist G (1983) Effects of low temperature on photosynthesis: a review. Plant Cell Env. 6:281-300.

Osmond CB, Björkman O, and Anderson DJ (1980) Physiological processes in plant ecology: toward a synthesis with *Atriplex*. Ecological Studies Series, Vol. 36, Springer-Verlag, Heidelberg.

Peet RK (1988) Forests of the Rocky Mountains. pp. 63-102. In Barbour MG and Billings DW (editors) North American terrestrial vegetation. Cambridge Univ. Press, Cambridge.

Rackham O (1975) Temperatures of plant communities as measured by pyrometric and other methods. pp 477-496. In Evans EC, Bainbridge R, and Rackham O (editors) Light as an ecological factor II. Blackwell Scientific Publications, Oxford.

Richards JH and Bliss LC (1986) Winter water relations of a deciduous timberline conifer, *Larix lyalli* Parl. Oecologia 69:16-24.

Running SW (1976) Environmental control of leaf water conductance in conifers. Can. J. For. Res. 6:104-112.

Running SW (1980) Relating plant capacitance to the water relations of *Pinus contorta*. For. Ecol. Manag. 2:237-252.

Running SW and Reid CPP (1980) Soil temperature influences on root resistance of *Pinus contorta* seedlings. Plant Physiol. 65:635-640.

Rutter MR (1977) An ecophysiological field study of three Sierra conifers. PhD Thesis, Univ. California, Berkeley, CA.

Sakai A and Larcher W (1987) Frost Survival of Plants: Responses and Adaptations to Freezing Stress. Springer-Verlag, Heidelberg.

Salo DJ (1974) Factors affecting photosynthesis in Douglas-fir. PhD Thesis, Univ. Washington, Seattle, WA.

Sanford AP and Jarvis PG (1986) Stomatal responses to humidity in selected conifers. Tree Physiol. 2:105-114.

Schulze E-D (1982) Plant life forms and their carbon, water and nutrient relations. pp. 615-676. Physiological Plant Ecology II: Water Relations and Carbon Assimilation. Encyclopedia of Plant Physiology (New Series), Vol. 12B. Springer-Verlag, Heidelberg.

Schulze E-D (1986) Carbon dioxide and water vapor exchange in response to drought in the atmosphere and soil. Ann. Rev. Plant Physiol. 37:247-274.

Schulze E-D and Hall AE (1982) Stomatal responses, water loss and CO_2 assimilation rates of plants in contrasting environments. pp. 181-230. In Lange OL, Nobel PS, Osmond CB, Ziegler H (editors) Physiological Plant Ecology II: Water Relations and Carbon Assimilation. Encyclopedia of Plant Physiology (New Series), Vol. 12B. Springer-Verlag, Heidelberg.

Schulze E-D, Turner NC, Gollan T, and Shackel KA (1987) Stomatal responses to air humidity and to soil drought. pp. 311-321. In Zieger E, Farquhar GD, and Cowan IR (editors) Stomatal function. Stanford Univ. Press, Stanford, CA.

Sharkey TD and Ogawa T (1987) Stomatal responses to light. pp. 175-208. In Zieger E, Farquhar GD, and Cowan IR (editors) Stomatal function. Stanford Univ. Press, Stanford, CA.

Sharma ML (1984) Evapotranspiration from plant communities. Elsevier, Oxford.

Smith WK (1981) Temperature and water relation patterns in subalpine understory plants. Oecologia 48:353-359.

Smith WK (1985a) Montane forests. pp. 95-126. In Chabot BF and Mooney HA (editors) Physiological ecology of North American plant communities. Chapman and Hall, NY.

Smith WK (1985b) Environmental limitations on leaf conductance in central Rocky Mountain conifers, USA. pp. 95-101. In Turner H and Tranquillini W (editors) Establishment and tending of subalpine forest: Research and management. Eidg. Anst. forstl. Versuchsw. Ber. 270.

Smith WK and Carter GA (1988) Shoot structural effects on needle temperature and photosynthesis in conifers. Amer. J. Bot. 75:496-500.

Smith WK and Geller GN (1979) Plant transpiration at high elevation: theory, field measurements, and comparisons with desert plants. Oecologia 41:109-122.

Smith WK and McClean TM (1989) Adaptive importance of leaf water repellency, stomatal distribution, and gas exchange. Amer. J. Bot. 76:465-469.

Smith WK and Hollinger DY (1990) Stomatal behavior. In Lassoie JP and Hinckley TM (editors) Ecophysiology of forest trees: techniques and approaches. CRC Press, NY, (in press).

Smith WK and Nobel PS (1978a) Influence of seasonal changes in leaf morphology on water use efficiency for three desert broadleaf shrubs. Ecology 58:1033-1043.

Smith WK and Nobel PS (1978b) The influence of leaf temperature, soil water potential, and illumination on leaf morphology in *Encelia farinosa* Gray. Amer. J. Bot. 28:171-185.

Smith WK, Knapp AK, and Reiners WA (1989) Penumbral effects on sunlight penetration in plant communities. Ecology 3:1603-1609.

Smith WK, Young DR, Carter GA, Hadley JL, and McNaughton GM (1984) Autumn stomatal closure in six conifer species of the central Rocky Mountains. Oecologia 63:237-242.

Squire GR and Black CR (1981) Stomatal behavior in the field. pp. 223-246. In Zieger E, Farquhar GD, and Cowan IR (editors) Stomatal function. Stanford Univ. Press, Stanford, CA.

Szawiawski RK and Wierzbicki B (1978) Net photosynthetic rate of some coniferous species at diffuse high irradiance. Photosynthetica 12:412-417.

Tenhunen J, Pearcy RW, and Lange OL (1987) Diurnal variations in leaf conductance and gas exchange in natural environments. pp. 323-352. In Zieger E, Farquhar GD, and Cowan IR (editors) Stomatal function. Stanford Univ. Press, Stanford, CA.

Tesky RO, Hinckley TM, and Grier CC (1983) Effect of interruption of flow path on stomatal conductance of *Aibes anabilis*. J. Exp. Bot. 34:1251-1259.

Tesky RO, Hinckley TM, and Grier CC (1984) Temperature induced change in the water relations of *Aibes anabilis* (Dougl.) Forbes. Plant Physiol. 74:77-80.

Teskey RO, Fites JA, Samuelson LJ, and Bongarten BC (1987) Stomatal and nonstomatal limitations to net photosynthesis in *Pinus taeda* L. under different environmental conditions. Tree Physiol. 2:131-142.

Tilman D (1988) Plant strategies and the dynamics and structure of plant communities. Princeton Univ. Press, Princeton, NJ.

Tranquillini W (1959) Die stoffproduktion der Zirbe (*Pinus cembra* L.) an der Waldgrenze wahrend eines Jahres. Planta 54:107-151.

Tranquillini W (1976) Water relations and alpine timberline. pp. 473-491. In Lange OL, Kuppen L, and Schulze E-D (editors) Water and Plant Life. Ecological Studies 19. Springer-Verlag, Heidelberg.

Tranquillini W (1979) Physiological Ecology of the Alpine Timberline. Springer-Verlag, Heidelberg.

Tranquillini W (1982) Frost drought and its ecological significance. pp. 379-400. In Lange OL, Nobel PS, Osmond CB, and Ziegler H (editors) Physiological Plant Ecology II: Water Relations and Carbon Assimilation. Encyclopedia of Plant Physiology (New Series), Vol. 12B. Springer-Verlag, Heidelberg.

Waring RH and Schlesinger WH (1985) Forest ecosystems: concepts and management. Academic Press, NY.

Willmer CM (1983) Stomata. Longman, NY.

Winner WE and Phelps LB (editors) (1988) Responses of trees to air pollution: the role of branch studies. Western Conifers Research Cooperative, National Forest Response Program, USDA, USEPA, Boulder, CO.

Wong SC, Cowan IR, and Farquhar GD (1979) Stomatal conductance correlates with photosynthetic capacity. Nature 282:424-426.

Woodward FI (1983) The significance of interspecific differences in specific leaf area to the growth of selected herbaceous species from different altitudes. New Phytol. 95:313-323.

Woodward FI (1986) Ecophysiological studies on the shrub *Vaccinium myrtillus* L. taken from a wide altitudinal range. Oecologia 70:580-586.

Woodward FI, Körner C, and Crabtree RC (1986) The dynamics of leaf extension in plants with diverse altitudinal ranges I. Field observations on temperature responses at one altitude. Oecologia 70:222-226.

Young DR and Smith WK (1979) Influence of sunflecks on the temperature and water relations of two subalpine, understory congeners. Oecologia 43:195-20.

Young DR and Smith WK (1980) Influence of sunlight on photosynthesis, water relations, and leaf structure in the understory species *Arnica cordifolia*. Ecology 48:353-359.

Young DR and Smith WK (1982) Simulation studies of the influence of understory location on transpiration and photosynthesis of *Arnica cordifolia* on clear days. Ecology 63:1761-1770.

Young DR and Smith WK (1983) Effect of cloudcover on photosynthesis and transpiration in the subalpine understory species *Arnica latifolia*. Ecology 64:681-687.

Zelawski W and Niwinski Z (1966) Variability of some needle characteristic in Scots pine (*Pinus silvestris* L.) ecotypes grown in native conditions. Ekol. Polika Ser. A 14:301-311.

Zieger E, Farquhar GD, and Cowan IR (editors) (1987) Stomatal Function. Stanford Univ. Press, Stanford, CA.

Plate 9.1. (upper) In the high elevation spruce-fir forests, sunlight penetration and penumbral patches form a mosaic of full sun, low irradiance sunflecks, and deep shade that changes dramatically on an hourly and seasonal basis.

Plate 9.2. (lower) Periods of high irradiance and low windspeeds commonly result in very high evaporative demands on high elevation species. *Arnica cordifolia* is one of several subalpine herbs that routinely wilts during the prolonged sunpatches characteristic of coniferous forest understory environments.

Plate 10.1. (upper) Extreme wind and ice abrasion of leaf cuticles is characteristic of the timberline ecotone during winter.

Plate 10.2. (lower) As a result, severe winter desiccation can lead to high leaf mortality on already established trees. Snow coverage is crucial for new leaves on mature trees as well as for the establishment and survival of seedlings at this extreme upper limit of growth.

Plate 11.1. (upper) Island of Sierra Nevada pines on hydrothermally altered granite south of Reno, NV. Trees are surrounded by sagebrush vegetation. Pinyon-juniper woodland of the Virginia Mountains is in the foreground, and east-slope Sierra Nevada forest is in the background.

Plate 11.2 (lower) *Eriogonum robustum*, closely related to the Sierra Nevada species *E. lobbii*, is endemic to altered rock in the western Great Basin.

Plate 12.1. (upper) Creosote-bursage (*Larrea tridentata - Ambrosia dumosa*) vegetation type in the lowland Great Basin-Mojave Desert margins of eastern California.

Plate 12.2. (lower) Riparian zone within the lowland Mojave Desert of southern Nevada. Most of the riparian corridor is dominated by the exotic saltcedar (*Tamarix chinensis*) and arrow-weed (*Tessaria sericea*), with a single native cottonwood (*Populus fremontii*) in the center, and mesquite (*Prosopis juliflora*) along the edges.

Chapter 5

Ecophysiology of Great Basin and Sierra Nevada Vegetation on Contrasting Soils

Evan H. DeLucia and William H. Schlesinger

Introduction

A dramatic feature of the landscape in the western Great Basin desert is the occurrence of "tree islands" in the midst of sagebrush vegetation (Plate 11.1). Small stands of Sierra Nevada conifers, primarily *Pinus ponderosa* and *P. jeffreyi*, occur up to 60 km east of the eastern limits of the Sierra Nevada montane forest. These stands range from one to several hectares, are largely devoid of characteristic Great Basin species, and are restricted to outcrops of azonal soil derived from hydrothermally altered andesitic bedrock (Billings 1950). Soil derived from altered bedrock is light yellow and contrasts strongly with adjacent dark brown soil derived from unaltered andesite.

These disjunct Sierra Nevada communities provide an opportunity to examine the ecophysiological responses of dominant Great Basin and Sierra Nevada plants growing in a desert climate. We have used this system to explore the importance of drought tolerance and water- and nutrient-use efficiency in enabling Sierra Nevada and Great Basin plants to persist at different levels of resource availability. Additionally, the role of bedrock geology in defining the distribution of desert vegetation will be considered. Much of the research discussed in this chapter stems from the original description of these "tree islands" by W.D. Billings, and was undertaken with his collaboration.

Vegetation

Approximately 140 stands of Sierra Nevada pines on altered rock, ranging from one to several hectares, are located in the western Great Basin desert in Nevada and on the east slopes of the Sierra Nevada in California (Billings pers. comm.). Of the 15 sites used in our studies (Figure 5.1), 10 sites occur in sagebrush steppe and pinyon-juniper woodland (1,400-2,000 m above sea level) and 5 sites occur among high-elevation Sierra Nevada montane forest (1,750-2,100 m). Vegetation data are summarized for one site in each community type (Table 5.1).

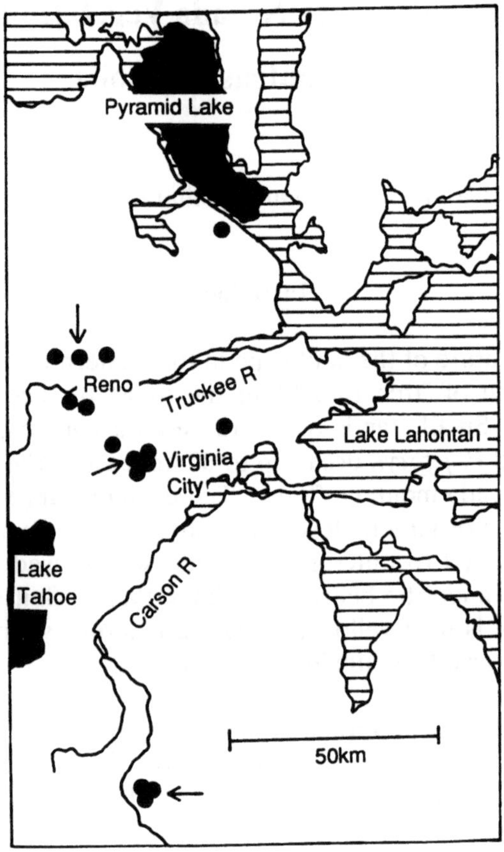

Figure 5.1. Western Nevada and eastern California showing the western shoreline of the Pleistocene Lake Lahontan (hatched), the extent of present-day lakes (black), and location of the study sites. Intensive study sites as described in Table 5.1 and by DeLucia et al. (1988) are circled. From Schlesinger et al. (1989).

In the Great Basin *Pinus ponderosa* and *P. jeffreyi* are dominant on altered rock of the Virginia Mountains and Peavine Mountain, whereas *Artemisia tridentata* and other desert shrubs are dominant on adjacent unaltered country rock (Table 5.1). At least 8 of the 15 taxa of Sierra Nevada conifers occur on altered sites at other locations in the Great Basin. Interspersed among widely spaced trees on the altered rock are a few endemics and a number of Sierra Nevada alpine plants and their congeners (Billings 1950). *Eriogonum robustum* (Plate 11.2), closely related to the Sierra Nevada species *E. lobbii*, and *Calyptridium umbellatum*, are two endemics on altered soil in the Great Basin. Total plant cover is uniformly lower on soils derived from altered compared to unaltered andesite. On the driest sites there is little floristic overlap with adjacent sagebrush vegetation; few of the typical desert species occur on altered rock to which the Sierra Nevada species are restricted. Moreover, the ecotone between communities on soils derived from altered and unaltered bedrock typically is defined sharply (Plate 11.1).

In the slightly more mesic pinyon-juniper woodlands of the western Great Basin ranges, scattered individuals of *Juniperus osteosperma*, *Pinus monophylla*, and *Amelanchier alnifolia* occur on altered rock in addition to the Sierra Nevada pines. As moisture availability further increases with higher elevation in the Sierra Nevada montane forest, the species composition on adjacent rock types becomes more similar (Alpine County, Table 5.1). *Pinus jeffreyi* as well as *Arctostaphylos patula* and other shrubs occur on both soils, although at greatly reduced densities on nutrient-poor altered soils. The pattern of increasing similarity of vegetation on these two soil types with increasing elevation and precipitation suggests that moisture availability is one important factor in maintaining these unique forested communities in sagebrush desert.

Soil Chemistry

Hydrothermal alteration probably occurred in the late Miocene (Calkins 1944, Gianella 1936). Rising hot water introduced elemental sulfur that combined with iron in the ferromagnesian minerals of the rock to form pyrite. As exposed altered rock eroded, pyrite oxidized to sulfuric acid leaching most of the base cations and leaving shallow, acidic soils of mostly alumino-silicates.

Soils derived from andesitic bedrock in the Great Basin are primarily Xerollic Haplargids, typical of desert brown soils, whereas altered rock

Table 5.1. Vegetation summary of selected research sites on altered and unaltered rock (from DeLucia et al. 1988). Density (plants ha^{-1}), dominance (Dom.) of trees (basal area, $m^2\ ha^{-1}$) or of shrubs (frequency, %), and importance value (%) are shown for all species in the tree-size class, and for the two species in the shrub-size class with the highest importance value. Importance value in the tree-size class was calculated as the mean of the relative density and relative dominance. Importance value for plants in the shrub-size class was calculated as the mean of the relative density and relative frequency.

Vegetation	Unaltered site			Altered Site		
	Density	Dom.	Importance Value	Density	Dom.	Importance Value
Alpine County[3]						
Trees (all species)						
Pinus jeffreyi	510	36.0	98	260	23.9	100
Abies concolor	10	<0.1	1	—[1]	—	—
Cercocarpus ledifolius	10	<0.1	1	—	—	—
Shrubs						
Arctostaphylos patula	50	2	6	150	6	17
Purshia tridentata	2300	42	72	—	—	—
Pinus seedlings[2]	600	10	18	1500	18	83
Other	100			50		
Total density	3580			1960		
Herb cover (%)	0.1			<0.1		
Virginia Mountains[3]						
Trees (all species)						
Pinus ponderosa	—	—	—	100	3.9	83
Pinus monophylla	40	6.2	100	10	0.2	6
Juniperus osteosperma	—	—	—	30	<0.1	11
Shrubs						
Artemisia tridentata	4200	68	30	—	—	—
Chrysothamnus parryi	350	10	4	5050	40	68
Eriogonum wrightii	9450	66	45	450	6	8
Pinus seedlings[2]	—	—	—	800	20	20
Other	2400			100		
Total density	16420			6540		
Herb cover (%)	3.4			0.1		

Table 5.1 continued

Vegetation	Unaltered site			Altered Site		
	Density	Dom.	Importance Value	Density	Dom.	Importance Value
Peavine Mountain[3]						
Trees (all species)						
Pinus ponderosa	—	—	—	70	3.3	58
Pinus jeffreyi	—	—	—	60	2.3	42
Shrubs						
Artemisia tridentata	6550	88	47	550	14	31
Tetradymia glabrata	6600	94	48	—	—	—
Eriogonum sp.	—	—	—	1200	18	51
Other	400			350		
Total density	13550			2230		
Herb cover (%)	1.2			0.2		

[1] Species was not present in the sample.

[2] *Pinus* seedlings were either *P. ponderosa* or *P. jeffreyi*.

[3] Site characteristics:

	Unaltered site	Altered site
Alpine County		
Elevation (m)	2088	2088
Slope (°)	14	21
Aspect (° from true N)	170	182
Virginia Mountains		
Elevation (m)	1615	1615
Slope (°)	16	15
Aspect (° from true N)	268	268
Peavine Mountain		
Elevation (m)	1710	1735
Slope (°)	15	19
Aspect (° from true N)	117	192

Table 5.2. Soil chemical properties on altered and unaltered parent materials in the Sierra Nevada and Great Basin, measured on 1:5 soil extract and HCO_3^--P extracted at pH 8.5 (from Schlesinger et al. 1989). All data are means with one SE in parentheses. Analysis of variance was used to distinguish main effects due to location and soil type; t-tests were used to distinguish differences between soil types within a location. Asterisks indicate significance at < .05.

Location		pH in 0.01 M $CaCl_2$	pH	HCO_3^-	SO_4^- (mg l^{-1})	Ca^{2+}	H_2O-P ($\mu g\ g^{-1}$)	HCO_3^--P	C (mg g^{-1})	N
Sierra Nevada			*	*	*	*	*			
Unaltered	(n = 5)	5.72	6.23	0.30	1.01	6.38	0.54	22.6	0.234	0.011
		(0.19)	(0.19)	(0.06)	(0.12)	(1.03)	(0.23)	(5.2)	(0.044)	(0.002)
Altered	(n = 5)	3.99	5.06	0.07	5.54	2.24	0.27	32.0	0.224	0.008
		(0.22)	(0.17)	(0.03)	(1.46)	(0.70)	(0.22)	(19.0)	(0.031)	(0.002)
Great Basin Pinyon Zone		*	*	*		*	*	*		
Unaltered	(n = 4)	6.19	6.70	0.56	1.21	4.06	1.14	23.3	0.142	0.008
		(0.24)	(0.13)	(0.21)	(0.23)	(1.10)	(0.33)	(6.1)	(0.043)	(0.002)
Altered	(n = 4)	3.66	4.81	0.04	6.46	0.98	0.02	2.9	0.107	0.005
		(0.12)	(0.18)	(0.02)	(2.61)	(0.43)	(0.02)	(1.7)	(0.042)	(0.002)
Sagebrush zone		*	*	*			*			
Unaltered	(n = 4)	6.55	7.00	0.45	0.84	6.07	2.81	31.4	0.078	0.006
		(0.17)	(0.13)	(0.11)	(0.22)	(2.15)	(1.25)	(4.4)	(0.012)	(0.001)
Altered	(n = 4)	3.74	4.69	0.03	19.86	2.64	0.16	5.6	0.115	0.009
		(0.15)	(0.23)	(0.02)	(12.81)	(1.55)	(0.16)	(2.9)	(0.015)	(0.002)
ANOVA for main effects										
Location		ns	ns	ns	ns	ns	ns	ns	*	ns
Soil		*	*	*	*	*	*	ns	ns	ns

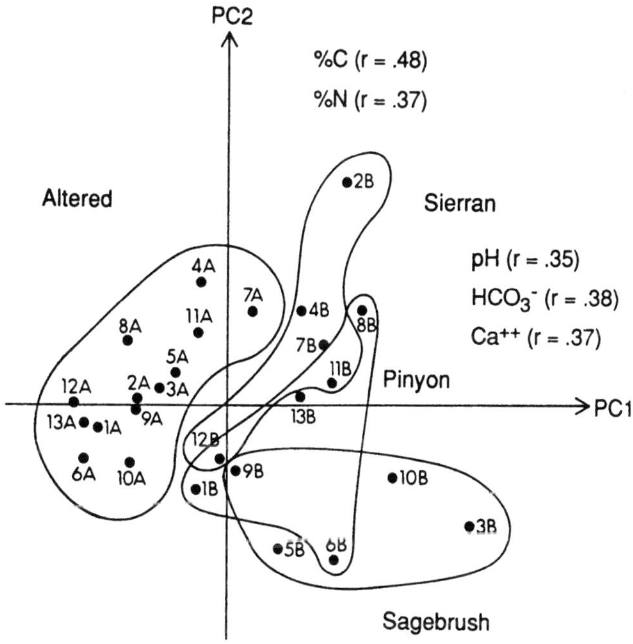

Figure 5.2. Principle component analysis of soil data. At each location a sample was analyzed for soils derived from adjacent hydrothermally altered andesite (A) and unaltered parent material (B). Soils derived from altered rock form one group whereas unaltered soils vary with dominant vegetation type. The first two axes explain 48% of the variance and are correlated most strongly with the variables indicated (Schlesinger et al. 1989).

formed shallow Lithic Entisols. Altered soils have low concentrations of base cations and phosphorus and are acidic (Table 5.2; Billings 1950; Salisbury 1954, 1964; Goldberg 1982, 1985; Schlesinger et al. 1989). In an analysis of 18 soil parameters, Schlesinger et al. (1989) found that altered soils from different sites share basic similarities, but adjacent unaltered soils vary depending on whether they are collected from forest vegetation in the Sierra Nevada, from pinyon-juniper woodland, or from sagebrush in the Great Basin (Figure 5.2, Table 5.2). As indicated by principle components analysis, the percentage of carbon and nitrogen is greater in soils from Sierra Nevada forest than from sagebrush vegetation, and the differences between soil types is least in the Sierra Nevada where abundant precipitation has had a modifying effect on soil chemistry (Jenny 1980). The principle components analysis indicates that pH, alkalinity, and calcium explain most of the variance among sites.

However, the biological effect of these differences is probably through the influence on phosphorus availability, which is reduced by precipitation with iron and aluminum minerals under the acidic conditions (Salisbury 1954).

Hypotheses

The presence of these unique forested ecosystems within desert sagebrush vegetation of strikingly different physiognomy and microclimate poses two important questions in plant ecology. First, the question of origin: how do these tree islands come to be? Second, the question of maintenance: how do Sierra Nevada trees and other species that are adapted to cool moist montane and alpine conditions persist in a desert climate? In this chapter we present and discuss data relevant to the maintenance of these disjunct forested communities, but first the question of origin warrants further discussion. Billings and Schlesinger proposed that these disjunct Sierra Nevada communities are relicts of a widespread occurrence of Sierra Nevada pines on many rock types during the cool moist climate of the late Pleistocene. As the climate dried, Sierra Nevada species remained only on the altered rocks because of the inability of the Great Basin taxa to invade and survive on these rocks. Although this hypothesis has not been tested explicitly, it is supported by the observation that disjunct Sierra Nevada communities do not occur east of the western shoreline of the Pleistocene Lake Lahontan (Figure 5.1).

A two-part hypothesis for the maintenance of disjunct Sierra Nevada communities in the Great Basin proposes that species from the infertile soil type (the Sierra Nevada taxa) are excluded from the more fertile soil by competition for water and conversely, species from the more fertile soil (sagebrush vegetation) are excluded from the less fertile soil by physiological intolerance of nutrient deficiency. There are many similarities in the physiognomy and dynamics of vegetation on soils derived from hydrothermally altered and serpentine rock, and a similar hypothesis has been proposed for the maintenance of distinct serpentine communities (Kruckeberg 1954, Whittaker 1954). The primary effect of altered soil on vegetation is, however, through the influence of low pH and phosphorus availability rather than the nickel and chromium toxicity of serpentine-derived soil. In addition to illustrating the role of edaphic factors in the local modification of climatically governed distribution of vegetation in the region, the altered andesite stands provide the opportunity to compare physiological attributes of two contrasting vegetation types under a similar climatic regime. Implicit in our hypothesis is the notion that an evolutionary compromise has occurred in the ability of perennial plants to adjust to low water versus low nutrient availability.

Water Relations

Snow is the dominant form of precipitation for east-slope Sierra Nevada forests (Barbour 1988) as well as western Great Basin steppe (West 1988). However, annual precipitation for the western Great Basin is often less than one third of that for east-slope forests (Chapter 2). Low annual precipitation of marked seasonality has selected for a high degree of drought tolerance in Great Basin vegetation suggesting the hypothesis that Sierra Nevada pines are excluded from the matrix of sagebrush vegetation by competition for water. This hypothesis raises several questions regarding the comparative water relations of Great Basin and Sierra Nevada vegetation. For example, is drought tolerance an attribute of certain physiognomic groups (trees versus shrubs) or is it a characteristic of Great Basin vegetation as a whole? It has been proposed that efficient utilization of water, expressed as water-use efficiency (WUE), is an attribute of drought tolerant desert vegetation (Fischer and Turner 1978). However, the compromises associated with high WUE (Cohen 1970, Orians and Solbrig 1977) raise questions about the relationship between drought tolerance and WUE during competition in a desert moisture regime. Furthermore, the maintenance of Sierra Nevada pines on altered rock in a desert climate suggests that water is more abundant on this rock type, possibly as a result of lower vegetation cover. These predictions have been tested in field and greenhouse experiments.

Soil Moisture

Although a number of reliable methods for measuring soil moisture are currently available (Fonteyn et al. 1987, Reginato and Nakayama 1988), inferences based on these techniques are limited by the physical placement of probes. In the case of deeply rooted trees it is unlikely that the location of probes will correspond with actual root distribution. Moreover, the depth of soil on altered sites is typically <15 cm, and large

Table 5.3. Predawn water potential measured in early and late summer (from DeLucia et al. 1989). At each site measurements were made on several dominant species growing on adjacent altered (A) and unaltered rock (U). The Alpine County site is in Sierran montane forest, the Virginia Mountains site is in pinyon-juniper woodland, and the Peavine Mountain and Ramsey Mine sites are in sagebrush steppe (n = 5 ± 1 SD).

Species and Location	Site	Predawn Water Potential June (mPa)	Predawn Water Potential August (mPa)	Seasonal Amplitude 1987 (mPa)	Seasonal Amplitude 1986[1] (mPa)
Alpine County					
Arctostaphylos patula	U	-0.96 ± 0.07	-2.74 ± 0.20	1.78	2.13
Arctostaphylos patula	A	-0.75 ± 0.12	-1.97 ± 0.25	1.22	1.70
Artemisia tridentata	U	-0.98 ± 0.07	-2.80 ± 0.33	1.82	—
Pinus jeffreyi	U	-1.02 ± 0.06	-1.50 ± 0.06	0.48	0.86
Pinus jeffreyi	A	-0.95 ± 0.10	-1.31 ± 0.04	0.36	0.67
Virginia Mountains					
Amelanchier alnifolia	U	-1.15 ± 0.17	-2.66 ± 0.59	1.51	2.47
Amelanchier alnifolia	A	-0.77 ± 0.21	-1.83 ± 0.22	1.06	1.05
Artemisia tridentata	U	-1.23 ± 0.64	-2.33 ± 0.09	1.10	3.45
Juniperus osteosperma	U	-1.27 ± 0.06	-2.45 ± 0.22	1.18	1.41
Pinus monophylla	U	-1.08 ± 0.10	-1.52 ± 0.03	0.44	0.81
Pinus ponderosa	A	-0.82 ± 0.18	-1.15 ± 0.09	0.33	0.13
Peavine Mountain					
Artemisia tridentata	U	-1.43 ± 0.37	-3.85 ± 0.56	2.42	2.27
Pinus jeffreyi	A	-0.80 ± 0.26	-0.84 ± 0.12	0.04	0.26
Pinus ponderosa	A	-0.65 ± 0.20	-1.06 ± 0.06	0.41	0.39
Purshia tridentata	U	-0.72 ± 0.20	-0.75 ± 0.10	0.03	0
Ramsey Mine					
Artemisia tridentata	U	-1.88 ± 0.21	-3.33 ± 0.91	1.45	—
Juniperus osteosperma	U	-1.75 ± 0.13	-2.61 ± 0.54	0.86	—
Juniperus osteosperma	A	-1.30 ± 0.20	-2.29 ± 0.54	0.99	—
Pinus ponderosa	A	-0.96 ± 0.08	-1.24 ± 0.11	0.28	—

[1]Calculated from data in DeLucia et al. (1988).

trees on these sites are rooted in deep cracks in the altered grus (decomposed bedrock). Because of these conditions, measurements of predawn water potential have been used to estimate soil moisture availability integrated over the rooting depth of representative plants (Ritchie and Hinckley 1975).

Soil moisture depletion in desert shrub communities is strongly correlated with plant cover (Bronson et al. 1976, Miller et al. 1982, Schlesinger et al. 1987). Although total plant cover and leaf area index were not measured, the density of woody and herbaceous vegetation on altered soils is substantially lower than on adjacent brown desert soils (DeLucia et al. 1988). Measurements of predawn water potential made during the summers of 1986 and 1987 indicate that moisture availability is greater on altered rock toward the end of summer. In the few cases where the same species grow on both soil types (*Arctostaphylos patula* and *Pinus jeffreyi* at Alpine County; *Amelanchier alnifolia*, *Juniperus osteosperma*, and *Pinus monophylla* at Virginia Mountains), plants on altered rock consistently have higher predawn water potential by late summer (Table 5.3). In a parallel study in eastern deciduous forest, oaks growing on serpentine and nonserpentine soils show higher predawn water potential, and higher water availability on the serpentine soils (Hull and Wood 1984). Similar to the situation on altered rock, vegetation cover is lower on serpentine than on adjacent soils. Seasonal variation in predawn water potential for Sierra Nevada pines is low compared to shrubs (Table 5.3). In early summer it is typically below -1.0 MPa for the Sierra Nevada pines but remains above -1.5 MPa for the duration of the summer. The shrubs on soils derived from unaltered rock begin the summer with comparable or even higher predawn water potential values, which decline to below -2.0 MPa, and below -4.0 MPa for *Artemisia*. In addition to the high water storage capacity of trees (Waring and Running 1978), the lower seasonal amplitude of predawn water potential may indicate a relatively stable moisture supply on altered rock.

By late summer, shrubs of low stature have substantially lower predawn water potential than Sierra Nevada pines, and values for the smaller Great Basin conifers, *Juniperus osteosperma* and *Pinus monophylla*, are intermediate. The relationship between size and predawn water potential may imply a hierarchy of rooting depths, with the Sierra Nevada pines being the deepest rooted followed by *Juniperus* and *P. monophylla*, and the Great Basin shrubs. A distinct decrease in predawn water potential in *Pinus ponderosa* on altered bedrock occurs for trees that are less than ca. 12 cm tall, corresponding in age to 2 to 4 years (Figure 5.3). These small seedlings and saplings have not penetrated the altered grus and are rooted in the upper 10 cm of soil. Once roots penetrate altered bedrock (trees greater than 10 cm high), a relatively stable water supply is assured. Measurements of volumetric soil moisture in the upper 10 cm also indicate a more rapid depletion of water on the unaltered soils in early summer (DeLucia et al. 1988). This is probably a result of high transpiration rates in late winter and early spring by the pervasive winter annual *Bromus tectorum* (see Chapter 6). *Bromus* is restricted largely to the sagebrush brown soils, although it will grow on altered soil in localized areas of litter accumulation.

Other physiological attributes of vegetation on the two soil types are also important in establishing seasonal differences in substrate water availability. Shrubs, particularly *Artemisia tridentata*, are dominant on the unaltered soil type. As a physiognomic group shrubs maintain substantially higher leaf conductances than the Sierra Nevada pines over a range of predawn water potential from -0.5 to less than -5 MPa (DeLucia et al. 1988) and consequently can extract water at lower levels of availability. In deeper volcanic soils, *Artemisia* uses and depletes soil moisture strongly to depths of 2 m (Black and Mack 1986; Campbell and Harris 1977; Sturges 1977). For many *Pinus* species complete stomatal closure occurs at water potentials above -2.0 MPa (Lassoie et al. 1985; Smith 1985). Stomatal closure at relatively high values of predawn water potential in Sierra Nevada conifers reduces the demand on soil moisture reserves on altered rock.

Drought Tolerance

Drought tolerance of adjacent Great Basin and Sierra Nevada plants was assessed by two methods. The response of photosynthesis to moisture availability was determined by measuring maximum daily net photosynthetic rates as a function of predawn water potential. Measurements were made in early and late summer at several research sites differing in total annual precipitation and offering a range in predawn water potential (DeLucia and Schlesinger 1990). Measurements were also made of its maximum seasonal amplitude. In so far as predawn water potential reflects the ability of perennial plants to continue extracting soil moisture by maintaining stomatal conductance at progressively decreasing availability, its amplitude provides an index of drought tolerance.

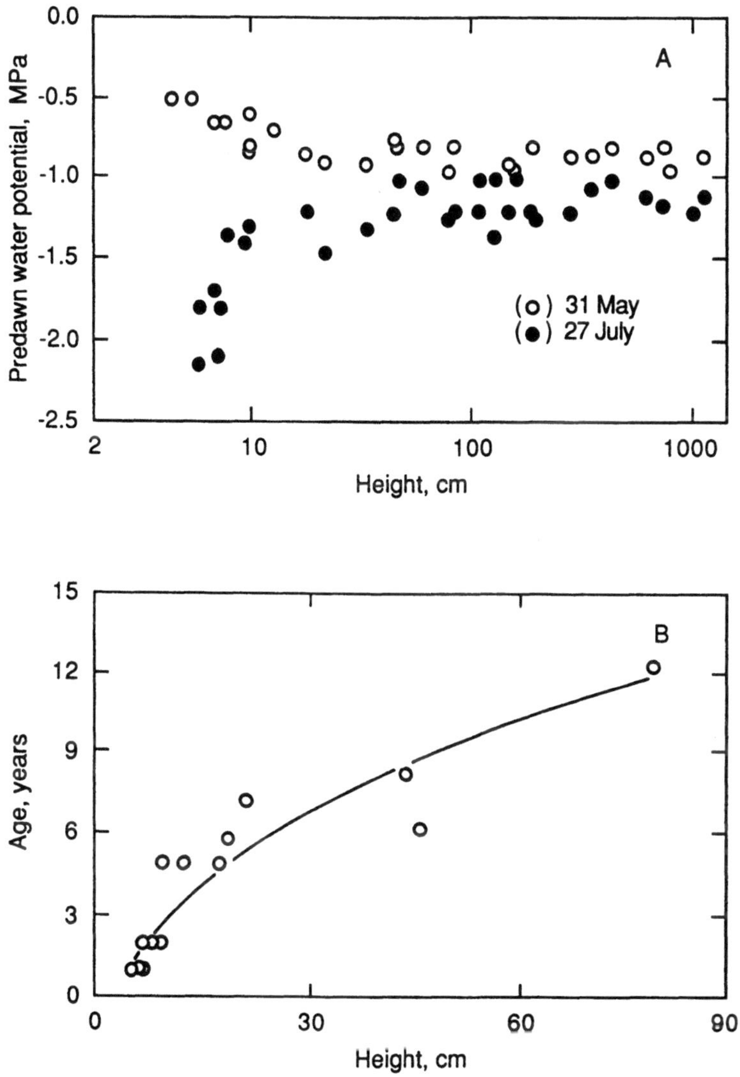

Figure 5.3. Predawn water potential of shoot-tips of *Pinus ponderosa* as a function of tree height (A), and the relationship between height and age (B). Measurements were made between 03:00 and 05:00 hours (solar) on the designated dates during 1987. Trees were growing on an altered site surrounded by sagebrush vegetation near Reno, Nevada. The site was gently sloping (slope: 11°; aspect: 215° magnetic), and soil depth ranged from 4 to 19 cm. Soil was underlain by altered grus. Following water potential measurements, several trees were cut or cored at the base for age determination.

Cessation of net photosynthesis occurs at substantially higher predawn water potential for Great Basin and Sierra Nevada trees than for shrubs. Water potentials greater than approximately -2.0 MPa allow positive rates of carbon assimilation for *Pinus ponderosa*, *P. jeffreyi*, and *P. monophylla* (Figure 5.4). *Juniperus osteosperma* maintains positive photosynthetic rates to ca. -3.0 MPa. Although the precise minimum water potential supporting net carbon assimilation for the shrubs cannot be determined from these data, shrubs, regardless of whether derived from Great Basin or Sierra Nevada flora, clearly have higher rates of photosynthesis at lower predawn water potential than the tree species. *Artemisia tridentata* exhibits positive carbon assimilation at below -4.5 MPa and is known to maintain gas exchange at even lower water potentials (Campbell and Harris 1977; DePuit and Caldwell 1973).

By late summer, shrubs have substantially lower predawn water potential than the *Pinus* species (Table 5.3); thus, the seasonal amplitude in water potential is greater for shrubs. Minimum predawn water potential and the magnitude of seasonal fluctuation for *Juniperus osteosperma* are intermediate between the shrubs (*Artemisia*, *Arctostaphylos*, *Amelanchier*, *Purshia*) and the pines. Based on these data and the relationship between photosynthesis and predawn water potential, we conclude that shrubs including Great Basin and Sierra Nevada species have the highest degree of drought tolerance whereas pines have the lowest. *Juniperus osteosperma* is intermediate in drought tolerance, though it is more similar to the pines than to the shrubs.

Our results confirm previous studies reporting complete stomatal closure for many western conifers at soil water potentials above -2.0 MPa (Lassoie et al. 1985; Lopushinsky 1969; Smith 1985; Drivas and Everett 1988) and at ca. -1.8 MPa for *Pinus ponderosa* (Lopushinsky and Klock 1974; Running 1976). These values provide a conservative estimate for a threshold predawn water potential supporting net carbon balance and growth for conifers. In contrast, vegetative growth in *Artemisia tridentata* continues at water potentials below -3.0 MPa, and reproductive growth and relatively high conductance values are observed below -6.0 MPa (Campbell and Harris 1977). It is surprising that physiognomic differences (trees versus shrubs) in the response to soil water deficit are more striking than correlations with native distribution (montane versus desert plants). Despite coexisting with *Artemisia* and other shrubs on the slopes of basin ranges, the Great Basin conifers, *Pinus monophylla* and *Juniperus osteosperma*, perform similarly to the Sierra Nevada pines. The same is true for shrubs. Whether Great Basin (*Artemisia*, *Purshia*,

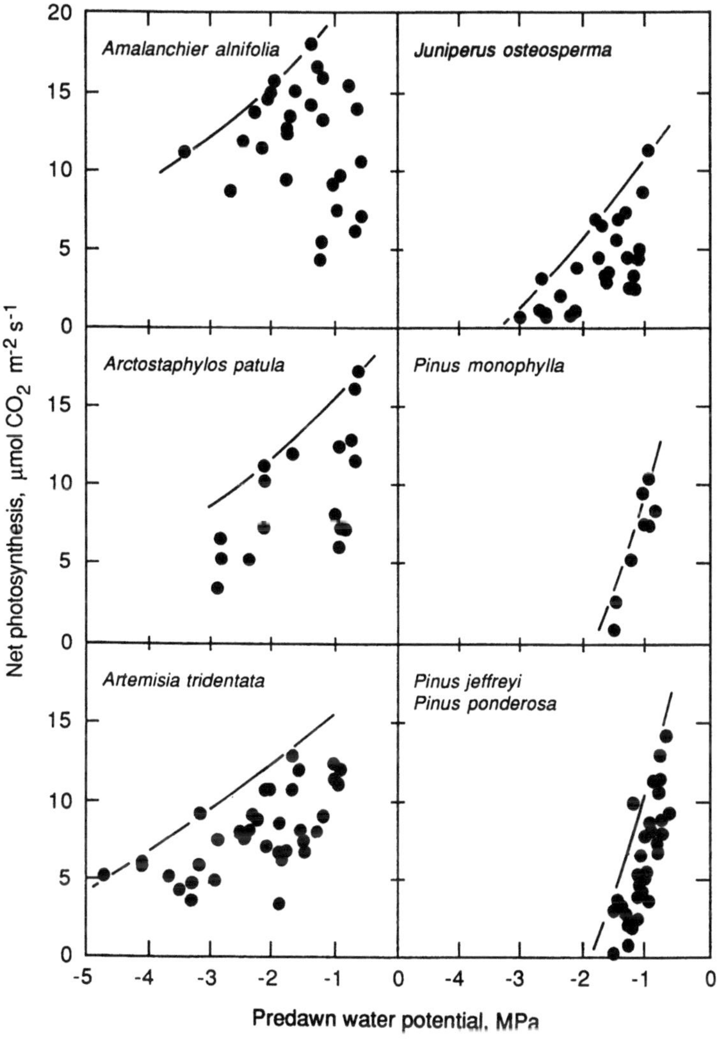

Figure 5.4. Boundary line plots of net photosynthesis versus predawn water potential for several Great Basin and Sierra Nevada shrubs and trees. *Amelanchier* and *Artemisia* are shrubs, and *Juniperus osteosperma* and *Pinus monophylla* are trees of the Great Basin. *Arctostaphylos* is a shrub and *Pinus ponderosa* and *P. jeffreyi* are trees of the Sierra Nevada. Values are maximum daily photosynthetic rates expressed on a one-sided leaf area basis. Variation in predawn water potential was generated by measuring plants at several locations and at different times during the summer. It is assumed that factors other than moisture availability limit photosynthesis for those points below the boundary lines (DeLucia and Schlesinger 1990).

Amelanchier) or Sierra Nevada (*Arctostaphylos*), shrubs have a higher degree of drought tolerance than trees. Higher drought tolerance of shrubs may be associated with a more shallow root distribution.

Given the physiological performance of the Great Basin conifers (*Pinus monophylla* and *Juniperus osteosperma*), it is unclear how these species coexist with *Artemisia* at higher elevations. *Pinus monophylla* seedlings are associated with *Artemisia* nurse plants during establishment (Everett et al. 1986). Nurse plants ameliorate microclimate but also compete with tree seedlings for water and nutrients (Drivas and Everett 1988). The ability of *P. monophylla* to survive long periods of negative carbon balance following stomatal closure induced by water stress may be a function of low growth and respiration rates. Furthermore, Great Basin trees may depend on unusually moist years for successful establishment. Relatively high values of predawn water potential in larger trees (Table 5.3) suggest that, after establishing roots at depth, competition for water between *Juniperus osteosperma* and *Pinus monophylla* and shrubs is reduced.

Water-Use Efficiency

Water-use efficiency (WUE) can be defined in ecosystem science as the ratio of net primary production to evapotranspiration and in plant physiology as the ratio of net photosynthesis to transpiration (Fischer and Turner 1978; Kramer 1983). In dry habitats, high WUEs have been observed in plants with widely different physiognomies (Field et al. 1983; Knapp and Smith 1987; Monson et al. 1986) and greater WUE in C_4 plants is well established (Osmond et al. 1982). However, contrary to expectation, the less drought tolerant Sierra Nevada and Great Basin conifers have higher WUE than the more drought tolerant shrubs (DeLucia et al. 1988; DeLucia and Heckathorn 1989; DeLucia and Schlesinger 1990).

We compared WUE of *Artemisia tridentata* and *Pinus ponderosa* during a drought cycle using gas exchange techniques (DeLucia and Heckathorn 1989). The WUEs for these species are similar at high water potentials but as water potential decreases, *Pinus ponderosa* consistently maintains higher WUE than *Artemisia tridentata* (Figure 5.5). Because of difficulty interpreting instantaneous values of WUE under field conditions, we have used carbon isotope composition of foliage (expressed as delta ^{13}C) as an index of seasonally integrated WUE (DeLucia et al. 1988).

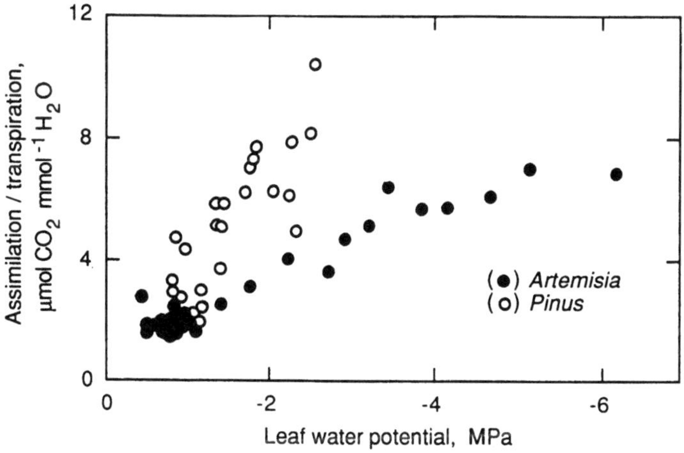

Figure 5.5. Water-use efficiency (A/T) as a function of bulk shoot water potential for potted seedlings of *Artemisia tridentata* and *Pinus ponderosa*. Water potential was manipulated by withholding irrigation (DeLucia and Heckathorn 1989).

The theoretical relationship between delta ^{13}C and WUE for C_3 plants was developed by Farquhar et al. (1982) and is used widely as an index of WUE (Ehleringer and Cooper 1988; O'Leary 1988). High WUE is associated with lower discrimination between $^{12}CO_2$ and $^{13}CO_2$ and is indicated by less negative values for delta ^{13}C (see Chapter 8). This relationship is based on the assumption that the foliages being compared experienced similar leaf-air water vapor gradients over their physiologically active lives.

Carbon isotope discrimination measured in foliage collected at the end of the 1986 growing season shows less negative delta ^{13}C values in *Pinus monophylla* and *Juniperus osteosperma* than for the Sierra Nevada pines (Figure 5.6). Shrubs have the highest maximum stomatal conductances and most negative delta ^{13}C values. These data indicate that shrubs, with highest drought tolerance have the lowest WUE, suggesting there may be a physiological cost that potentially compromises the ecological benefits of maintaining high WUE (Cohen 1970; Cowan 1982).

Ecological Implications

The patterns of water use intrinsic to these species suggest that seedlings of Sierra Nevada pines cannot effectively compete with the

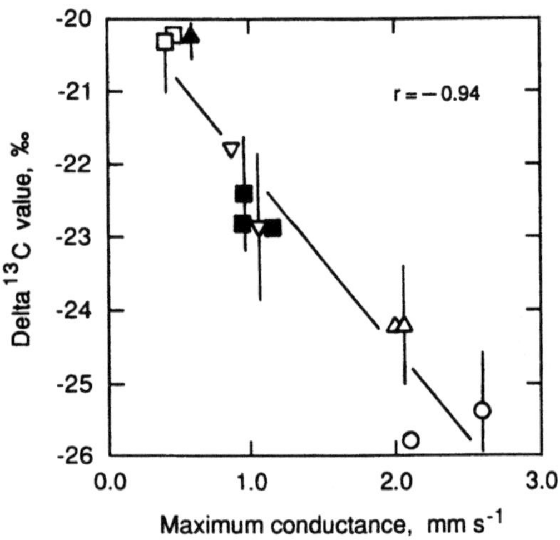

Figure 5.6. Foliage delta ^{13}C versus maximum stomatal conductance for several Great Basin and Sierra Nevada species. Isotopic composition was measured in cellulose extracted from current-year foliage. (o) *Amalenchier alnifolia*; (o) *Artemisia tridentata*; (△) *Juniperus osteosperma*; (□) *Pinus monophylla*; (■) *Pinus jeffreyi*; (▽) *Pinus ponderosa*. Open and closed symbols designate plants on unaltered and altered rock, respectively (see Table 5.1). Each point is a mean of 5 measurements and the largest standard deviation for each species is shown. The curve was fitted by least-squares regression (delta ^{13}C = 2.47 x conductance + 19.6, r^2 = 0.89) (DeLucia et al. 1988).

suite of shrub species in desert soils of the Great Basin. Using a modified de Wit-replacement design, Shainsky and Radosevich (1986) found *Pinus ponderosa* seedlings are less competitive for water and have lower relative growth rates when grown with *Arctostaphylos patula* than when grown in monoculture. Similarly, White and Newton (1989) observed significant reductions in stem volume in *Pinus ponderosa* and *Pseudotsuga menziesii* when grown with *Arctostaphylos viscida* and various herbaceous plants.

Soil moisture in the western Great Basin is recharged mainly during the winter, and seedlings of *Pinus ponderosa* and *P. jeffreyi* germinate early in the spring when water availability is high. On unaltered soil rapid growth of annual and perennial grasses in early spring, followed by high rates of water use by shrub species, depletes soil water to a level unavailable to the Sierra Nevada conifer seedlings by June. Interference from the grass *Calamagrostis rubescens* causes substantial growth reductions in *P. ponderosa* seedlings, as well as enhancing small size classes in the population structure of tree seedlings (Petersen 1988). In addition to increasing the frequency of fire in sagebrush steppe (Billings pers. comm.), the introduced annual *Bromus tectorum* also may alter significantly the patterns of seedling establishment by exploiting surface moisture reserves early in the season. Rapid depletion of surface soil moisture by *Bromus trinii* plays a similar important role in restricting the establishment of evergreen chaparral species in adjacent grasslands in southern California (Davis and Mooney 1985). Stands of *Pinus ponderosa* and *P. jeffreyi* appear to be excluded from adjacent sagebrush steppe on brown unaltered soils by competition for water.

Nutrient Relations

Few Great Basin plants occur on soil derived from altered rock, and Sierra Nevada plants thus persist in an exceptionally dry climate without competition for water with the more drought tolerant Great Basin shrubs. The question remains: why are Great Basin plants excluded from soils derived from altered rock? Schlesinger et al. (1989) proposed that as a result of evolution in a region characterized by circumneutral soils of relatively high nutrient content, Great Basin vegetation is intolerant of the low nutrient and acidic conditions of altered substrate. Dominant Great Basin and Sierra Nevada plants share two life history traits of stress tolerant plants (Grime 1979); they are long-lived woody perennials that range from evergreen to semi-deciduous. It seems, however, that physiological adaptations to drought may be disadvantageous in low nutrient situations and vice versa. Other studies have reported a physiological compromise between nutrient and water utilization in evergreen shrubs. Field et al. (1983) found that instantaneous values of water-use efficiency and nitrogen-use efficiency were inversely correlated for several chaparral species. Sierra Nevada "tree islands" provide an opportunity to compare nutrient use in adjacent Great Basin and Sierra Nevada plants.

Foliar Nutrients and Nutrient Amendments

The most pronounced differences in chemistry for soils derived from altered and adjacent unaltered andesite are low pH, phosphorus, and base cations in altered soils (Schlesinger et al. 1989). In a similar situation in the Sierra Madre of northwestern Mexico, Goldberg (1985) found that seedlings of the deciduous tree *Lysiloma divaricata* survive on altered soil only when pH was increased by the addition of $CaCO_3$. The biological effect of raising pH was probably related to an increase in the availability of phosphorus. Billings (1950) reported the appearance of phosphorus-deficiency symptoms in agronomic species grown in altered soils. These studies suggest that, in addition to drought tolerance, the ability to acquire phosphorus plays a fundamental role in ecologically differentiating dominant Great Basin and Sierra Nevada plants.

In examining the foliage nutrient contents from field-collected tissue, Schlesinger et al. (1989) found that for the same species growing on both soil types, mean foliar concentrations of nitrogen, phosphorus, and calcium are consistently lower for plants growing on altered rock, although differences are generally small. Pronounced differences in foliar nutrient contents between species are evident. The Sierra Nevada and Great Basin conifers have relatively low tissue nitrogen and phosphorus, and high specific leaf mass compared to the shrubs (Figures 5.7 and 5.8). The Sierra Nevada shrub *Arctostaphylos patula* is an exception. Although this species has a relatively broad flat leaf, it is extremely sclerophyllous and has physical and chemical properties similar to conifers. Low tissue nutrient contents and high specific leaf mass are characteristic of the long-lived foliage produced by many conifers and presumably represent an adaptation to low nutrient conditions (Chabot and Hicks 1982). In evergreens this response may maximize net carbon gain per unit of foliar nutrient content over the life of the leaf (Orians and Solbrig 1977, Schlesinger and Chabot 1977).

Results from nutrient amendment experiments support the inference that phosphorus is the primary limitation to plant growth on altered rock. When provided with a complete fertilizer, agronomic species grew as well on altered soil as in a fertilized inert medium (Billings 1950). However, withholding phosphorus and nitrogen from the fertilizer caused approximately 98 and 81% reductions in final biomass, respectively. In similar experiments conducted with native species grown under controlled conditions, the response to nutrient additions and soil type was highly species specific (DeLucia et al. 1989).

Total biomass for *Pinus monophylla, P. ponderosa*, and *P. jeffreyi* is 10 to 28% lower when grown in altered soil. This compares with greater than 90% reductions for *Artemisia tridentata* and *Bromus tectorum* (DeLucia

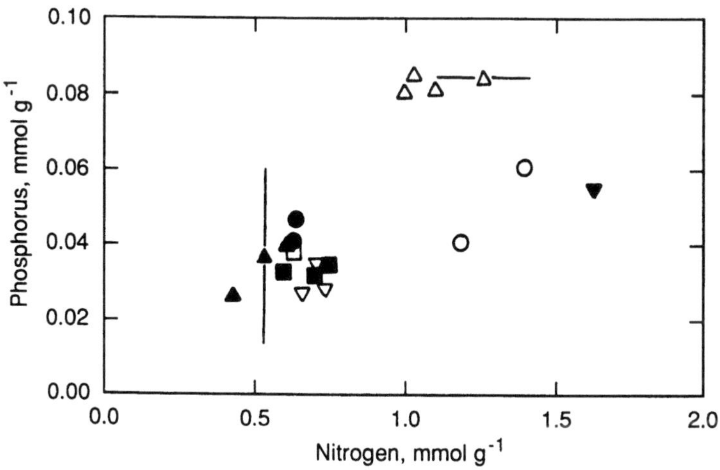

Figure 5.7. Foliage N and P contents for several Great Basin and Sierra Nevada species. Each point is a mean of 5 different plants; (o) *Amalanchier alnifolia*; (o) *Arctostaphylos patula*; (Δ) *Artemisia tridentata*; (▲) *Juniperus osteosperma*; (□) *Pinus monophylla*; (■) *Pinus jeffreyi*; (▽) *Pinus ponderosa*; (▼) *Purshia tridentata*. Replicate samples for each species were taken from populations at different elevations which received different mean annual precipitation (DeLucia and Schlesinger 1990).

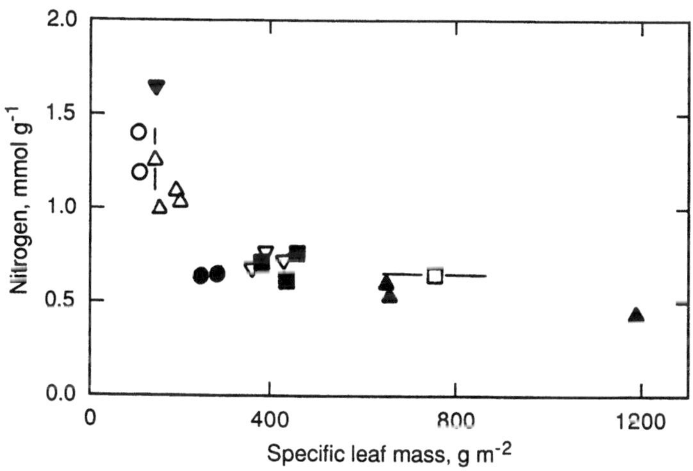

Figure 5.8. Leaf N content versus leaf specific mass for several Great Basin and Sierra Nevada species. Sampling strategy and plant indentification is as shown in Figure 5.7 (DeLucia and Schlesinger 1990).

et al. 1989). Simultaneous additions of nitrogen and phosphorus cause an increase in total dry mass, relative growth rate, and net photosynthesis, and when added separately, phosphorus has a greater effect than nitrogen (DeLucia et al. 1989). Substantially greater sensitivity of *Artemisia tridentata* to substrate pH and phosphorus compared to the Sierra Nevada pines, supports the hypothesis that dominant Great Basin vegetation is excluded from altered soil by intolerance to nutrient deficient conditions (Billings 1950). The effect of limited phosphorus on physiological performance of these species is complex, however.

Photosynthesis and stomatal conductance are lower in pines grown in altered soil (DeLucia et al. 1989). Net photosynthesis is correlated positively with foliar nitrogen, and the addition of this nutrient significantly increases photosynthetic rates and growth of *Pinus ponderosa*. Although phosphorus deficiency can inhibit photosynthesis directly (Brooks 1986; Foyer and Spencer 1986; Morrison and Batten 1986), addition of phosphorus does not cause an increase in CO_2 exchange in Sierra Nevada conifers. The primary effect of phosphorus deficiency on growth of Great Basin and Sierra Nevada plants appears to be its influence on leaf area development. Fertilizing with phosphorus has no effect on net photosynthesis but substantially increases leaf area. Phosphorus deficiency can directly affect leaf area expansion, possibly by decreasing hydraulic conductivity of the root system thereby reducing turgor in expanding foliage (Radin and Eidenbock 1984).

Leaf nitrogen is a major determinant of photosynthetic capacity (Evans 1983, 1989; Field and Mooney 1983; Gulmon and Chu 1981; Natr 1975) and further differentiates the physiological performance of Great Basin and Sierra Nevada trees and shrubs. Maximum photosynthetic rates for different plants are correlated significantly with leaf nitrogen (Figure 5.9a). Great Basin shrubs have higher foliage nitrogen contents and photosynthetic rates than the Great Basin and Sierra Nevada conifers. As is the case for chemical and physical properties of foliage, physiological performance for *Arctostaphylos patula*, as indicated by the photosynthesis-nitrogen relationship, is more similar to the conifers than shrubs. A weak positive correlation between foliage phosphorus contents and photosynthesis is evident, however, this relationship probably results from correlation of foliage nitrogen and phosphorus contents (Figure 5.7) rather than from a functional control of carbon assimilation by phosphorus.

The photosynthesis-nitrogen relationship indicates basic differences in patterns of resource partitioning and allocation and consequently in the physiological and ecological attributes of these contrasting groups of plants. The long-lived foliage of conifers has low nutrient content resulting in low photosynthetic and growth rates. These attributes are part of a suite of responses that are often seen in conditions of low nutrient availability (Chapin 1980, Chapin and Kedrowski 1983, Chabot and Hicks 1982). They may enable conifers to persist on nutrient-deficient altered soil and at higher elevations in the Sierra Nevada, where mineralization is reduced by low soil temperature. Rapid growth rates associated with high foliage nitrogen and high photosynthetic rates in the shrubs may ensure stronger competitive ability for water and nutrients during relatively short periods of abundant water availability in the Great Basin desert.

Nutrient-Use Efficiency

As part of the working hypotheses explaining vegetation mosaics associated with altered rock, we proposed that certain Sierra Nevada plant species formed unique forest ecosystems on nutrient-deficient soils in an arid climate because they had higher nutrient-use efficiency (NUE) than their potential competitors in the Great Basin. Testing this hypothesis depends on the operational definition of NUE. Biomass produced per unit of nutrient taken up, or the inverse of tissue nutrient concentration, provides one index of NUE. However, for perennial plants this definition of NUE may be inadequate. Vitousek (1982) observed that in long-lived perennial plants, nutrients are recycled internally to varying degrees prior to abscission of leaves and other structures, and are reused to produce additional biomass. In this case he suggests that NUE should be defined as net primary production per unit nutrient lost in leaf litter, where nutrient losses are an index of uptake since they must be replenished from the soil. Using this definition, Vitousek (1982) concluded that NUE is related inversely to nutrient availability in forest ecosystems.

We have explored several methods of assessing NUE in adjacent Great Basin and Sierra Nevada plants. In field investigations NUE was estimated as the ability of plants to reabsorb nutrients (nitrogen and phosphorus) prior to leaf abscission (Schlesinger et al. 1989). This index, designated NUE_{reab}, is calculated as the difference in nutrient concentration between mature foliage and litterfall divided by the

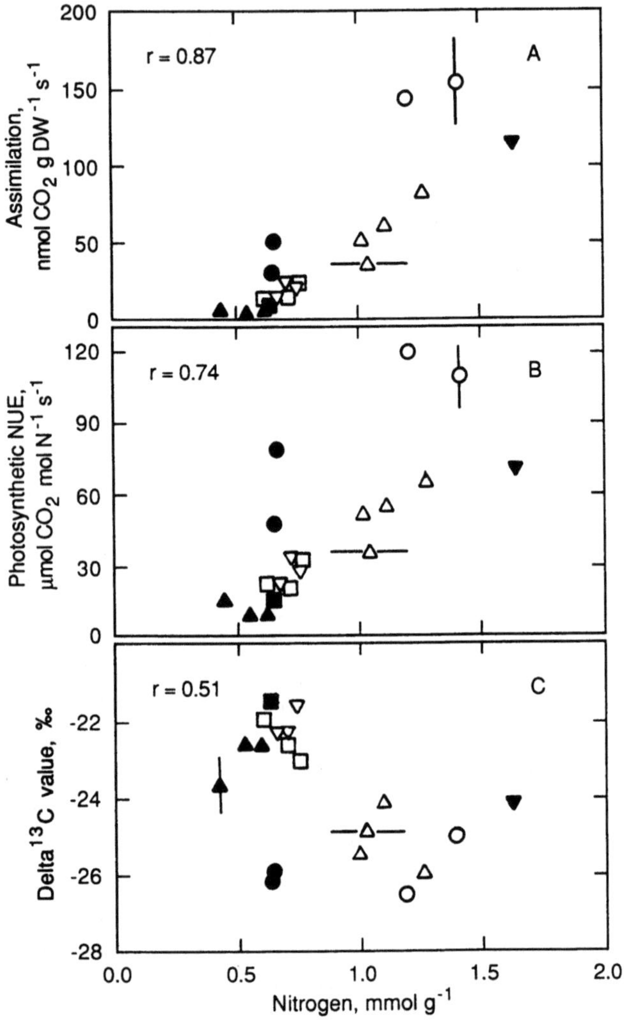

Figure 5.9. A, Maximum net photosynthesis; B, photosynthetic nitrogen-use efficiency; and C, foliage delta ^{13}C versus foliage N content. Plants and sampling strategy was as in Figure 5.7. The largest standard deviation for the dependent and independent variables is shown (DeLucia and Schlesinger 1990).

concentration in mature foliage. When adjusted for seasonal changes in specific leaf mass (g m^{-2}), NUE_{reab} indicates the fractional reabsorption of foliar nutrients, which Vitousek (1982) found is higher in forests on nutrient-limited substrate. As an alternative, recent physiological studies define instantaneous values of nutrient-use efficiency as the rate

Table 5.4. Fractional nutrient reabsorption for species occurring in many of the field sites (from Schlesinger et al. 1989). All data are given ± 1 SE. Analysis of variance was used to distinguish main effects due to location; t-tests were used to distinguish significance (< 0.05) between soil types within a location.

Location and Parent Material	Nitrogen			Phosphorus		
	Pinus ponderosa	*Pinus monophylla*	*Artemisia tridentata*	*Pinus ponderosa*	*Pinus monophylla*	*Artemisia tridentata*
Sierra Nevada						
unaltered	0.68 (0.06)	0.35 (0.10)	0.40 (0.04)	0.62 (0.08)	0.50 (0.09)	0.37 (0.08)
altered	0.63 (0.03)	—	—	0.54 (0.07)	—	—
Great Basin Pinyon Zone						
unaltered	—	0.48 (0.04)	0.56 (0.03)	—	0.60 (0.02)	0.44 (0.06)
altered	0.58 (0.07)	0.49 (0.04)	—	0.57 (0.05)	0.64 (0.03)	—
Sagebrush Zone						
unaltered	—	—	0.53 (0.05)	—	—	0.26 (0.08)
altered	0.61 (0.03)	—	—	0.62 (0.05)	—	—
ANOVA for main effect						
Location	ns[1]	ns	*	ns	ns	ns

[1] Not significant, or significant *.

of net photosynthesis per unit of tissue nutrient content, usually nitrogen or phosphorus (Field and Mooney 1983; Reich and Schoettle 1988; Sage and Pearcy 1987). Such physiological indices of NUE provide a mechanistic component to the more ecologically complete definition of NUE that includes NUE_{reab} and whole plant patterns of nutrient allocation and relative growth rate (Chapin 1980).

Fractional reabsorption of nitrogen and phosphorus are higher for *Pinus ponderosa* than for *Artemisia tridentata* (Table 5.4). In a literature review Chapin and Kedrowski (1983) calculated mean fractional reabsorptions

Table 5.5. Relative growth rate, specific absorption rate, and nutrient-use efficiency for seedlings grown in the Duke University Phytotron (from Schlesinger et al. 1989).

Species and Soil	Relative Growth Rate[1] ($mg\ g^{-1}\ d^{-1}$)	Final Biomass[1] (g)	Specific Absorption Rate ($mg\ g^{-1}$) N	P	Ca	Nutrient Use Efficiency ($mg\ mg^{-1}$) N	P	Ca
Pinus ponderosa								
Altered	13.8[a]	0.81[a]	19.0	2.0	2.0	114.2	1092	1082
Unaltered	24.0[a]	1.13[a]	11.7	1.7	6.1	175.5	1239	336
P. monophylla								
Altered	7.4[b]	0.37[b]	20.8	5.6	-	139.2	516	-
Unaltered	10.8[b]	0.46[b]	46.4	3.8	8.5	60.1	135	328
Artemisia tridentata								
Altered	53.9[c]	0.01[c]	46.6	-	-	55.7	-	-
Unaltered	64.2[c]	0.27[d]	21.1	-	-	99.6	-	-

[1]data followed by the same superscript are not significantly different in ANOVA, followed by Tukey's range test (< 0.05).

of 0.52 for nitrogen and 0.43 for phosphorus. *Pinus ponderosa* is generally higher and *Artemisia tridentata* lower than these mean values. However, among *P. ponderosa* occurring in different habitats there is no indication of greater NUE_{reab} in the low nutrient soils on altered bedrock. Similarly, Birk and Vitousek (1986) found little acclimation in NUE_{reab} among *P. taeda* occurring along a nitrogen fertility gradient, and limited acclimation in NUE_{reab} is seen in other species of Great Basin and Sierra Nevada trees and shrubs (Schlesinger et al. 1989).

There are no consistent trends in specific absorption rate (mg nutrient in plant/g root dry weight) of nitrogen or phosphorus for seedlings of *Pinus ponderosa, P. monophylla*, and *Artemisia tridentata* grown under controlled growth chamber conditions (Table 5.5). However, growth based nutrient-use efficiency for nitrogen is consistently higher for the two pines than for *Artemisia tridentata*. Conifers generally show higher reabsorption efficiency than broad-leaf species (Vitousek 1982; Waring and Schlesinger 1985). It seems that evolution in soils with relatively high available phosphorus has not selected for high nutrient reabsorption efficiency in desert shrubs, although *Artemisia* shows specialized mechanisms to enhance phosphorus uptake in calcareous soils (Jurinak et al. 1986).

A very different picture emerges when Great Basin and Sierra Nevada plants are compared using a physiological index of nutrient-use efficiency based on photosynthetic nitrogen-use efficiency (NUE_{ps}) (DeLucia and Schlesinger 1990). NUE_{ps} for different species is correlated positively with foliage nitrogen contents (Figure 5.9b). Compared to the shrubs, Sierra Nevada and Great Basin conifers have low foliage nitrogen and NUE_{ps}. *Purshia tridentata*, a nitrogen fixing species (Virginia and Delwiche 1982), has the highest foliage nitrogen and *Amelanchier alnifolia* the highest NUE_{ps}. Although *Arctostaphylos patula* has nitrogen contents similar to the conifers, this species has values of NUE_{ps} similar to the shrubs (Figure 5.10b). There is no consistent relationship between photosynthetic phosphorus-use efficiency and phosphorus contents. However, high foliage phosphorus and low photosynthetic phosphorus-use efficiency for *Artemisia tridentata* may be ecologically significant in that they suggest a high phosphorus requirement for this species.

Ecological Implications

Reich and Schoettle (1988) found a strong correlation between instantaneous photosynthetic measurements of NUE and growth estimates of NUE for *Pinus strobus* grown under a variety of climatic and edaphic conditions,. Our data suggest high NUE_{ps} in shrubs of the Great Basin Desert, whereas data for NUE_{reab} and growth per unit nutrient uptake suggest the opposite. This raises questions about the relative merits of these indices and their ecological interpretation. A complication arises in interspecific comparisons of instantaneous measurements of NUE for plants with different leaf durations. In the case of evergreen foliage, an appropriate index of NUE should include carbon fixed per unit nitrogen or phosphorus over the life of the leaf to accommodate use of reabsorbed nutrients (Vitousek 1982). Assuming the nutrient content and growth based estimates of NUE provide a more complete ecological assessment of nutrient use, our data support the hypothesis of greater tolerance to low nutrient conditions in Sierra Nevada vegetation.

The physiologically based indices of NUE are not without merit since they provide a means of examining allocation and utilization of nitrogen and other nutrients at the functional level of the leaf. NUE_{ps} is one of many components determining patterns of whole plant nutrient use. For species shown in Figure 5.9b there is a positive relationship between NUE_{ps} and foliage nitrogen, as reported by Field and Mooney (1986) in a survey of 21 species ranging from desert annuals to evergreen trees and shrubs. This relationship illustrates the functional dependence of photosynthesis on nitrogen content and partitioning and leaf structure (Evans 1989).

Shrubs, with the possible exception of *Arctostaphylos patula*, probably invest a high proportion of leaf nitrogen in various components of the photosynthetic apparatus enabling these species to maintain high maximum photosynthetic rates and NUE_{ps}. In contrast, sclerophyllous conifer foliage has high specific leaf mass, and low nitrogen, low NUE_{ps} and low maximum photosynthetic rates. Field and Mooney (1986) proposed that a high proportion of leaf nitrogen is invested in non-photosynthetic components in sclerophyllous leaves- components perhaps necessary for leaf longevity. Medina (cited in Field and Mooney 1986) suggested that low NUE_{ps} in this type of foliage results from high mesophyll resistance to CO_2 diffusion associated with relatively impermeable cell walls. However, less negative delta ^{13}C values (Figure 5.9c) and relatively steep slopes for the photosynthesis versus conductance relationship in conifers (DeLucia unpublished) indicate that stomatal limitations predominate for conifers. Various attributes of dominant Sierra Nevada trees, including leaf longevity, slow relative growth rates, and high NUE in growth, support the hypothesis that Sierra Nevada trees are better adapted to low nutrient conditions than shrubs.

Physiognomic Relationships

As a physiognomic group Great Basin and Sierra Nevada shrubs possess many physiological characteristics that distinguish them from trees. Shrubs examined here have a higher degree of drought tolerance and lower WUE than trees (Table 5.6), and, with the exception of *Arctostaphylos patula*, shrubs also possess higher foliage nutrient contents and NUE_{ps}. When gas exchange parameters are calculated on a projected leaf area basis, the maximum values for leaf conductance and rates of net photosynthesis are higher for shrubs than for the trees (DeLucia et al. 1988).

Fisher and Turner (1978) proposed that in dry habitats natural selection should favor plants with high WUE. The opposite appears to be the case for dominant Great Basin and Sierra Nevada plants; low WUE is associated with drought tolerance in the shrubs (Table 5.6) suggesting a possible ecological cost associated with the maintenance of high WUE (Cohen 1970, Cowan 1982). Because of the non-linear relationship between photosynthesis and transpiration (Cowan 1982, Raschke 1979) and the additional resistances to diffusion of CO_2 associated with the mesophyll (Nobel 1983), stomatal closure causes a proportionately greater decrease in transpiration than photosynthesis thereby increasing WUE. High WUE for trees is maintained by operating at low seasonal conductances. However, this also reduces absolute net photosynthetic rates therefore reducing growth and possibly competitive ability. High WUE also can be achieved by increasing the intrinsic photosynthetic capacity at low leaf nitrogen as illustrated by C_4 photosynthesis. Increasing photosynthetic capacity in C_3 plants often involves an increase in nitrogen allocation to photosynthetic enzymes, which, depending on the nature of the photosynthesis-nitrogen relationship, may decrease nitrogen-use efficiency (Field and Mooney 1986).

The relationship between low WUE and drought tolerance in shrubs raises several questions regarding the ecological significance of WUE.

Table 5.6. Summary of some physiological and growth characteristics for several Great Basin and Sierra Nevada trees and shrubs. Growth forms are: Sh, shrub; Mt, multistem tree; Tr, single-stem tree. Community type is either Great Basin (GB), which includes shrub steppe and pinyon-juniper woodland or Sierra Nevada montane forest (SN). (+) and (-) indicate a high or low degree, respectively, for each physiological parameter. The rankings for drought tolerance are based on the photosynthesis versus water potential relationship and the seasonal amplitude of predawn water potentials. Water-use efficiency (WUE) is based on carbon isotope discrimination. Photosynthetic nitrogen-use efficiency (PNUE) is based on gas exchange data. Growth-based nutrient-use efficiency (NUE growth) and fractional reabsorption of P prior to leaf abscission were determined from data in Schlesinger et al. (1989).

Species	Growth Form	Community Type	Drought Tolerance	WUE	PNUE	NUE (Growth)	Fractional Reabsorption
Amelanchier alnifolia	Sh	GB	+++	--	+++	[1]	[1]
Artemisia tridentata	Sh	GB	+++	-	++	-	-
Purshia tridentata	Sh	GB	[1]	-	++	[1]	[1]
Arctostaphylos patula	Sh	SN	++	--	++	[1]	-
Juniperus osteosperma	Mt	GB	-/+	++	--	[1]	++
Pinus monophylla	Mt	GB	-	+++	-	++	++
Pinus jeffreyi	Tr	SN	-	+++	-	[1]	++
Pinus ponderosa	Tr	SN	-	+++	-	+++	++

[1] insufficient data.

Given the potential costs of maintaining high WUE, namely lower rates of photosynthesis and growth, Cohen (1970) argues that selection for high WUE would be advantageous only if conserved soil moisture is available for uptake later in the season. Although *Artemisia tridentata* can develop root systems up to 2 m deep, a high proportion of its roots are in the upper 0.5 m where there is intense competition with grasses and other shrubs for water (Caldwell et al. 1985; Campell and Harris 1977; Eissenstat and Caldwell 1988; Sturges 1977). It is therefore unlikely that surface moisture conserved by *Artemisia* would be available later in the growing season. Instead, moisture would be lost by evaporation or to a competitor. The relatively small magnitude of seasonal change in predawn water potential for mature pines suggests that these species are deeply rooted and largely free of competition for water with the shallow-rooted shrub species. Under these conditions conservative use of water (high WUE) would lengthen the period favorable for net carbon assimilation and more closely match growth with low rates of nutrient availability on nutrient-poor sites (Chapin 1980).

In our studies plants with high WUE have low physiologically based NUE_{ps}. However, WUE is positively correlated with growth based measurements of NUE such as fractional reabsorption from senescing tissue and the ratio of biomass production to nutrient content (Table 5.6). Field et al. (1983) suggest an inverse relationship between WUE and NUE_{ps} for several evergreen species that have different degrees of drought tolerance. This tradeoff is based on the relationship of WUE and NUE_{ps} to intercellular CO_2 concentration (C_i); as C_i increases WUE decreases and NUE_{ps} increases (Field et al. 1983). The balance between WUE and NUE_{ps} probably represents a functional compromise between maximizing net photosynthesis (high leaf N, high conductances, low specific leaf mass) and physiological and morphological adaptations to drought (low leaf N, low conductances, high specific leaf mass) (Orians and Solbrig 1977), rather than a compromise between nutrient and water acquisition per se. For Sierra Nevada trees high WUE, high growth-based NUE, and relatively low growth rates represent adaptations to low nutrient availability. In this case high WUE represents a conservative ecophysiological strategy that is detrimental under conditions of intense competition for water. For more shallowly rooted shrubs, the combination of an "exploitive" use of water (low WUE) and high degree of drought tolerance is a successful combination of physiological characteristics in water-limited but relatively high nutrient habitats.

Acknowledgements

We thank JL Nachlinger, WT Peterjohn, SM Billings, and SA Heckathorn for assistance with fieldwork and the Desert Research Institute, Reno, Nevada, for providing logistical support. This project was funded by NSF grant BRS 85-04859 (Ecology Program) to WD Billings and WH Schlesinger. Additional support was provided by a USDA Competitive Research Grant (Forest Biology Program) and a University of Illinois Research Board grant to EH DeLucia.

References

Barbour MG (1988) California upland forest. pp. 131-164. In Barbour MG and Billings WD (editors) North American Terrestrial Vegetation. Cambridge Univ. Press, NY.

Billings WD (1950) Vegetation and plant growth as affected by chemically altered rocks in the western Great Basin. Ecology 31:62-74.

Birk EM and Vitousek PM (1986) Nitrogen availability and nitrogen use efficiency in loblolly pine stands. Ecology 67:69-79.

Black RA and Mack RN (1986) Mount St. Helens ash: recreating its effects on the steppe environment and ecophysiology. Ecology 67:1289-1302.

Branson FA, Miller RF, and McQueen IS (1976) Moisture relationships in twelve northern desert shrub communities near Grand Junction, Colorado. Ecology 57:1104-1124.

Brooks A (1986) Effects of phosphorus nutrition on ribulose-1,5-bisphosphate carboxylase activation, photosynthetic quantum yield and amounts of some Calvin cycle metabolites in spinach leaves. Aust. J. Plant Physiol. 13:221-237.

Caldwell MM, Eissenstat DM, Richards JH, and Allen MF (1985) Competition for phosphorus: Differential uptake from dual-isotope-labeled soil interspaces between shrub and grass. Science 229:384-386.

Calkins FC (1944) Outline of the geology of the Comstock Lode district, Nevada. 35 pp. US Geological Survey Washington, DC.

Campbell GA and Harris GA (1977) Water relations and water use patterns for *Artemisia tridentata* Nutt. in wet and dry years. Ecology 58:652-659.

Chabot BF and Hicks DJ (1982) The ecology of leaf life spans. Ann. Rev. Ecol. Sys. 13:229-259.

Chapin III FS (1980) The mineral nutrition of wild plants. Ann. Rev. Ecol. Sys. 11:233-260.

Chapin III FS and Kedrowski RA (1983) Seasonal changes in nitrogen and phosphorus fractions and autumn retranslocation in evergreen and deciduous taiga trees. Ecology 64:376-391.

Cohen D (1970) The expected efficiency of water utilization in plants under different competition and selection regimes. Israel J. Bot. 19:50-54.

Cowan IR (1982) Regulation of water use in relation to carbon gain in higher plants. pp. 589-613. In Lange OL, Nobel PS, Osmond CB, and Ziegler H (editors) Physiological Plant Ecology II: Water Relations and Carbon Assimilation. Encyclopedia of Plant Physiology (New Series), Vol. 12B. Springer-Verlag, Heidelberg.

Davis SD and Mooney HA (1985) Comparative water relations of adjacent California shrub and grassland communities. Oecologia 66:522-529.

DeLucia EH and Heckathorn SA (1989) The effect of soil drought on water-use efficiency in contrasting Great Basin desert and Sierran montane species. Plant Cell Environ. 12:935-940.

DeLucia EH and Schlesinger WH (1990) Water-use efficiency and drought tolerance in adjacent Great Basin and Sierran plants. Ecology, in press.

DeLucia EH, Schlesinger WH, and Billings WD (1988) Water relations and the maintenance of Sierran conifers on hydrothermally altered rock. Ecology 69:303-311.

DeLucia EH, Schlesinger WH, and Billings WD (1989) Edaphic limitations to growth and photosynthesis in Sierran and Great Basin vegetation. Oecologia 78:184-190.

DePuit EJ and Caldwell MM (1973) Seasonal pattern of net photosynthesis of *Artemisia tridentata*. Amer. J. Bot. 60:426-435.

Drivas EP and Evertt RL (1988) Water relations characteristics of competing singleleaf pinyon seedlings and sagebrush nurse plants. For. Ecol. Manag. 23:27-37.

Ehleringer JR and Cooper TA (1988) Correlations between carbon isotope ratio and microhabitat in desert plants. Oecologia 76:562-566.

Eissenstat DM and Caldwell MM (1988) Competitive ability is linked to rates of water extraction: a field study of two aridland tussock grasses. Oecologia 75:1-7.

Evans JR (1983) Nitrogen and photosynthesis in the flag leaf of wheat (*Triticum aestivum* L.). Plant Physiol. 72:297-302.

Evans JR (1989) Photosynthesis and nitrogen relationships in leaves of C_3 plants. Oecologia 78:9-19.

Everett RL, Koniak S and Budy J (1986) Pinyon seedling distribution among soil microsites. USDA Forest Service Intermountain Research Station Research Paper INT-363.

Farquhar GD, O'Leary MH and Berry JA (1982) On the relationship between carbon isotope discrimination and the intercellular carbon dioxide concentration in leaves. Aust. J. Plant Physiol. 9:121-137.

Field C and Mooney HA (1983) Leaf age and seasonal effects of light, water, and nitrogen use efficiency in a California shrub. Oecologia 56:348-355.

Field C and Mooney HA (1986) The photosynthesis-nitrogen relationship in wild plants. pp. 25-55. In Givnish TJ (editor) On the Economy of Plant Form and Function. Cambridge Univ. Press, NY.

Field C, Merino J, and Mooney HA (1983) Compromises between water-use efficiency and nitrogen-use efficiency in five species of California evergreens. Oecologia 60:384-389.

Fischer RA and Turner NC (1978) Plant productivity in the arid and semiarid zones. Ann. Rev. Plant Physiol. 29:277-317.

Fonteyn PJ, Schlesinger WH, and Marion GM (1987) Accuracy of soil thermocouple hygrometer measurements in desert ecosystems. Ecology 68:1121-1124.

Foyer C and Spencer C (1986) The relationship between phosphate status and photosynthesis in leaves: effects of orthophosphate distribution, photosynthesis and assimilate partitioning. Planta 167:369-375.

Gianella VP (1936) Geology of the Silver City district on the southern portion of the Comstock Lode, Nevada. Univ. Nev. Bull. 30(9):1-108.

Goldberg DE (1982) The distribution of evergreen and deciduous trees relative to soil type: an example from the Sierra Madre, Mexico, and a general model. Ecology 63:942-951.

Goldberg DE (1985) Effects of soil pH, competition, and seed predation on the distributions of two tree species. Ecology 66:503-511.

Grime JP (1979) Plant Strategies and Vegetation Processes. John Wiley and Sons, NY.

Gulmon SL and Chu CC (1981) The effects of light and nitrogen on photosynthesis, leaf characteristics, and dry matter allocation in the chaparral shrub, *Diplacus aurantiacus*. Oecologia 49:207-212.

Hull JC and Wood SG (1984) Water relations of oak species on and adjacent to a Maryland serpentine soil. Amer. Midl. Nat. 112:224-234.

Jenny H (1980) The Soil Resource. Ecological Studies Vol. 37. Springer-Verlag, NY.

Jurinak JJ, Dudley LM, Allen MF, and Knight WG (1986) The role of calcium oxalate in the availability of phosphorus in soils of semiarid regions: A thermodynamic study. Soil Science 142:255-261.

Knapp AK and Smith WK (1987) Stomatal and photosynthetic responses during sun/shade transitions in subalpine plants. Oecologia 74:62-67.

Kramer PJ (1983) Water Relations of Plants. Academic Press, NY.

Kruckeberg AR (1954) The ecology of serpentine soils. III. Plant species in relation to serpentine soils. Ecology 35:267-274.

Lassoie JP, Hinckley TM, and Grier CC (1985) Coniferous forests of the Pacific Northwest. pp. 127-161. In Chabot BF and Mooney HA (editors) Physiological Ecology of North American Plant Communities. Chapman Hall, NY.

Lopushinsky W (1969) Stomatal closure in conifer seedlings in response to leaf moisture stress. Bot. Gaz. 130:250-263.

Lopushinsky W and Klock GO (1974) Transpiration of conifer seedlings in relation to soil water potential. Forest Science 20:181-186.

Miller RF, Branson FA, McQueen IS, and Snyder CT (1982) Water relations in soils as related to plant communities in Ruby Valley, Nevada. J. Range Manag. 34:462-468.

Monson RK, Sackschewsky MR, and Williams GW III (1986) Field measurements of photosynthesis, water-use efficiency, and growth in *Andropogon smithii* (C_3) and *Bouteloua gracilis* (C_4) in the Colorado shortgrass steppe. Oecologia 68:400-409.

Morrison JIL and Batten GD (1986) Regulation of mesophyll photosynthesis in intact wheat leaves by cytoplasmic phosphate concentrations. Planta 168:200-206.

Natr L (1975) Influence of mineral nutrition on photosynthesis and the use of assimilates. pp. 537-556. In Cooper JP (editor) Photosynthesis and Productivity in Different Environments. Cambridge Univ. Press, Cambridge.

Nobel PS (1983) Biophysical Plant Physiology and Ecology. WH Freeman and Company, NY.

O'Leary MH (1988) Carbon isotopes in photosynthesis. BioScience 38:328-336.

Orians GH and Solbrig OT (1977) A cost-income model of leaves and roots with special reference to arid and semiarid areas. Amer. Nat. 111:677-690.

Osmond CB, Winter K, and Ziegler H (1982) Functional significance of different pathways of CO_2 fixation in photosynthesis. pp. 479-547. In Lange OL, Nobel PS, Osmond CB, and Ziegler H (editors) Physiological Plant Ecology II: Water Relations and Carbon Assimilation. Encyclopedia of Plant Physiology (New Series), Vol. 12B. Springer-Verlag, Heidelberg.

Petersen TD (1988) Effects of interference from *Calamagrostis rubescens* on size distributions in stands of *Pinus ponderosa*. J. Appl. Ecol. 25:265-272.

Radin JW and Eidenbock MP (1984) Hydraulic conductance as a factor limiting expansion of phosphorus-deficient cotton plants. Plant Physiol. 76:392-394.

Raschke K (1979) Movements of stomata. pp. 383-441. In Haupt W and Feinleib ME (editors) Physiology of Movements. Encyclopedia of Plant Physiology (New Series), Vol. 7. Springer-Verlag, Heidelberg.

Reginato RJ and Nakayama FS (1988) Neutron probe calibration based on plastic transfer standards. Soil Science 145:381-384.

Reich PB and Schoettle AW (1988) Role of phosphorus and nitrogen in photosynthetic and whole plant carbon gain and nutrient use efficiency in eastern white pine. Oecologia 77:25-33.

Ritchie GA and Hinckley TM (1975) The pressure chamber as an instrument for ecological research. Adv. Ecol. Res. 9:166-254.

Running SW (1976) Environmental control of leaf water conductance in conifers. Can. J. Forest Res. 6:104-112.

Sage RF and Pearcy RW (1987) The nitrogen use efficiency of C_3 and C_4 plants. II. Leaf nitrogen effects on the gas exchange characteristics of *Chenopodium album* (L.) and *Amaranthus retroflexus* (L.). Plant Physiol. 84:959-963.

Salisbury FB (1954) Some chemical and biological investigations of materials derived from hydrothermally altered rock in Utah. Soil Science 78:277-294.

Salisbury FB (1964) Soil formation and vegetation on hydrothermally altered rock material in Utah. Ecology 45:1-9.

Schlesinger WH and Chabot BF (1977) The use of water and minerals by evergreen and deciduous shrubs in Okefenokee swamp. Bot. Gaz. 138:490-497.

Schlesinger WH, DeLucia EH, and Billings WD (1989) Nutrient-use efficiency of woody plants on contrasting soils in the western Great Basin, Nevada. Ecology 70:105-113.

Schlesinger WH, Fonteyn PJ, and Marion GM (1987) Soil moisture content and plant transpiration in the Chihuahuan desert of New Mexico. J. Arid Env. 12:119-126.

Shainsky LJ and Radosevich SR (1986) Growth and water relations of *Pinus ponderosa* seedlings in competitive regimes with *Arctostaphylos patula* seedlings. J. Appl. Ecol. 23:957-966.

Smith WK (1985) Western montane forests. pp. 95-126. In Chabot BF and Mooney HA (editors) Physiological Ecology of North American Plant Communities. Chapman and Hall, NY.

Sturges DL (1977) Soil water withdrawal and root characteristics of big sagebrush. Amer. Mid. Nat. 98:257-274.

Virginia RA and Delwiche CC (1982) Natural ^{15}N abundance of presumed N_2-fixing and non-N_2-fixing plants from selected ecosystems. Oecologia 54:317-325.

Vitousek PM (1982) Nutrient cycling and nutrient use efficiency. Amer. Nat. 119:553-572.

Waring RH and Running SW (1978) Sapwood water storage: its contribution to transpiration and effect upon water conductance through the stems of old-growth Douglas-fir. Plant Cell Environ. 1:131-140.

Waring RH and Schlesinger WH (1985) Forest Ecosystems. Academic Press, NY.

West NE (1988) Intermountain deserts, shrub steppes, and woodland. pp. 209-231. In Barbour MG and Billings WD (editors) North American Terrestrial Vegetation. Cambridge Univ. Press, NY.

White DE and Newton M (1989) Competitive interactions of whiteleaf manzanita, herbs, Douglas-fir, and Ponderosa pine in southwest Oregon. Can. J. For. Res. 19:232-238.

Whittaker RH (1954) The ecology of serpentine soils. IV. The vegetational response to serpentine soils. Ecology 35:275-288.

Chapter 6
Ecophysiology of Plants in the Intermountain Lowlands

Stanley D. Smith and Robert S. Nowak

Introduction

The Great Basin of Billings (1951) and the Great Basin Division of Cronquist et al. (1972) are major subdivisions of the Basin and Range Province (Fenneman 1931), a region which extends from Washington to Mexico and encompasses well over one million square kilometers. The Great Basin is approximately triangular in shape, and lies between the Sierra Nevada and Cascade ranges to the west and the Wasatch-Rockies to the east, and thus is termed also the Intermountain Region (Cronquist et al. 1972). The northern and southern boundaries are not defined as clearly, being generally demarcated by the Columbia-Snake drainages to the north and the Colorado River to the south (Billings 1951). These northern and southern boundaries are determined as much by plant physiognomic and floristic changes as by topographic barriers.

Despite its name, the Great Basin Province is not a single large basin, but instead is a series of approximately one hundred basins separated by relatively parallel, north-south oriented mountain ranges of fault-block origin. Most of these basins exhibit internal drainage, resulting in extensive playa lakes or salt flats throughout the region (Chapter 7; Young and Evans 1986). However, sloping bajadas (alluvial fans) can occupy up to 75% of the valley areas (Shreve 1942).

Essentially all of the lowland basins of the region are characterized by great expanses of small-statured, shrubby vegetation which surround

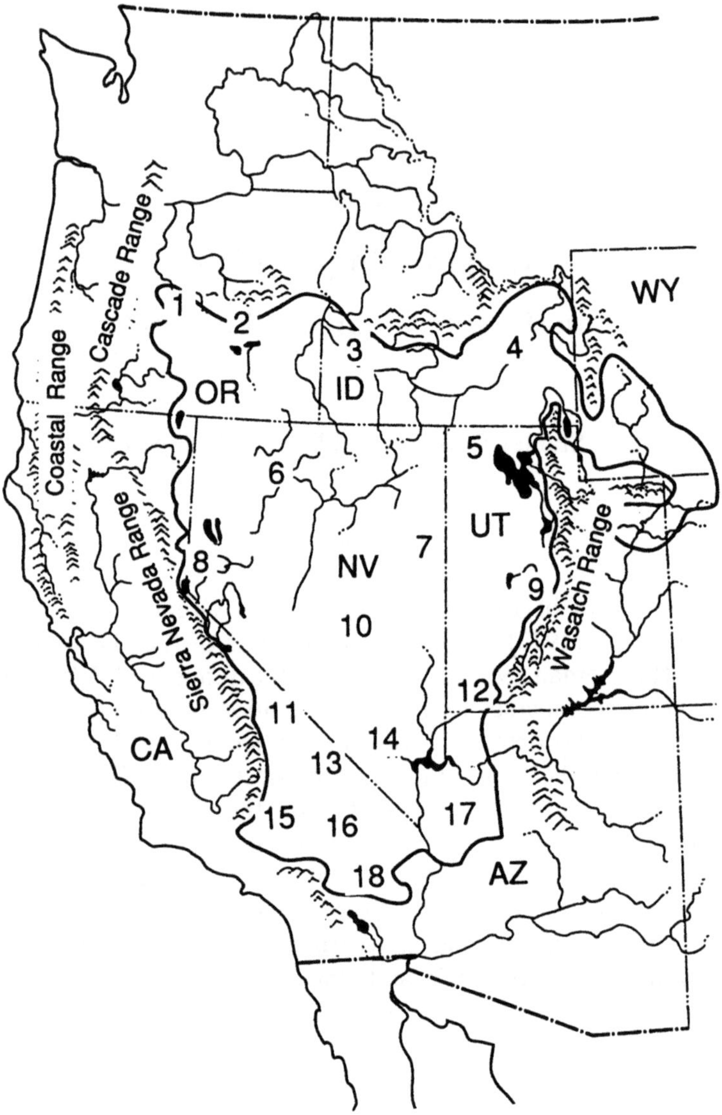

Figure 6.1. Map of the Intermountain Region, showing the Great Basin and Mojave Deserts. Representative climate stations are: 1, Bend, OR; 2, Burns, OR; 3, Boise, ID; 4, Idaho Falls, ID; 5, Curlew Valley, UT; 6, Winnemucca, NV; 7, Ely, NV; 8, Reno, NV; 9, Fillmore, UT; 10, Tonopah, NV; 11, Bishop, CA; 12, St. George, UT; 13, Furnace Creek (Death Valley), CA; 14, Las Vegas, NV; 15, Mojave, CA; 16, Barstow, CA; 17, Kingman, AZ; and 18, Twentynine Palms, CA.

"islands" of woodland vegetation on the higher mountain ranges. The dominant shrubs are microphyllous, and are principally members of the Asteraceae, Chenopodiaceae, and the Zygophyllaceae (Billings 1949). In the southern part of the region, the bajadas are dominated primarily by *Larrea tridentata* (creosotebush) and its associates. These communities are replaced along an irregular boundary extending across southern Utah, Nevada, and California by *Artemisia tridentata* (sagebrush) and *Atriplex confertifolia* (shadscale) vegetation (Billings 1949). This is also the assumed boundary between the Great Basin and Mojave Deserts (Shreve 1942; Jaeger 1957; Beatley 1975).

The Basin and Range Province extends further south to include the Sonoran and Chihuahuan Deserts. Although we will utilize the literature from those deserts, this chapter will focus on the Great Basin and Mojave Deserts which initially were recognized, mapped and described by Billings (1949, 1951) in his treatment of the Intermountain Region of the western United States. Furthermore, these two deserts share consistencies in climate and topography that separate them from the more southerly deserts, which exhibit a warmer, more monsoonal climate and an increased number of external drainage basins, particularly in the Sonoran Desert.

Deserts are considered to represent one of the more extreme habitats for plant life on earth. Plants must possess unique adaptations in order to survive these stressful environments. Noy-Meir (1979/80) advanced an "Autecological hypothesis" that states: "In deserts, the dynamics of each population are determined mainly by its independent reaction to the environment, in particular to the highly intermittent availability of water." Using such an approach, we will review current information related to how the climate and physical conditions of the region influence the water relations, photosynthesis, carbon balance, growth habits, and survival of Great Basin plants and plant assemblages. Some recent reviews of the physiological ecology of desert plants overlap with our discussion of the Intermountain lowlands, and we refer the reader to Caldwell (1985), Ehleringer (1985b), and Smith and Nobel (1986). Finally, it should be pointed out that despite the stressful conditions under which desert vegetation of the region must survive and grow, a moderately high species diversity often can be found in undisturbed plant communities. This suggests that a variety of successful adaptations and plant growth forms have evolved to cope with these stressful conditions. An additional goal of this review will be to explore the nature of these adaptations and growth habits, and to assess their prevalence in native desert communities. Our goal is a more thorough understanding of the ecology and physiology of Great Basin lowland vegetation, which may facilitate predictions of vegetation response to future human impacts or to long-term climatic change.

Climate and Substrates

Desert regions are characterized by sparse vegetation cover and climatic extremes, with annual precipitation typically below 200 mm, potential evapotranspiration (PET) greater than 2,000 mm, and a wide diurnal and seasonal variation in surface temperatures (Noy-Meir 1973). The Great Basin-Mojave region contains the driest habitats in North America (Chapter 2), with precipitation averaging 50-300 mm per year. Abundant sunshine and low humidity result in high PET, with annual lake evaporation ranging from 1,100 mm in the north to over 2,000 mm in the south (Flaschka et al. 1987). The precipitation regime for the region exhibits a winter maximum and a summer minimum, with the summers being almost rainless at many locations (Billings 1949).

On a worldwide basis, the arid and semiarid Intermountain Region may be considered to be relatively well watered and benign when compared to more subtropical deserts such as the central Sahara, where precipitation is less than 100 mm and PET is greater than 5,000 mm. Nonetheless, both of these regions share a dry climate in which evaporation from surface soils is a continuous process that often overtakes the episodic occurrence of precipitation events. Thus, most recharge events are of potentially short duration (Bailey 1981). Seasonal precipitation patterns in the Intermountain Region are illustrated in Figure 6.2. Annual precipitation is lowest in the southwestern part of the region, which also happens to be the area of lowest average elevation. Total precipitation increases as one goes north, east, or south. This increase is a result of large increases in elevation of the drainage basins from south to north (Beatley 1975) and to spatial and seasonal airmass dynamics (Neilson 1987). Furthermore, the percentage of precipitation that falls in the summer "growing season" is lowest along the western edge of the Intermountain Region, i.e., immediately east of the Sierra Nevada and Cascade ranges, the primary rainshadow for the Great Basin and Mojave Deserts (Figure 6.2). Farther east or south, monsoonal precipitation sources become increasingly more important as a percentage of annual moisture input.

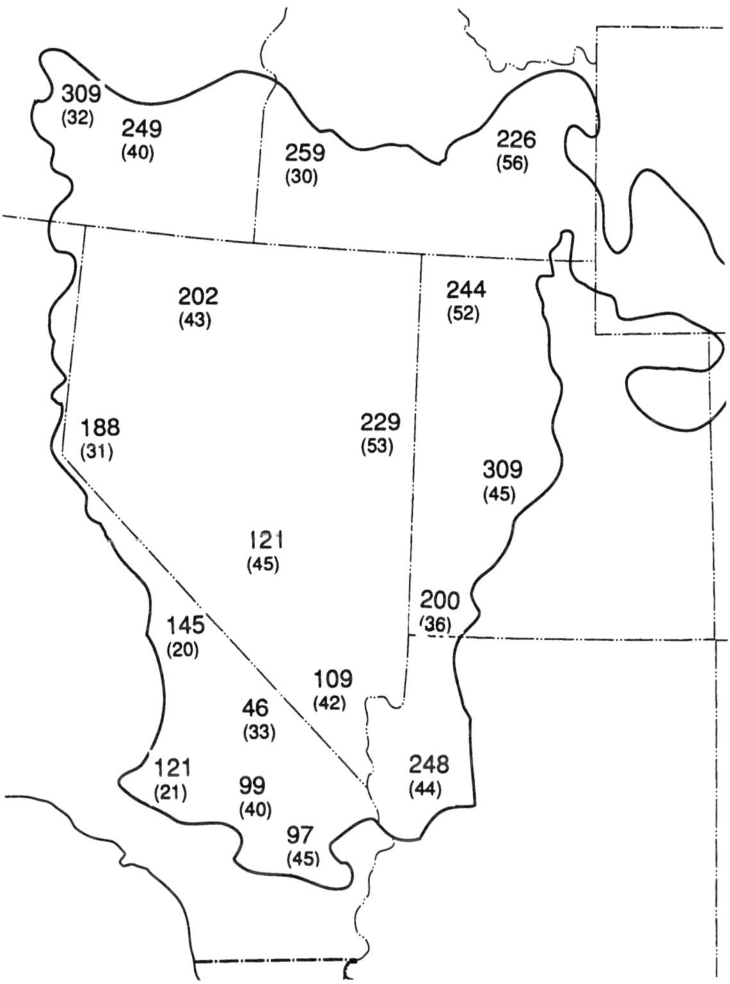

Figure 6.2. Precipitation patterns at various climate stations in the Intermountain Region (see Figure 6.1). Data from each station are total annual precipitation (mm) and the percentage of annual precipitation that occurs in the April to September "potential growing season" (figure in parentheses below total precipitation number).

Analysis of climate diagrams from different stations in the region (see Walter et al. 1975; Caldwell 1985) indicates that even regions with relatively abundant summer rainfall exist at a moisture deficit during that time of the year. For winter precipitation regions, a moisture surplus usually occurs in the winter and early spring, but the moisture deficit in the warm season is significantly more pronounced. Although summer precipitation tends to be highly variable on a year to year basis in desert regions, particularly at the most arid stations (Ehleringer 1985b), it is relatively low in the Intermountain Region, ranging from 20-30% of total annual precipitation in the Great Basin to 30-40% in the Mojave Desert (Bell 1979). Most recharge events are of potentially short duration (Bailey 1981), and are quickly lost from surface soils via evapotranspiration.

Temperature can also pose strong limitations to plant growth and survival in the Intermountain Region. In most of the higher latitudes or elevations of the region there is a wide seasonal fluctuation in air, surface, and soil temperatures. For example, Caldwell (1985) has noted that cold desert shrubs must endure temperatures ranging between -40° C to in excess of 40° C over the course of the year. On a diurnal basis, temperatures also can vary widely between daytime maxima and nighttime minima. This high diurnal variability is a result of the typically clear conditions and low humidity of the region, allowing rapid solar heating during the day followed by rapid surface cooling via longwave radiation loss at night. In addition, the characteristic basin and range topography can result in substantial nocturnal cold air drainage during certain times of the year (e.g., Beatley 1974a).

The primary climatic demarcation between the Great Basin and Mojave Deserts is based on temperature, with the Great Basin being considered a "cold desert" and the Mojave a "hot desert." Indeed, the northern extension of *Larrea tridentata*, and thus the northern extension of the Mojave Desert, has been suggested to be limited primarily by low winter temperatures (Billings 1951; Beatley 1974a, 1975). Figure 6.3 shows that the length of the frost-free season throughout the basins of the Intermountain Region increases substantially in the southern part of the region. Some stations, such as Furnace Creek in Death Valley, may have entire years without a single freezing event, although this is rare. In the more northern and higher elevation stations freezing conditions predominate for the majority of the year, with frost-free seasons as short as 80-90 days in the northern and central Great Basin.

The interplay between seasonal patterns of precipitation and temperature have important implications for the physiological ecology of native desert taxa. In the Great Basin winter precipitation may not be utilized during much of the year because cold temperatures reduce the

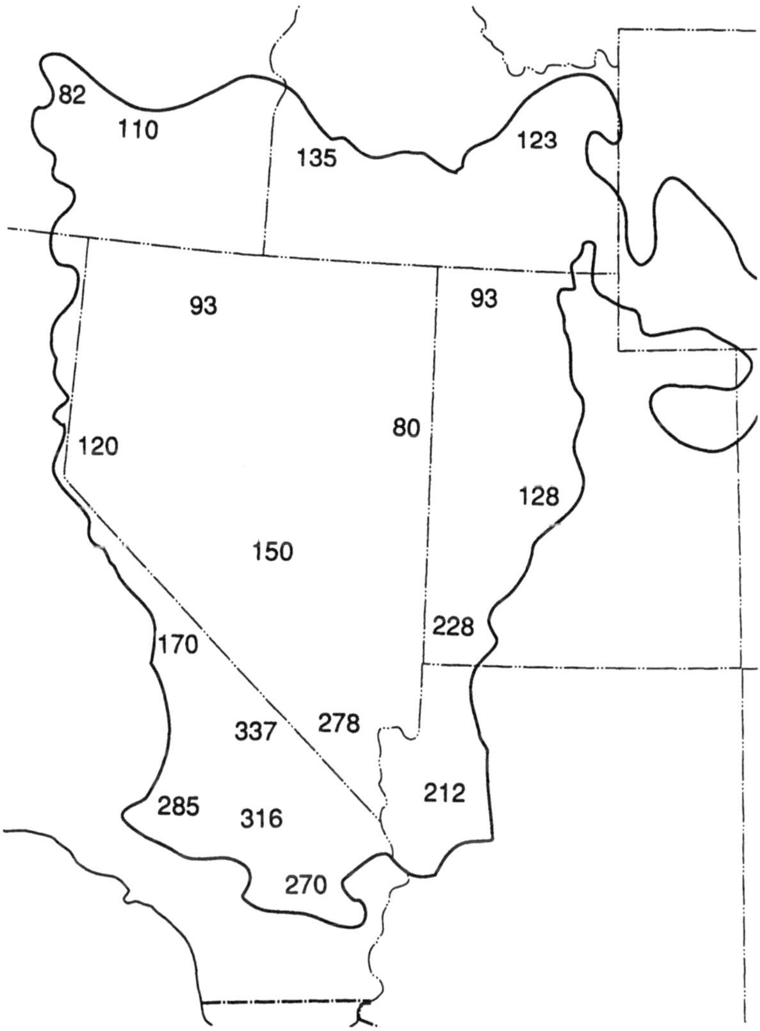

Figure 6.3. Mean length of the frost-free season, in days, at various climate stations in the Intermountain Region (see Figure 6.1).

physiological activities of many perennial species, yet moisture still can be lost via sublimation and evaporation. Even after plants become active, they can be exposed to physiological drought due to high solar radiation and moderate air temperatures occurring in concert with cold or frozen soils (Caldwell 1985). By the time air and soil temperatures become elevated in late spring, many sites may have entered the "water deficit" portion of the year, resulting in a potentially narrow window of favorable conditions for plant photosynthesis and growth.

Soil substrate plays a very important role in plant water and ion relations, and thus in species distribution and vegetation zonation within the Intermountain Region. Much of the mosaic structure that is observed in the vegetation of many basins of the region reflects physical and chemical patterns in substrate. The western Great Basin is primarily underlain by volcanic rocks, whereas the eastern Great Basin is underlain more by sedimentary rocks (Fiero 1986). However, a mixture of rock types can be found in almost every sub-region of the Great Basin. It has been shown that vegetation composition can be dramatically different between igneous and limestone substrates in the Chihuahuan and Sonoran Deserts (Wentworth 1981; Aide and Van Auken 1985), and that these differences are related to differences in soil nutrient balance (Wentworth and Davidson 1987).

On bajadas in the region, the relative age of geomorphic surfaces can influence the degree to which caliche (calcium carbonate deposition layers) has formed in subsurface soils (Gile et al. 1966), which in turn can have an influence on plant water relations by potentially limiting the depth to which roots can penetrate (Shreve and Mallery 1933; Schlesinger 1985). Caliche layers become increasingly indurated with age, which appears to result in a long-term successional change in the vegetation of bajadas toward a more open, xeric community (J. McAuliffe, unpublished data). Within individual closed drainage basins, substrate is often the primary determinant of vegetation zonation, since only plants capable of tolerating high salinities or low soil oxygen can exist on the fine-textured soils of the basin floor. Indeed, the evergreen dominants of each major desert, *Artemisia tridentata* in the Great Basin and *Larrea tridentata* in the Mojave, are thought to be excluded from fine-textured and poorly-drained soils due to their high oxygen requirements for root growth (Lunt et al. 1973).

Vegetation Structure

Shreve (1942) was probably the first plant ecologist to classify the vegetation associations of the Basin and Range Province, while the comprehensive studies of Billings (1945, 1949, 1951) and Fautin (1946) were the first to characterize the vegetation of the Intermountain Region in relation to climate, substrate, and important species assemblages. Billings (1951) identified three broad vegetation zones in the Great Basin: (1) the creosotebush zone; (2) the shadscale zone; and (3) the sagebrush-grass zone. Billings defined the creosotebush zone as the southern extension of the Great Basin from southwestern Utah to southeastern California, an area that is considered presently to be the Mojave Desert. He defined the shadscale zone as the region of bajadas and foothills of western Nevada, extending from the Amargosa Desert to the Lahontan Desert, and another large region in the lower valleys of western Utah. The sagebrush-grass zone occupied the remainder of the Great Basin, primarily the basins and foothills of northern and eastern Nevada, northern Utah, southern Idaho, and eastern Oregon that occur above about 1,500 m elevation. Utilizing the climatic model of Thornthwaite (1948), Billings determined that the creosote-shadscale transition is largely a function of winter temperatures excluding *Larrea* from the colder Great Basin. Within the "cold desert," the shadscale zone is in the arid western troughs of the region, whereas the sagebrush-grass zone occupies the more semiarid, higher elevation basins of the northern and eastern Great Basin.

Numerous studies have been conducted in the past several decades in an effort to quantify vegetation structure in the Intermountain Region, and these have been summarized by Cronquist et al. (1972), MacMahon (1979, 1987) and West (1987). These reviews suggest that there may be up to six primary vegetation associations or formations that are fairly repeatable across the landscape: (1) creosote-bursage scrub; (2) mixed Mojave shrubland; (3) blackbrush; (4) saltbush-greasewood; (5) Great Basin sagebrush; and (6) sagebrush steppe. We will describe briefly each of these formations, stressing the important species that are discussed in more detail in subsequent sections of this chapter.

The creosote-bursage (*Larrea tridentata-Ambrosia dumosa*) association dominates the bajadas and lower foothills of the Mojave Desert (Plate 12.1). This association often has few perennial subordinates present, except in the small washes that dissect the bajadas, where *Acacia* spp., *Hymenoclea salsola* and other shrubs become important. *Larrea* and *Ambrosia* tend to drop out as the bajadas grade into finer textured soils of valley bottoms, where they are replaced by *Atriplex* spp. and other halophytic shrubs. Where upper bajadas become rockier, steeper, or higher in elevation, the mixed Mojave shrub assemblage becomes dominant. This community includes *Yucca brevifolia*, *Yucca schidigera*,

Ephedra nevadensis, *Grayia spinosa*, widely scattered *Larrea*, and often a diverse assemblage of cacti. Major washes in the Mojave Desert support extensive stands of *Prosopis* spp. (mesquite), *Chilopsis linearis* (desert willow) and *Acacia* spp. Along more perennial water courses, or adjacent to springs, riparian zones of *Populus fremontii* (cottonwood), *Salix* spp. (willow) and *Prosopis* spp. dominate, although many of the riparian and spring areas of the Mojave Desert (Plate 12.2) have been invaded extensively and overtaken by the exotic *Tamarix* spp. (saltcedar).

The blackbrush (*Coleogyne ramosissima*) vegetation type usually occurs as an elevational belt on nonsaline sites in the southern Intermountain Region (Plate 13.1). It replaces *Larrea* at higher elevations on bajadas in much of the Mojave Desert, but also can intergrade with *Artemisia* in the southern Great Basin. *Coleogyne* often can be found in association with *Yucca brevifolia* in the Mojave Desert. Although it appears widespread in the region, *Coleogyne* may be largely confined to limestone soils underlain by a caliche layer (Beatley 1975).

The saltbush-greasewood (*Atriplex confertifolia-Sarcobatus vermiculatus*) association of West (1987) is essentially analogous to Billings' (1949) shadscale zone, only it covers not only the *A. confertifolia* dominated bajadas in the arid basins of western Nevada and Utah (Plate 13.2), but also the valley bottoms that are dominated primarily by *S. vermiculatus*. Although *A. confertifolia* often dominates the bajadas in this formation, other important sub-dominants can include *Artemisia spinescens*, *Ceratoides lanata*, *Grayia spinosa*, and the grasses *Oryzopsis hymenoides* and *Sitanion hystrix*. Associates of *Sarcobatus* in bottomlands include *Allenrolfea* sp., *Salicornia* sp., *Suaeda* sp., and the grasses *Distichlis stricta* and *Sporobolus airoides*. [Note that a recent revision of the perennial Triticeae of North America recommends different taxonomic identification for many of the grass species discussed in this chapter: see Barkworth and Dewey (1985) for a synopsis of the revisions.]

The Great Basin sagebrush (*Artemisia tridentata*) formation occurs in the central Great Basin and on the northern Colorado Plateau, and is considered transitional between the shadscale and sagebrush steppe zones (West 1987). In this region, *A. tridentata* dominates in almost monospecific fashion, with few perennial grasses present. This formation has, however, been quite susceptible to invasion by exotic annual grasses, particularly *Bromus tectorum*. The sagebrush steppe (Plate 14.1) covers most of the northern part of the Intermountain Region, and is considered the largest of the North American semidesert vegetation types

(West 1983). This region is characterized by *Artemisia tridentata* plants of larger stature and higher cover than in the Great Basin sagebrush, and they occur in association with a higher cover of perennial bunchgrasses such as *Agropyron*, *Stipa*, *Festuca*, and *Elymus* (Plate 14.2). Shrub subordinates include *Purshia tridentata*, *Chrysothamnus* spp., *Ephedra viridis*, and *Tetradymia* spp., depending on site. Within the sagebrush region as a whole, five taxa of sagebrush have been identified that segregate along environmental gradients according to microhabitat requirements (West et al. 1978). At lower elevations, *Artemisia tridentata tridentata* occurs on moister lowlands with *A. tridentata wyomingensis* dominating the upper bajadas (Barker & McKell 1983). At intermediate elevations *A. tridentata tridentata* again dominates the lowlands, but *A. nova* occurs on the bajadas. At the highest elevations, *A. arbuscula* dominates on dry sites and *A. tridentata vaseyana* on moist sites.

The age and space distribution of perennial plants in the Intermountain Region is of considerable interest to physiological ecologists who attempt to ascertain the importance of specific plant adaptations for growth and survival. Most of the studies conducted on the dominant desert taxa of the region suggest that germination and survival are rare events, resulting in even aged, often clonal, stands in many habitats (Barbour 1969; Stutz 1978; Jordan and Nobel 1979; Vasek 1980). Additionally, recruitment of most perennial species is limited largely to protected microsites beneath established perennial plants (McAuliffe 1988).

Desert perennials have been observed to exhibit aggregated (clumped), random, or regular spatial distribution patterns. A number of quantitative analyses, primarily using distance-based measurements, suggest that many mature desert perennials are distributed regularly (Yeaton and Cody 1976; Phillips and MacMahon 1981; Cody 1986), whereas younger or small-statured plants tend toward clumping (Fonteyn and Mahall 1981; Phillips and MacMahon 1981). Shrub species on the Nevada Test Site are aggregated into "shrub clumps" separated by bare areas of desert pavement (Wallace et al. 1980b). Spatial pattern in *Larrea tridentata* apparently changes along climatic gradients (Barbour 1969), ranging from aggregated populations in mesic habitats to regular distribution on xeric sites (Woodell et al. 1969). Many Mojave Desert perennials exhibit positive interactions between species and negative interactions with conspecifics, which Cody (1986) attributed to differential interspecific compatibility of root systems. Fonteyn and Mahall (1981) concluded that newly established individuals are usually clumped, but tend toward regularity through time due to progressive thinning. Thus, only long-lived populations may exhibit a regular distribution. The determination of true "individuals" may influence greatly spatial pattern analyses; for example, Smith and Ludwig (1978) found individual rosettes of *Yucca elata* to exhibit a highly clumped distribution, whereas clones of vegetatively propagated rosettes were distributed regularly.

Virtually all of the basins of the Intermountain Region have been exposed to grazing pressure by domestic livestock over the past 100 or more years. In many areas this pressure has been intensive and constant, resulting in the decimation of many stands of native bunchgrasses and their replacement by *Artemisia*, *Chrysothamnus*, and other invasive shrubs (Young et al. 1975). Domestic livestock appear to reduce native bunchgrasses not only due to their direct "clipping" effect, but also due to surface disturbance and trampling, which can eliminate surface cryptogamic crusts (Jeffries and Klopatek 1987), reduce soil mycorrhizal fungi (Bethenfalvay and Dakessian 1984; Doerr et al. 1984), and cause soil compaction. Although comparative studies have clearly shown sites with no grazing history to have more grass and shrub cover than on intensively grazed sites (e.g., Young et al. 1975; Jeffries and Klopatek 1987), "recovery" studies have not provided as consistent results. For example, some studies have shown sites that are relieved from grazing pressure for a decade or more to exhibit dramatic increases in both perennial grass and shrub cover (Anderson and Holte 1981; Bock et al. 1984; Rasmussen and Brotherson 1986), whereas other sites have shown no such increases (Rice and Westoby 1978; West et al. 1984; Jeffries and Klopatek 1987). In the latter cases, degradation may have been so complete that succession back to the pristine condition may have been no longer possible. In many degraded areas, especially in the sagebrush steppe, the exotic bunchgrasses, *Agropyron cristatum* and *A. desertorum* (crested wheatgrass), have been planted in extensive, monospecific stands to restore productivity for livestock forage. These grazing tolerant species are adapted to the Intermountain climate and can be considered as naturalized species in the region.

Significant portions of the sagebrush-grass rangelands in the Great Basin are degraded to the condition that they produce less than 50% of their potential (Evans and Young 1984). The elimination of many native grasses resulted in the rapid expansion of several annual invaders in the early 1900s. Foremost among these was *Bromus tectorum* (cheatgrass), which now is distributed widely from the saltbush-greasewood vegetation type through the sagebrush zones (where it is most abundant), extending into the montane coniferous forests of the Great Basin (Morrow and Stahlman 1984). *Bromus tectorum* is now the most important plant species of most disturbed steppe communities, and is the dominant plant on more than 41 million ha in the Intermountain Region (Mack 1981). In the transition zone between the Great Basin and Mojave Deserts, *B. tectorum* intergrades with, and then gives way to *B. rubens*, which is frequently the dominant winter annual of the blackbrush and creosote-bursage community types (Beatley 1966). *Bromus*, particularly *B. tectorum*, appears to attain rapid dominance because it can germinate successfully over a wide range of temperature and moisture conditions (Thill et al. 1979) and quickly establishes an extensive fibrous root system that allows it to out-compete seedlings of native perennial grasses (Harris 1967) and other annual forbs (Young and Evans 1973) effectively.

Fires are infrequent in native, undisturbed desert vegetation, especially in the driest areas. When they do occur, desert wildfires have been shown to cause complete or near-complete mortality of the dominant shrubs of the Intermountain Region, such as *Artemisia tridentata* (Young and Evans 1978; Cluff et al. 1983), *Coleogyne ramosissima* (Callison et al. 1985), and *Larrea tridentata* (O'Leary and Minnich 1981; Brown and Minnich 1986). Both species of *Bromus* strongly promote fire. Invasion of disturbed sites by *Bromus* may thus "predispose" native shrub vegetation to recurring wildfires and cyclic environmental degradation (Young and Evans 1978).

Morphology and Phenology

Desert plant species exhibit many different morphological and phenological patterns, which accomodate the highly unpredictable conditions of their environment. Desert plants have been classified along an "adaptation continuum," from drought avoidance at one extreme to drought tolerance at the other (Levitt 1980). Shantz (1927) proposed four categories of desert plants based on their seasonal responses to water stress: (1) drought-escapers, which include the annual vascular plants and poikilohydric cryptogams; (2) drought-evaders, which include the perennial grasses and forbs and the drought-deciduous shrubs; (3) drought-endurers, which include the succulents and phreatophytes; and (4) drought-resistors, which include the evergreen shrubs and trees and

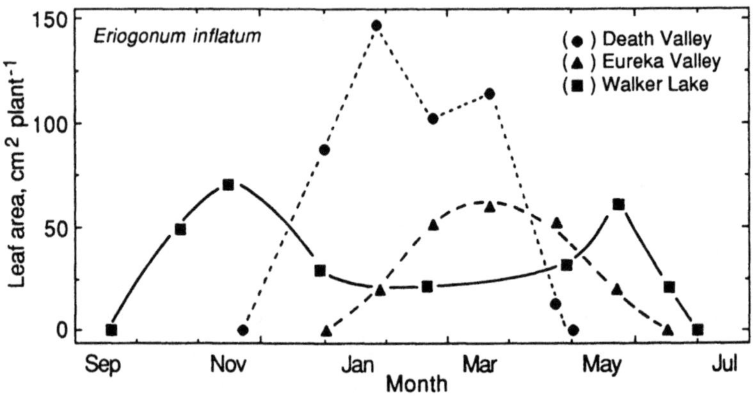

Figure 6.4. Seasonal leaf area of *Eriogonum inflatum* plants from Death Valley, Eureka Valley, and Walker Lake for a single growth cycle (redrawn from Smith and Osmond 1987).

the winter-deciduous, summer-active shrubs. Although all of the above morphological strategies are represented in Great Basin vegetation, the primary perennial dominants of the region are evergreen shrubs. For example, *Artemisia tridentata*, *Atriplex confertifolia*, *Larrea tridentata*, *Coleogyne ramosissima*, *Ceratoides lanata*, and *Yucca brevifolia* are all species that never drop all their leaves and can exhibit net growth in response to spring, summer or fall precipitation (Ackerman et al. 1980).

Phenological development of different growth forms appears to be sequentially offset in most Great Basin communities (Smith and Nobel 1986). Ephemeral plants, both annuals and perennial forbs, grow and flower in response to snowmelt or winter-spring rainfall, and then senesce by early summer. Perennial grasses survive longer, but enter dormancy after seed set in midsummer. Drought-deciduous shrubs also enter partial to complete dormancy by midsummer, as do most succulents. Only evergreen shrubs and phreatophytes are capable of maintaining moderate activity during the hot, dry summer months. Phenology of individual species also can be offset in different climatic regions of the Great Basin. For example, seasonal leaf development in *Eriogonum inflatum*, a perennial forb with a basal leaf rosette, is shown for three contrasting sites in its distribution (Figure 6.4). In the Great Basin Desert south of Reno, *E. inflatum* produced a rosette in the fall and did not exhibit leaf senescence until mid summer. In contrast, *E. inflatum* populations in Death Valley had an accelerated developmental cycle, with complete leaf senescence in late spring. An intermediate, Mojave Desert site showed an intermediate phenological pattern. Extended rosette life is attributed to the cooler temperatures and higher soil moisture prevailing in the more northern sites, particularly in the spring-summer transition period (Smith and Osmond 1987).

In the Great Basin, the start of the "growing season" is determined by: (1) the amount of winter-spring precipitation; and (2) increasing air and soil temperatures. For example, Turner and Randall (1987) determined that regression models required both precipitation and air temperature data to give accurate predictions of first leafing and flowering in shrubs from southern Nevada. The importance of moisture in driving phenology patterns in the Great Basin is illustrated by the study of Caldwell et al. (1977a), who compared the activity and growth schedules of two sympatric shrub species that differed in photosynthetic pathway: *Ceratoides lanata* (C_3) and *Atriplex confertifolia* (C_4). Although *Ceratoides* was active somewhat earlier and *Atriplex* later, the two species had remarkably similar timing of photosynthetic activity and growth even though C_4 plants are documented to operate more efficiently at high temperatures (Berry 1975). The two species also exhibit an almost identical progression of phenophases when averaged over multiple growing seasons (West and Gasto 1978) and similar timing of root extension (Fernandez and Caldwell 1975). The coincidence of timing of phenology and metabolic activity in these two shrubs from a winter precipitation region contrasts with the staggered activity patterns found in more humid regions or in dry regions with a bimodal rainfall regime, with C_3 plants most active during the cool season and C_4 plants in the warm season.

All perennial species of the region (except phreatophytes) undergo at least partial dormancy in the hot, dry summer months. In many species, this dormancy appears to be the result of an interaction of high temperatures and low soil moisture (Ackerman et al. 1980). In an analysis of 19 Mojave and Sonoran Desert shrub and subshrub species, Comstock et al. (1988) found a 32-94% reduction in total photosynthetic area from maximal canopy development in the spring to mid-August peak dormancy. No woody shrub was completely without green surface area, although in several species this surface area was reduced to green, leafless twigs. A comparison of leaf area index as a function of declining water potential in three shrub species from a single site in southern Nevada illustrates interspecific differences in summer deciduousness (Figure 6.5). *Grayia spinosa* and *Hymenoclea salsola* are drought-deciduous

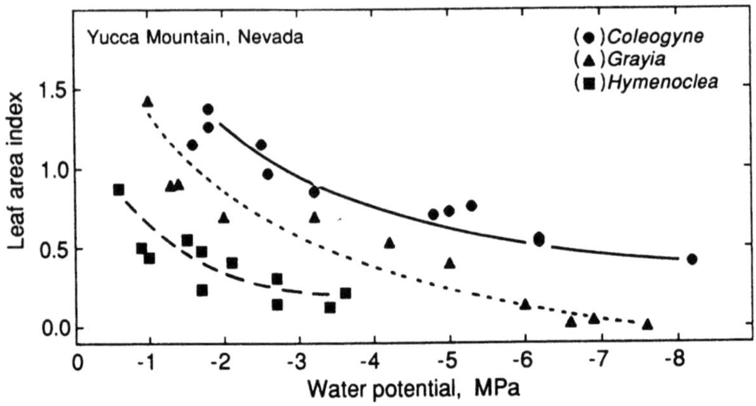

Figure 6.5. Leaf area index of *Coleogyne ramosissima*, *Grayia spinosa*, and *Hymenoclea salsola* as a function of seasonally declining predawn water potential in a transitional Great Basin-Mojave Desert community near Yucca Mountain, Nevada (unpublished data of S.D. Smith).

shrubs that became completely deciduous in midsummer when predawn plant water potentials fell below -5 to -6 MPa, whereas *Coleogyne ramosissima*, an evergreen, maintained a leaf canopy throughout the summer at predawn water potentials down to -8 MPa. However, a continuum of responses still can be seen between the three species (e.g., the evergreen *Coleogyne* still lost over 50% of its leaf area during the summer dry season). Similar to *Coleogyne*, the evergreen *Artemisia tridentata* exhibits high leaf abscission during the summer, but this occurs during a canopy turnover from large, ephemeral spring leaves to smaller, perennial leaves that survive both the summer dry season and winter freezing temperatures (Miller and Shultz 1987). The phenology of the shrubs of the Nevada Test Site have been studied in considerable detail (Bamberg et al. 1975; Strojan et al. 1979; Ackerman et al. 1980; Turner and Randall 1987). The vegetation of the Region is transitional between the Great Basin and Mojave Deserts, and is composed of primarily evergreen shrubs such as *Artemisia*, *Atriplex*, *Coleogyne* and *Larrea*. Drought-deciduous shrubs are also important, including *Ambrosia*, *Grayia*, *Hymenoclea* and *Lycium*. Reproductive phenologies vary considerably. *Atriplex*, *Coleogyne*, *Hymenoclea* and *Ephedra* are obligate spring flowering shrubs, whereas *Artemisia* always flowers in the fall. *Ambrosia*, *Larrea*, and *Lycium* facultatively flower and set seed in the spring, summer, or fall, depending on seasonality of precipitation. Regardless of growth habit and reproductive phenology, all species exhibit maximum canopy development in response to winter or spring rains, then enter relative states of dormancy in the dry summer period. Recovery from dormancy depends on the timing and magnitude of late summer or fall rains. Most species require at least 2 cm of rainfall to reinitiate growth, whereas *Grayia* does not break dormancy in the summer, regardless of moisture conditions. *Atriplex confertifolia* in northern Utah also does not break summer dormancy in response to summer rainfall (Hodgkinson et al. 1978). Species that exhibit active growth in the fall then enter at least partial dormancy with the onset of freezing temperatures in the winter.

Many desert shrubs exhibit pronounced seasonal changes in leaf morphology, which have important implications for leaf energy budgets and resulting leaf temperatures. The polymorphic behavior of large, glabrous leaves that maximize absorption of solar radiation in the cool, moist winter-spring months and that are replaced by small, highly reflective leaves in the hot, dry summer is considered most adaptive. Some shrubs, such as the evergreen *Atriplex hymenelytra* (Mooney et al. 1977b) and the drought-deciduous *Encelia farinosa* (Smith and Nobel 1978, Ehleringer 1982) approach this pattern. For example, leaf reflectance increases (i.e., absorptance decreases) in both of the above species as plant hydration level decreases and also as air temperature increases into the summer (Figure 6.6). This increased reflectance is due to the progressive accumulation of surface salts in *Atriplex* (Sinclair and Thomas 1970; Mooney et al. 1977b) and pubescence in *Encelia* (Ehleringer 1982).

In addition to becoming more reflectant, leaves become smaller, thicker, and may be more vertically oriented in the summer months. In a controlled growth chamber study, Smith and Nobel (1978) found irradiance level, air temperature, and leaf water potential to have an influence on leaf morphological properties in *E. farinosa* (Table 6.1). Their experiments indicated that leaf size is influenced primarily by leaf water potential, leaf thickness by irradiance, and leaf absorptance by irradiance. To test the adaptive nature of these morphological changes in the field, Ehleringer (1985a) monitored leaf temperature in *E. farinosa* throughout a developmental cycle. He found that the large, glabrous winter leaves maintain leaf temperatures above air temperature during a majority of the day, whereas the small, pubescent spring/summer leaves maintained leaf temperatures that were 2-4° C below air temperature during the day (Figure 6.7). Since the upper temperature tolerance of photosynthesis averages about 50° C in the desert shrubs of the region

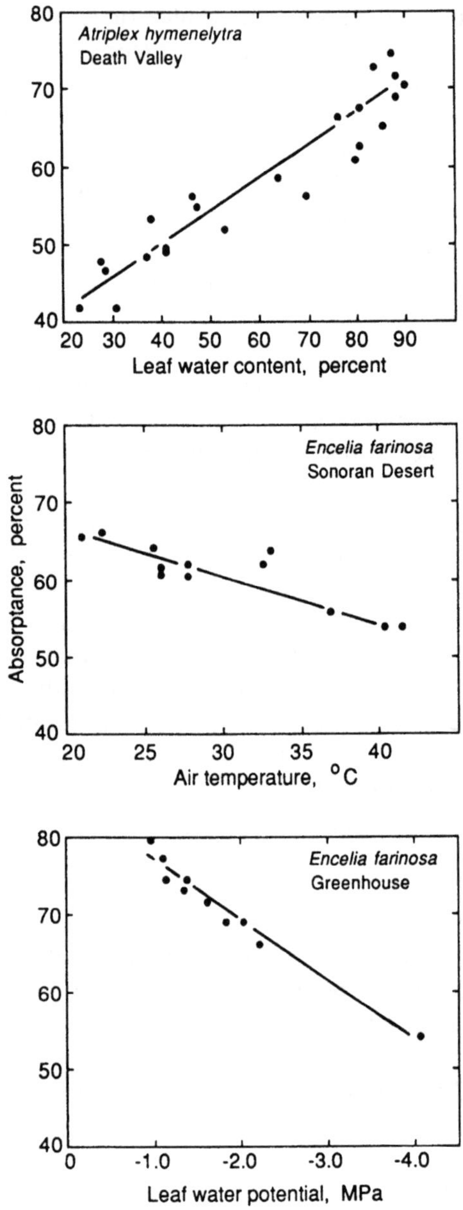

Figure 6.6. Leaf absorptance as a function of leaf water content in *Atriplex hymenelytra*, air temperature in *Encelia farinosa*, and leaf water potential in *E. farinosa*. Data are redrawn from Mooney et al. (1977b) for *Atriplex* and Ehleringer (1982) for *Encelia*.

Table 6.1. Effects of various environmental treatments on leaf morphological properties in *Encelia farinosa*. Plants were grown in growth chambers in the eight possible combinations of the environmental parameters listed below. Mean values are derived across varying conditions of light, temperature, and water potential, with only the listed parameter constant (data from Smith and Nobel 1978).

Parameter	Leaf Length (cm)	Leaf Thickness (μm)	Leaf Absorptance (%)
Low irradiance (300 μmol $m^{-2}s^{-1}$)	3.83	230	67
High irradiance (2,050 μmol $m^{-2}s^{-1}$)	2.80	280	48
Low temperature (22° C)	3.60	252	59
High temperature (35° C)	3.03	258	56
Low water potential (-1.0 MPa)	2.45	255	55
High water potential (-0.1 MPa)	4.18	254	59

(Downton et al. 1984), and since small temperature differences can have large consequences for transpiration, the temperature differentials appear adaptive for *E. farinosa* in the summer months.

Annual or 'ephemeral' species are common in most of the Intermountain Region and can be placed into two broad groups based on seasonal activity patterns (Went 1948; Beatley 1974b): (1) winter annuals that germinate during the cool season; and (2) summer annuals that germinate in response to monsoonal rains. Additionally, some species germinate in early fall and again in late winter or early spring without intervening seed production (Burk 1982). Winter annuals have a variable life cycle duration, with peak activity in spring, and are exclusively of the C_3 photosynthetic pathway. Summer-active annuals are characterized by the C_4 pathway (Mulroy and Rundel 1977) and are almost exclusively confined to the Mojave Desert. In both groups, the

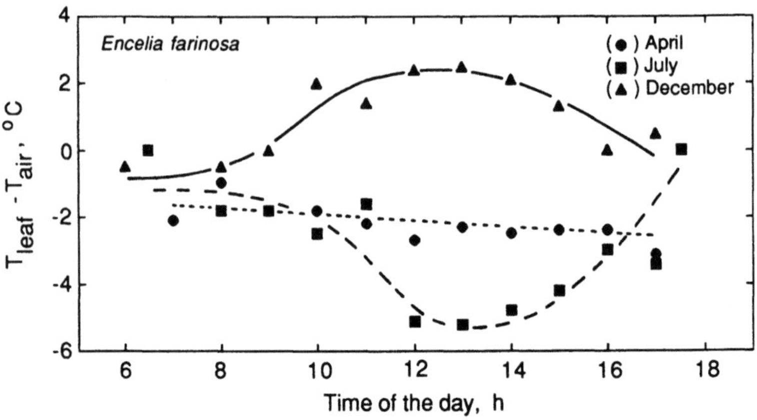

Figure 6.7. Diurnal trends in leaf-to-air temperature differential for a Sonoran Desert population of *Encelia farinosa* in December, April, and July. Data are redrawn from Ehleringer (1985a).

minimum precipitation requirement for seed germination is about 25 mm (Went 1949; Beatley 1974b). This minimum moisture threshold apparently supplies sufficient soil moisture to ensure completion of the life cycle in the event that no further rainfall occurs (Noy-Meir 1973). Not surprisingly, many desert annuals have accelerated life cycles that ensure seed set in as little as six weeks after germination. Other species commit a portion of their resources to nonstructural storage (Clark and Burk 1980), thus lengthening the potential growing season and significantly increasing seed set in the event that additional soil moisture becomes available during the life cycle.

Desert annuals often germinate at densities that cannot be supported by existing soil moisture reserves. High mortality in winter annuals often occurs during the early spring, at the time that a shift from slow vegetative growth to rapid stem elongation and a many-fold increase in plant volume is observed (Klikoff 1966; Beatley 1967). Some desert ephemerals also appear to be capable of facultative perennation, and to thus assume a biennial habit under favorable conditions (Beatley 1970). In a comparative study of two chromosomal races of the ephemeral *Machaeranthera gracilis*, Monson and Szarek (1981) found a desert race to have an accelerated life cycle compared with a foothills race under identical glasshouse conditions, but also to have the capacity to assume the biennial growth habit in the field during favorable years. Flexibility of life cycles is apparently one of the primary attributes that has allowed many introduced species (e.g., *Bromus*, *Salsola*, and *Schismus*) to obtain wide dominance on many degraded rangelands of the Great Basin. Most of these introduced ephemerals have been shown to exhibit high phenotypic plasticity, a lack of seed dormancy, and successful germination under a variety of conditions (Burk 1982; Mack and Pyke 1983).

In contrast to desert annuals, woody perennials appear to germinate primarily in response to summer rainfall; therefore, desert perennials only germinate in certain favorable years in winter rainfall deserts (Went 1979). This often results in clearly defined age groups among shrub populations. The best studied species for reproductive biology and germination behavior is *Larrea tridentata*. *Larrea* appears to suffer very high predispersal reproductive attrition, primarily due to abortion and insect predation during and after anthesis, but still attains a moderately high seed output (Boyd and Brum 1983a). Most seed dispersal occurs in the summer, and there is no significant carry-over of viable seeds between reproductive seasons (Boyd and Brum 1983b). *Larrea* requires high temperatures for germination (Barbour 1968), resulting in seedlings being found only after summer rains (Sheps 1973; Beatley 1974a). Xylem-ring counts of seedlings and small plants confirm that germination and establishment in *Larrea* only occurs in response to heavy, late summer rainfall (Boyd and Brum 1983b), which may be a relatively uncommon event in many parts of its range in the Mojave Desert. Unfortunately, few analogous studies have been conducted on other common shrub taxa of the Intermountain Region.

The root systems of some desert perennials, and certainly most phreatophytes, can be quite deep, but the root systems of most desert xerophytes are shallow (Barbour 1981). The root systems of *Larrea tridentata* have been reported to predominate in the top 10-25 cm of the soil (Chew and Chew 1965), whereas *Atriplex confertifolia* roots may extend to below 80 cm (Caldwell and Fernandez 1975). Root/shoot ratios for desert perennials show considerable variation among species, ranging from 0.15 to 8.0 (Barbour 1973). Representative root/shoot ratios for plants from the Intermountain Region are given in Table 6.2. *Larrea tridentata* has been reported to have root/shoot ratios ranging from 0.19 for glasshouse-grown seedlings to 1.24 in the Mojave Desert. In ten species from the northern Mojave Desert, Wallace et al. (1974a) observed a mean root/shoot ratio of 0.92. In contrast, Caldwell et al. (1977a) reported very high root/shoot ratios for cold desert dominants in northern Utah (Table 6.2). This may be due to the fact that, in the cold desert environment, the root system remains active most of the year as opposed to a very short period of activity for aboveground growth (Caldwell and Fernandez 1975). This results in a distinct cline occurring from high root/shoot ratios in the cold Great Basin Desert to intermediate in the transitional Mojave Desert to low root/shoot ratios in the hot Sonoran Desert.

In contrast to perennials, annual plants have extremely low root/shoot ratios (Table 6.2). Forseth et al. (1984) observed a mean root/shoot ratio of 0.13 for 13 winter annuals and 8 summer annuals from Death Valley, and Bell et al. (1979) found roots to comprise only 12-22% of total biomass in 8 species from southern Nevada. They also observed root allocation to be insensitive to differences in soil moisture. Obviously the patterns of biomass allocation between aboveground and below ground parts differ substantially among growth forms and climates, making it a questionable practice to identify generalized root/shoot allocation strategies for desert plants (see Barbour 1973). The role of root systems in the water relations, nutrient uptake, and plant-plant interactions of Great Basin vegetation is reviewed in Chapter 7.

Water and Ion Relations

Seasonal water relations of desert perennials in the Intermountain Region are a function of short- and long-term climatic conditions and soil structure, which in turn reflect seasonal changes in soil moisture storage. Throughout the region, the winter (October to March) is considered to be a soil moisture recharge period, whereas the summer is one of moisture deficit (Moore et al. 1972; Cable 1977). Campbell and Harris (1977) compared the seasonal water relations of *Artemisia tridentata* in wet and dry years (the wet year had about 2.5 fold higher precipitation than in the dry year), and found substantially lower leaf water potentials from June through September in the dry year (Figure 6.8). The wet year exhibited enhanced precipitation primarily in the winter recharge period, so differences in leaf water potential in the summer appear to be a function of an earlier depletion of stored soil moisture by *Artemisia* in the dry year.

Soil structural characteristics such as texture and depth also can be important in determining soil moisture storage and thus the availability of soil moisture in the dry season. For example, *Haplopappus cooperi* exhibits substantially lower summer water potentials on a bench surface

Table 6.2. Representative root/shoot ratios for desert plants of the Basin and Range Province.

Growth Form	Species	Desert	Root/ Shoot	Reference
Shrub	*Artemisia tridentata* (community level)	Great Basin	0.4-1.8	1
	Ambrosia dumosa	Mojave	1.16	2
	" "	Sonoran	0.64	3
	Atriplex canescens	Mojave	0.67	2
	Atriplex confertifolia	Great Basin	4.09	4
	" "	Mojave	0.43	2
	Ceratoides lanata	Great Basin	6.77	4
	Ephedra nevadensis	Mojave	0.83	2
	" "	Sonoran	1.22	3
	Grayia spinosa	Mojave	0.70	2
	Hymenoclea salsola	Sonoran	0.72	3
	Larrea tridentata	Mojave	1.24	2
	" "	Sonoran	0.33	3
	" "	(Glasshouse)	0.19	5
	Lycium spp.	Mojave	0.8-1.6	2
Annual	*Camissonia brevipes*	Mojave	0.15	6
	Camissonia claviformis	Mojave	0.08	7
	Geraea canescens	Mojave	0.11	7
	Plantago insularis	Mojave	0.06	6
	" "	Mojave	0.13	7
	Schismus arabicus	Mojave	0.04	6

References: 1, Pearson (1965); 2, Wallace et al. (1974a); 3, Garcia-Moya and McKell (1970); 4, Caldwell et al. (1977a); 5, Barbour et al. (1974); 6, Bell et al. (1979); 7, Forseth et al. (1984).

underlain by an impermeable caliche layer at ca. 50 cm than in an adjacent wash with gravelly soils and no caliche layer (Figure 6.8). Caliche layers have lower water holding capacity than do uncemented soil masses, and they are often an impermeable barrier to downward penetration of plant roots (Shreve and Mallery 1933; Gile et al. 1966). The maximum rooting depths of many desert species (Fernandez and Caldwell 1975; Sturges 1977; Wallace et al. 1980a) can be deeper than the depth of the caliche layer. Thus, widespread caliche formation in bajada soils, particularly in warmer regions, may be important in determining the species composition and the low and open character of the vegetation (Shreve and Mallery 1933).

Root growth and soil water extraction commence in most Great Basin perennials by early spring in the upper soil layers prior to the initiation of aboveground vegetative growth (Caldwell 1985). Root growth and water extraction then progressively shift downward (and outward) in the soil profile as surface soils increase in temperature and become depleted of moisture (Fernandez and Caldwell 1975; Sturges 1977). Finally, in late summer, root systems are either dormant (in shallow soils) or exhibit root extension at low water potentials in the deeper soil layers only. The ability of some species to initiate root extension at low temperatures may be a competitive advantage in some communities. For example, studies indicate that *Bromus tectorum* roots grow well at temperatures as low as 3° C, whereas the bunchgrass *Agropyron spicatum* grows very slowly at temperatures below 8-10° C (Harris 1967; Harris and Wilson 1970). By producing a finely branched root system with many root hairs earlier in the growing season, *Bromus* seems to be capable of gaining competitive advantage over *Agropyron* and other native bunchgrasses by effectively utilizing and depleting moisture reserves throughout the soil profile prior to the breaking of dormancy in *Agropyron* (Hulbert 1955; Harris 1967). Similarly, the introduced half-shrub *Kochia prostrata* appears to be competitively superior to *Artemisia tridentata* in many cold desert steppe sites via higher early spring transpiration rates and rapid depletion of soil moisture reserves (Romo and Haferkamp 1988).

Seasonal patterns and magnitudes of plant water potentials are also a function of atmospheric moisture demand. This is best illustrated with native riparian or phreatophytic plants that have a perennial water supply, so that seasonal or diurnal reductions are primarily a function of atmospheric stress. For example, *Betula occidentalis*, an obligate riparian species from the eastern Sierra Nevada, exhibits significantly lower predawn and midday water potentials during the summer at a low elevation (1,400 m) than at a high elevation (2,500 m) (Figure 6.8). Midsummer maximum temperatures were 5-7° C warmer at the lower elevation site, resulting in much higher vapor pressure deficits at that site.

Most of the available data suggest that two principal adjustments in plant cell water relations can occur. Decreases in osmotic potential

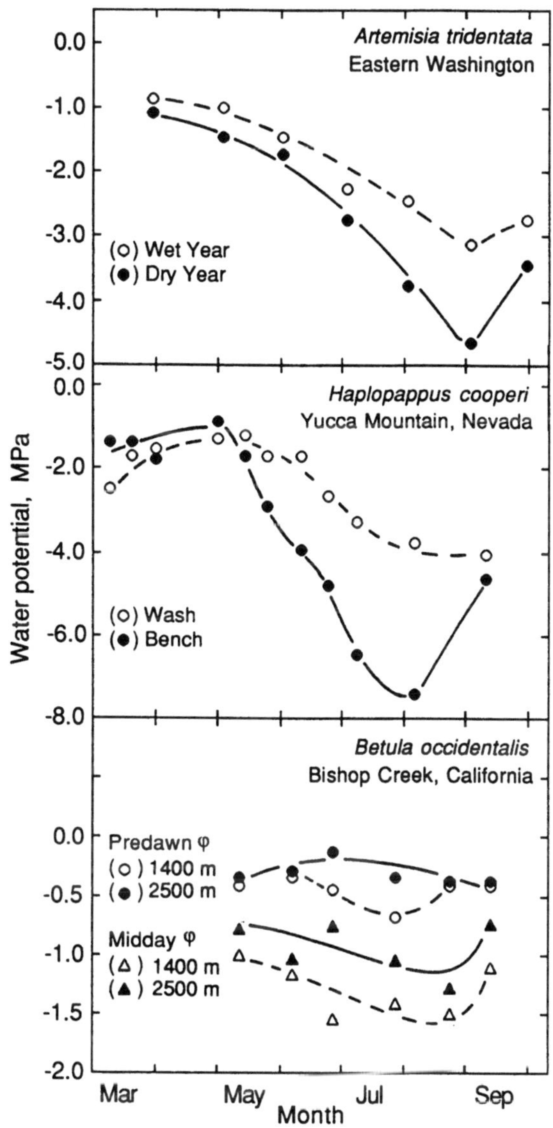

Figure 6.8. Seasonal predawn water potential in *Artemisia tridentata* in wet and dry years in eastern Washington; in *Haplopappus cooperi* in a wash and an adjacent bench in southern Nevada; and predawn and midday water potential of *Betula occidentalis* at 1,400 m elevation and 2,500 m in eastern California. *Artemisia* data are redrawn from Campbell and Harris (1977); *Haplopappus* and *Betula* data unpublished from S.D. Smith.

through accumulation of solutes or reduction in cell size may parallel decreases in total plant water potential, thus maintaining positive turgor pressure. Additionally, changes in cell wall elasticity can affect the magnitude of change in cell turgor pressure with respect to changes in water content. Unfortunately, studies that have investigated the cellular water relations of plants from the Great Basin region are few. Several field studies (Bennert and Mooney 1979; Monson and Smith 1982) have indicated that drought-tolerant evergreens maintain relatively low osmotic potentials throughout the year, whereas drought-avoiding ephemerals maintain high osmotic potentials. Drought-evading deciduous shrubs and drought-enduring deep-rooted trees exhibit intermediate osmotic potentials (Monson and Smith 1982). The capacity for osmoregulation also has been noted in most perennials that have been studied. Seasonal fluctuations in osmotic potential have been observed to range from 0.3-0.8 MPa in four Sonoran Desert woody perennials (Monson and Smith 1982) to 3.0 MPa in *Artemisia tridentata* (Campbell and Harris 1977) and *Yucca brevifolia* (Smith et al. 1983). In addition to seasonal fluctuations, Bennert and Mooney (1979) found osmotic potential to vary diurnally in *Larrea tridentata* and *Atriplex hymenelytra*. Some desert annuals also may exhibit seasonal osmoregulation (Forseth et al. 1984), and so may not be strict drought-avoiders in the conventional sense.

Seasonal water availability is important in determining vegetation structure in the shrublands and grasslands of the Intermountain Region. This has rarely been demonstrated quantitatively, but comparative studies of the subspecies of *Artemisia tridentata* are instructive. *A. tridentata* ssp. *tridentata* tends to occur in more mesic microsites of the cold desert steppe, whereas *A. tridentata* ssp. *wyomingensis* often forms monospecific stands on the bajadas of the region. *A. tridentata wyomingensis* has been observed to exhibit lower plant water potential and lower leaf conductance, and thus potentially to experience greater stress than contiguous populations of *A. tridentata tridentata* over much of the growth season (Barker and McKell 1983), although they have been observed to exhibit similar leaf water potentials when exposed to similar soil moisture conditions (Shumar and Anderson 1986). No comparative water stress experiments have been conducted on these subspecies under controlled conditions, but *A. tridentata wyomingensis* appears to be capable of surviving on more xeric sites by producing a much smaller aboveground canopy and having more rapid and longer root growth (Shumar and Anderson 1986; Welch and Jacobson 1988). Future

Table 6.3. Water potentials of desert perennials and soil following selective removal experiments conducted in the Intermountain Region. Predawn plant water potentials were measured in late July for *Ambrosia dumosa* and *Larrea tridentata* (Fonteyn and Mahall 1981), during the spring for *Encelia farinosa* (Ehleringer 1984). Soil water potential in the rooting zone was measured under *Hilaria rigida* during mid-September (Robberecht et al. 1983).

Species	No Neighbors Removed	Intraspecific Neighbors Removed	Interspecific Neighbors Removed	All Neighbors Removed
	Water potential (MPa)			
Ambrosia dumosa	-5.02	-4.98	-4.24	-3.99
Encelia farinosa	-3.30	-2.60	—	—
Larrea tridentata	-5.47	-5.01	-5.13	-4.72
Hilaria rigida	-4.89	-2.28	—	—

studies on the *A. tridentata* complex may yield further differences in adaptive morphology and water relations parameters that may aid in an explanation of its distribution in the Great Basin.

Competition between desert plants for limiting soil moisture reserves has been suspected for some time (Cohen 1970), and has been inferred from findings of regular spatial distributions in many desert species. However, until recently hypotheses concerning regular distributions in desert plants were not tested with controlled field experiments. Results of recent selective removal experiments (Table 6.3) indicate that removal of both intra- and interspecific competitors results in increases in plant water potential, particularly in dry seasons (Fonteyn and Mahall 1981, Robberecht et al. 1983, Ehleringer 1984). This alleviation of water stress has resulted in substantial, in some cases several-fold, increases in both aboveground and belowground productivity (Robberecht et al 1983, Ehleringer 1984, Nobel and Franco 1986), at least during the short-term experiments that were conducted. Caldwell, Richards, and colleagues have been conducting more long-term competition studies with Great Basin shrubs and grasses, as described in Chapter 7.

The importance of soil water availability and plant water relations on the fitness of individual species, and thus their distribution along environmental gradients, is illustrated effectively by the distribution of sexes within several species of Great Basin plants. In an analysis of five dioecious, wind pollinated species of diverse ecological habit (e.g., *Atriplex confertifolia, Distichlis spicata*), Freeman et al. (1976) found a dichotomous distribution to occur between male and female plants, with males more abundant on xeric microsites and females segregated in moister microhabitats. Males were assumed to be more tolerant of dry sites because each wind pollinated species studied shed pollen in the spring, prior to the summer dry season. In contrast, females of each species studied have greater moisture requirements due to the fact that they set seed throughout the summer. The authors concluded that these dioecious species, each of which can occur in monospecific stands, may maximize reproductive potential when males occupy marginal sites, and minimize competition with females on more optimum sites.

Shrublands of the Great Basin, particularly in salt desert regions, tend to exhibit striking examples of patterned mineral distribution in surface soils (Fautin 1946; Charley and West 1977; Doescher et al. 1984). Under typical soil conditions of low availability of nutrient ions (NO_3^- and $H_2PO_4^-$) and high levels of non-nutrient ions (Cl^- and HCO_3^-), the plant processes of ion uptake, transport, and excretion no doubt contribute to these patterns (Charley and West 1977; Osmond 1979). In general, the widespread upland evergreens of the region, *Artemisia tridentata* and *Larrea tridentata*, are intolerant of high soil salinity (Wallace et al. 1973). Intermediate species that are found on both saline and non-saline sites include *Agropyron* sp., *Atriplex confertifolia, Ceratoides lanata*, and *Chrysothamnus* spp. (Branson et al. 1976; Miller et al. 1982). Species that occur preferentially on saline sites include *Distichlis* spp. and *Sarcobatus vermiculatus*. *Sarcobatus* has been found to occur in soils with near-surface exchangeable sodium percentages of up to 50% (Eddleman and Romo 1987) and to tolerate a concentration of 1 M NaCl (-4.4 MPa) in solution culture.

Although desert vegetation in the region may exhibit striking distributional patterns along soil chemical gradients, the salinity tolerance of Great Basin plants is quite variable. Salt accumulation and metabolic processing in the tissues of desert halophytes is basically no different than that of halophytic species from other regions (Caldwell 1974). Desert shrubs can contain very high concentrations of inorganic ions (Wallace et al. 1973, 1974b), which may account at least partially for the low tissue water potentials that occur in response to soils drying to below -5 MPa in the dry season. These shrubs appear to regulate effectively ionic concentration within their cells and tissues by accumulating high concentrations of salt in the vacuole rather than cytoplasm. Increased succulence of plant tissues on saline sites also appears to play a role in this process. Effective salt excreting glands are found in several important halophytic genera, including *Atriplex* (Mozafar and Goodin 1970) and *Tamarix* (Thomson et al. 1970). A great deal of research has been conducted on the physiology of halophytes, particularly with regard to compartmentalization of solutes and function of salt glands, which has been reviewed by Caldwell (1974) and Osmond (1979) in the context of the physiology of desert halophytes.

Stomatal Conductance and Plant Water Use

In addition to soil moisture availability and osmotic regulation, the water relations of plants are influenced by stomatal control of transpiration and plant water use. Stomatal responses in leaves of *Agropyron dasystachyum* (streambank wheatgrass) to step changes in leaf microclimatic factors are typical of those for many Great Basin plants. Conductance increases rapidly at low irradiance, and maximum conductance occurs at photosynthetic photon flux densities (PPFD) between 33% and 50% full solar irradiance (Figure 6.9a). Stomates close as the air becomes dryer (Figure 6.9b) or as both leaf temperature and vapor gradient increase concomitantly (Figure 6.9d). If vapor gradient is kept constant as leaf temperature increases, then conductance increases (Figure 6.9c). Conductance is relatively constant over a wide range of intercellular CO_2 partial pressures between 150 and 350 μbar, but decreases rapidly above 500 μbar (Figure 6.9e). These stomatal responses are very similar to those found for leaves (Caldwell et al. 1983; Werk and Ehleringer 1985; Nowak and Caldwell 1986; Osmond et al. 1987b; Smith and Osmond 1987; Nowak et al. 1988), excised shoots (Davis et al. 1985), and whole plants (DePuit and Caldwell 1975a) of many species. In contrast to leaves, conductance of stems increases with PPFD over a wider range of irradiance (Nowak and Caldwell 1986), especially at high vapor gradients (Osmond et al. 1987b), and is less responsive than leaf conductance to concomitant changes in tissue temperature and vapor gradient at temperatures above 20° C.

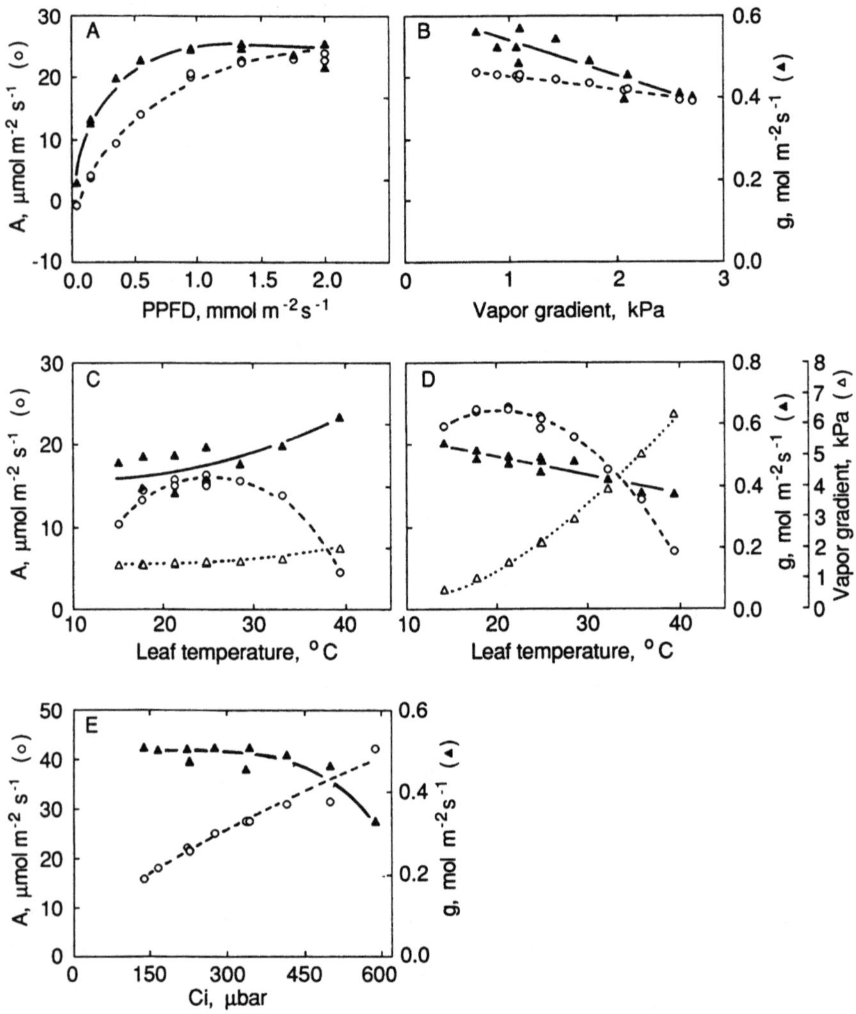

Figure 6.9. Changes in leaf conductance (stomatal plus boundary layer) and assimilation rate for *Agropyron dasystachyum* in response to step changes in: A, Photosynthetic photon flux density; B, Leaf-to-air vapor gradient; C, Leaf temperature; D, Leaf temperature concomitant with vapor gradient; and E, Intercellular CO_2. Data shown are typical for those from 2-7 sets of measurements, except for E which was not replicated. Unless indicated, all other microclimatic factors were constant. Values for constant microclimatic factors were PPFD at 2.0 mol m^{-2} s^{-1}; leaf temperature at 25° C; vapor gradient at 1.8 kPa for A and C, 1.1 kPa for E; and ambient CO_2 partial pressure at 297 μbar (350 ppm).

Cold soil temperatures reduce leaf conductance, but freezing air temperatures at night do not inhibit stomatal opening the following day. Leaves of Great Basin plants grown at cold soil temperatures have lower conductances than those of plants grown at warmer soil temperatures (Anderson and McNaughton 1973). However, decreasing daytime temperatures in the fall and cold nighttime air temperatures in the spring do not decrease conductance of leaves from two bunchgrass species (Nowak and Caldwell 1984a).

Leaf conductance (Campbell and Harris 1977; Frank 1983; Comstock and Ehleringer 1984; Smith and Osmond 1987; Nowak et al. 1988) as well as whole-plant conductance (DePuit and Caldwell 1975a; Caldwell et al. 1977a) decrease with declining plant and soil water status in the short- and long-term , but are not affected by leaf aging. For some of these studies, the long-term decrease in conductance was not confounded by leaf age because similar-aged foliage was used throughout the studies. Furthermore, independent measurements of conductance as leaves aged did not show any marked changes in conductance (Frank 1981; Nowak and Caldwell 1986). However, many of these measurements, which were made at optimal conditions for photosynthesis, may not be representative of the microclimate that plants experience, especially during summertime drought periods. For example, whole-plant conductance for the shrub *Gutierrezia sarothrae* decreases with increased plant water stress when measured near the temperature optimum for photosynthesis, but is relatively constant at higher temperatures that are more typical of midsummer (DePuit and Caldwell 1975a).

Leaf conductance for many species peaks before solar noon, especially when water availability is low and vapor gradients between leaf and air are high. *Agropyron desertorum* (crested wheatgrass) and *Gutierrezia sarothrae* show midday stomatal closure throughout the growing season, regardless of water status (DePuit and Caldwell 1975a; Nowak et al. 1988). For sagebrush and *Eriogonum inflatum*, midday decreases in conductance only occur later in the growing season when plants experience greater water stress and higher vapor gradients (DePuit and Caldwell 1973; Smith and Osmond 1987). Under conditions of severe water stress, leaf conductance of sagebrush and crested wheatgrass peaks early in the morning, then slowly declines for the rest of the day (Black and Mack 1986; Nowak et al. 1988).

Control of the diurnal pattern of conductance is influenced by both soil water availability and stomatal responses to vapor gradients. Midday stomatal closure of Basin wildrye leaves does not occur when soil water availability is high, even though vapor gradients are also high (J.E. Anderson, R.S. Nowak, and N.L. Toft, unpublished data). On the other hand, when vapor gradients decrease during the day, conductance of crested wheatgrass leaves increases, regardless of soil water availability (Nowak et al. 1988). If temperature is held constant during the day, midday decreases in conductance of sagebrush plants only occur later in the year when plants are water stressed, suggesting a direct influence of water stress on conductance (DePuit and Caldwell 1973). Although we are beginning to understand the relationships between conductance and both water stress and vapor pressure of the air (Gollan et al. 1985; Schulze 1986; Ball et al. 1987), much more research is needed to discover the roles and mechanisms that determine how these two factors regulate the diurnal and seasonal patterns of conductance.

Evapotranspiration (ET), the sum of plant water use and soil surface evaporation, is essentially equivalent to precipitation for most vegetation in the Intermountain lowlands on a yearly basis, with the possible exception of plant communities dominated by *Bromus tectorum* (cheatgrass). Native sagebrush steppe communities, individual plants, and monoculture stands of Great Basin species remove all available moisture from the soil profile (Rickard 1967; Campbell and Harris 1977; Anderson et al. 1987). In two salt-desert shrub communities, ET and precipitation are the same, with approximately half of the yearly water loss by transpiration and half by evaporation and sublimation (Caldwell et al. 1977a). In mixed Mojave shrublands, ET may not exactly equal precipitation in any one year, but over a several year period, the average values are very similar (Lane et al. 1984). In contrast, cheatgrass communities may represent one of the few communities in the Intermountain lowlands where ET is substantially less than precipitation. Soil water extraction by cheatgrass-dominated communities is not as deep as the native plant community in wet (Cline et al. 1977) or dry (R.S. Nowak and R.J. Tausch, unpublished data) years. Stand density of cheatgrass monocultures can affect the rate of soil water depletion, but not the depth nor the total amount of water extracted by the end of the growing season (Hulbert 1955). The extent that untapped water deep in the soil profile accumulates under cheatgrass communities or contributes to groundwater recharge is unknown.

The timing and depth of soil moisture extraction also vary among other species in the Great Basin. Soil water extraction by crested wheatgrass is more rapid than by *Agropyron spicatum* (bluebunch wheatgrass)

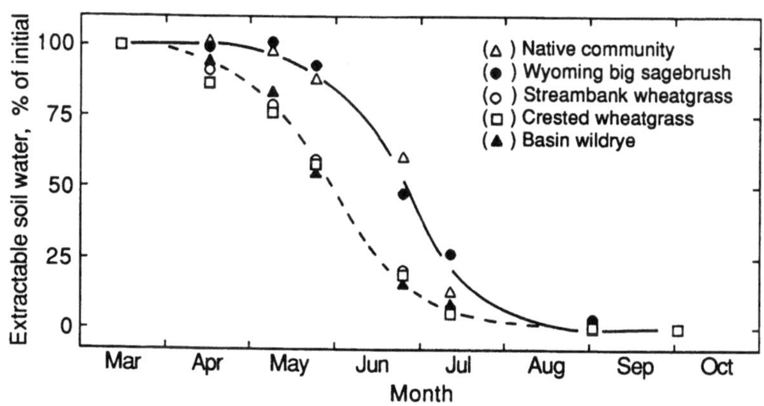

Figure 6.10. Depletion of extractable soil water during the 1986 growing season for a native sagebrush community and monoculture stands of 4 species. Extractable soil water is expressed relative to that in the soil profile at the beginning of the growing season. Redrawn from Anderson et al. (1987).

(Eissenstat and Caldwell 1988a). The maximum rate of soil water extraction by three grass species (crested wheatgrass, streambank wheatgrass, and Basin wildrye) and sagebrush are similar, but the rapid depletion of soil moisture during the spring occurs two weeks later in sagebrush monocultures and the sagebrush-dominated, native community (Figure 6.10). Soil moisture extraction near sagebrush plants is, however, more rapid during the spring than near other shrubs (Rickard 1967; Miller 1988) and deeper in the soil profile than for bluebunch wheatgrass (Shown et al. 1972; Thorgeirsson 1985). As indicated by the authors of these studies, when and where each species extracts water may affect their competitive relationships.

Environmental factors such as grazing, fertilization, and microenvironment affect the rate of soil water depletion. Soil moisture is depleted more rapidly in ungrazed stands of crested wheatgrass than in grazed stands, but the total amount of water extracted by the end of the growing season is similar for both treatments (Wraith et al. 1987). Fertilization also increases the rate of water use by crested wheatgrass (Williams et al. 1979), but not the total amount of water extracted (Power and Alessi 1970). The light-colored ash from the eruption of Mt. St. Helen's in 1980 altered the microclimate of sagebrush plants and increased transpirational demand (Black and Mack 1986). Consequently, plant water use increased during the first growing season after ash fall, and when coupled with an inhibitory effect by the ash on water infiltration, led to an earlier depletion of soil moisture in the subsequent year.

Few attempts have been made to model plant water use in the Great Basin using physiological data. Hydrologic water balance models such as SPUR (Simulation of Production and Utilization of Rangelands) (Wight and Skiles 1987) often do not include leaf conductance relationships, but use traditional, weather-based calculations of potential ET that are reduced by crop coefficients. In contrast, Caldwell et al. (1977a) relied on physiological, whole-plant measurements of transpiration to estimate seasonal water use in salt-desert shrub communities. A canopy layer model for the Portuguese macchia that incorporates leaf and canopy energy balance as well as functional relationships of conductance, helps scale leaf conductance measurements up to canopy water use and provides information on the adaptive significance of canopy development and architecture (Caldwell et al. 1986).

Linking stomatal conductance to canopy or regional water use by plants is a formidable task because of the enormous differences in scale (Jarvis and McNaughton 1986). Stomatal control at small scales (individual stomata or leaves) is greater than at larger scales (canopy or region); minute changes in leaf conductance in one part of the canopy may increase leaf transpiration in that part, but the concomitant increase in vapor pressure downwind will decrease minutely conductance and transpiration of leaves there. Thus, the net effect at the canopy level is no change in canopy transpiration, simply a redistribution within the canopy of where the water is lost. Nonetheless, we can place some limits on plant water use. With the possible exception of cheatgrass-dominated vegetation, many of the species and communities are capable of extracting all the available soil water from the soil profile in one growing season, regardless of a high initial soil moisture content (Anderson et al. 1987) or high precipitation (Campbell and Harris 1977).

Photosynthesis and Plant Carbon Gain

Photosynthetic rates of most cold-desert species are saturated at less than full solar irradiance, but many hot-desert species are not. Assimilation rate for leaves of the cold-desert grass *Agropyron dasystachyum* increases with increasing PPFD and is light-saturated at approximately 75% full solar irradiance (Figure 6.9a). Responses of assimilation to irradiance for mature leaves and stems as well as whole plants of other desert species (Frank and Barker 1976; Caldwell et al. 1983; Werk and Ehleringer 1985; Nowak and Caldwell 1986; Osmond et al. 1987b) are similar to *A. dasystachyum*. Responses for senescent leaves are even lower and reach light saturation at less than 50% full solar irradiance (Nowak and Caldwell 1986). Unlike most other cold-desert plants, assimilation rate of excised shoots from the shrub *Chrysothamnus nauseosus* is not light-saturated at full solar irradiance (Davis et al. 1985). This lack of photosynthetic saturation at full solar irradiance is also more characteristic of hot-desert species, including those found in the Mojave Desert (Toft and Pearcy 1982; Ehleringer 1985b).

Many desert plants increase light harvesting by solar tracking or by producing dense canopies. Solar tracking, or diaheliotropism, is common in hot-desert species and increases incident solar radiation flux over the diurnal period by 30% or more (Ehleringer 1985b). The widely-spaced bunchgrasses of the Great Basin, which have high foliage density, harvest more light than natural stands of Great Plains grasslands (Caldwell et al. 1983). Whole-plant measurements of assimilation show that light saturation increases with the size of the bunchgrass. Although the size of plants needed to harvest the same amount of light differs among species, larger plants harvest more light.

Photosynthesis of *A. dasystachyum* leaves decreases linearly with increasing vapor gradients (Figure 6.9b), but the response to leaf temperature is curvilinear (Figure 6.9c, 6.9d). Most desert species have very similar relationships between temperature and assimilation (Ehleringer 1985b; Werk and Ehleringer 1985; Nowak and Caldwell 1986; Sage and Sharkey 1987; Nowak et al. 1988). Plants in the Intermountain lowlands are adapted generally to a temperature range that corresponds with their period of most active growth. The temperature optimum near 25° C and high temperature compensation point above 40° C for *A. dasystachyum* is typical for cold-desert species as well as winter-active species from the Mojave Desert (Ehleringer 1985b). However, temperature optima above 30° C are found in many summer-active, Mojave Desert species. Changes in the temperature optimum for photosynthesis as leaves age (Nowak and Caldwell 1986), during the growing season (DePuit and Caldwell 1973; Depuit and Caldwell 1975a), or under different controlled-environment conditions (Toft and Pearcy 1982) are often less than 5° C. However, temperature acclimation between 5° and 10° C has been reported for *Larrea tridentata* (Mooney et al. 1977a), *Yucca brevifolia* (Smith et al. 1983), branches of sagebrush plants growing at high elevation (Mooney et al. 1966), and other Great Basin species (Chabot and Billings 1972). Assimilation rate for *A. dasystachyum* leaves increases almost linearly between intercellular CO_2 partial pressures of 150 to 600 μbar (Figure 6.9e). Increases in photosynthesis over this same range of intercellular CO_2 partial pressures are also nearly linear for 2 other species native to the Great Basin (Osmond et al. 1987b; Sage and Sharkey 1987). For the cold-desert forb *Scrophularia desertorum*, however, the response of assimilation to CO_2 was temperature dependent and assimilation was insensitive to O_2 below 6° C, which was much lower than cultivated species (Sage and Sharkey 1987).

Both C_3 and C_4 plants are found in the Intermountain lowlands, but the C_4 photosynthetic pathway is relatively rare. Although the C_4 pathway generally is considered to be superior to the C_3 pathway for desert plants, the lack of significant midsummer precipitation in the Great Basin restricts growth of most C_3 and C_4 plants to the cooler spring months when soil moisture is readily available. Carbon assimilation by two C_4 *Atriplex* species is tolerant of cold temperatures (Caldwell et al. 1977b). Carbon balance of the C_3 cold-desert shrub *Ceratoides lanata* is also strikingly similar to that of the C_4 shrub *Atriplex confertifolia* (Caldwell et al. 1977a; Caldwell 1985). These two species co-exist in the saltbush vegetation zone and have very similar photosynthetic characteristics, diurnal patterns of photosynthesis, and total carbon gain for the year. Daily carbon gain of the C_4 shrub tends to peak slightly later than the C_3 shrub, but daily carbon gain of the two species are nearly identical during the cold spring months when the C_3 species *C. lanata* would be expected to have greater carbon gain. *A. confertifolia* also maintains higher daily carbon gain during late-summer than *C. lanata*, but this carbon gain only accounts for 18% of the yearly carbon gain.

As with conductance, assimilation decreases with both short- and long-term decreases in plant or soil water status for leaves and whole plants of many species (DePuit and Caldwell 1973; Detling and Klikoff 1973; DePuit and Caldwell 1975a; Caldwell et al. 1977a; Frank 1983; Ehleringer 1985b; Osmond et al. 1987b; Nowak et al. 1988). Stomatal limitation of CO_2 diffusion as well as reduction in photosynthetic capacity cause the decrease in leaf assimilation rates (Comstock and Ehleringer 1984). As soils dry out, the decreased assimilation rate of leaves also restricts canopy carbon gain for drought-deciduous species (Comstock and Ehleringer 1986). Unlike many species, maximum assimilation rates of the shrub *Gutierrezia sarothrae* occur under mild water stress conditions (DePuit and Caldwell 1975a). Although decreases in photosynthetic capability under drought conditions might be slowed by osmotic adjustment (Ehleringer 1985b) or the maintenance of cell volume (Berkowitz and Kroll 1988), we do not know if either of these mechanisms occurs in *Gutierrezia sarothrae*.

Photosynthetic rates decline as leaves age. It may be difficult to distinguish the effects of aging on leaf assimilation rates from those of shading (Field 1983; Field and Mooney 1983) because these two phenomena often occur simultaneously for many species, especially for bunchgrass species. New leaf growth overtops older leaf tissue and older leaves are generally found in the lowest light environment within the canopy (Caldwell et al. 1983). For two bunchgrass species, assimilation rates of all cohorts of leaves declined through time, but clipping treatments that also allowed light penetration into the canopy delayed the senescence of some older leaves (Nowak and Caldwell 1984b). Thus, decreased assimilation rates with leaf aging also may be the result of shade acclimation.

Diurnal patterns of photosynthesis are influenced by both water availability and leaf microclimate. Under cool, spring microclimatic conditions, photosynthetic rates of crested wheatgrass leaves are relatively constant during the middle of the photoperiod when soil water availability is high, but photosynthesis peaks before noon when soil water is depleted (Figure 6.11). However, photosynthesis peaks early in the morning when leaves experience a hot, summer microclimate, regardless of soil water status. On the other hand, the diurnal peak in photosynthesis for sagebrush shifts from near noon to early morning only when plants become water stressed, regardless of leaf microclimate (DePuit and Caldwell 1973). Under prevailing ambient conditions, photosynthesis of shrub and bunchgrass species generally peaks before or near noon, and the diurnal peak tends to occur earlier in the day when plants experience greater water stress (DePuit and Caldwell 1975a; Caldwell et al. 1977a). Although these effects of water availability and leaf microclimate on diurnal patterns of photosynthesis influence total daily carbon gain, water use for the entire photoperiod often changes by a similar amount (Nowak et al. 1988).

Carbon fixation by green stems occurs in many plants in the Great Basin and can make substantial contributions to whole plant carbon gain. On an equivalent area basis, assimilation rates of stems can be substantially below those of leaves (Wiebe et al. 1974; DePuit and Caldwell 1975b),

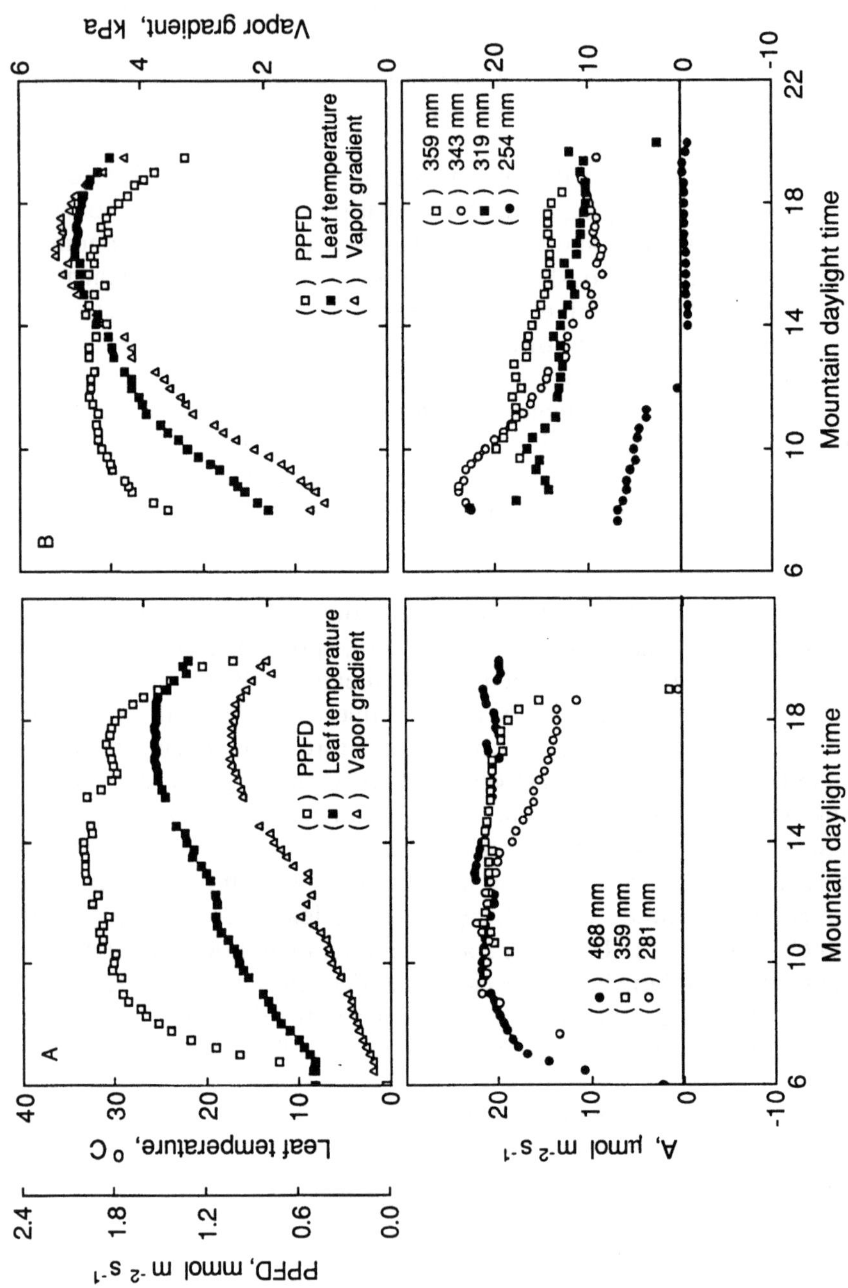
A
B
PPFD, mmol m-2 s-1
Leaf temperature, o C
Vapor gradient, kPa
A, μmol m-2 s-1
(□) PPFD
(■) Leaf temperature
(△) Vapor gradient
(●) 468 mm
(□) 359 mm
(○) 281 mm
(□) 359 mm
(○) 343 mm
(■) 319 mm
(●) 254 mm
Mountain daylight time
Mountain daylight time

30-60% that of leaf rates (Osmond et al. 1987b; Comstock and Ehleringer 1988), or nearly the same as leaf blades (Nowak and Caldwell 1984b). The potential contribution of stems to total plant carbon gain of two bunchgrass species was less than 25% before culm elongation, but was 75% or more after (Caldwell et al. 1981). Photosynthetic carbon gain of stems is also very important for grazed plants, especially in late-summer when stem area can be 50% or more of the total green foliage area (Caldwell et al. 1981; Nowak and Caldwell 1986).

Green stems of many desert species have greater water-use efficiency than leaves (Ehleringer et al. 1987). Green stems of *Eriogonum inflatum* exhibit greater water-use efficiency when measured either by gas exchange or with carbon isotope ratios (Figure 6.12). The greater water-use efficiency of *Eriogonum* stems is due primarily to their vertical orientation (Smith and Osmond 1987) and lower stomatal conductance (Osmond et al. 1987b), but additional water savings may accrue from the fixation of CO_2 that accumulates within the hollow stems. When coupled with the greater drought tolerance of stems (Smith and Osmond 1987; Comstock and Ehleringer 1988), the greater water-use efficiency of stems allows carbon gain under severe drought conditions near the end of the growing season.

Photosynthesis by inflorescences can contribute a large fraction of the carbon needed for reproduction (Bazzaz and Carlson 1979; Bazzaz et al. 1979), but few measurements have been conducted on plants in the Intermountain lowlands. Assimilation rates of inflorescences from two bunchgrass species are nearly equivalent to those for leaves on a projected area basis (Nowak and Caldwell 1984b), and carbon gain by inflorescences comprises up to 15% of total plant carbon gain during mid-June (Caldwell et al. 1981). However, photosynthetic carbon gain by flowers of *Encelia farinosa* is probably more typical: gross photosynthetic rates are nearly equivalent to dark respiration rates and at best offset maintenance respiration costs of the flower (Werk and Ehleringer 1983).

Figure 6.11. (left) Springtime (A) and summertime (B) diurnal leaf microclimatic conditions (upper frame) and assimilation rate (lower frame) of *Agropyron desertorum* as a function of the total amount of water in the soil profile, given for each successive day in a field dry-down experiment. Redrawn from Nowak et al. (1988).

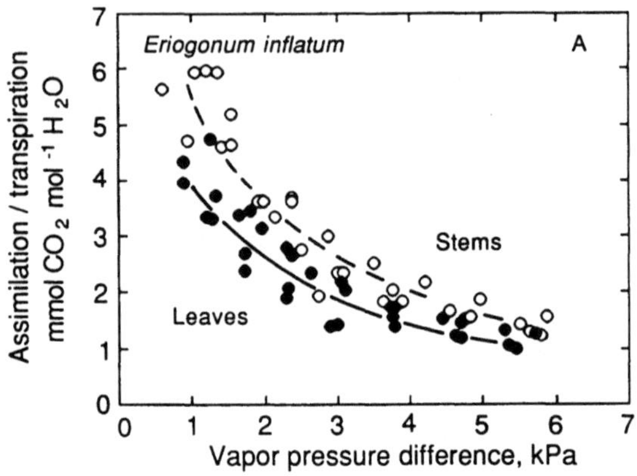

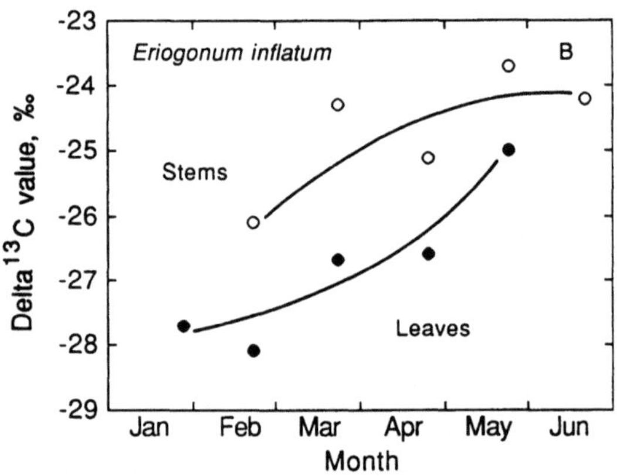

Figure 6.12. Instantaneous assimilation/transpiration ratios (redrawn from Osmond et al. 1987b), and delta [13] C values (redrawn from Smith and Osmond 1987) for leaves and stems of *Eriogonum inflatum* from Death Valley populations.

A large proportion of cold-desert species have green leaf tissue that persists during the winter. These "winter-green" leaves could have functions analogous to evergreen leaves: storage systems for scarce nutrients or opportunistic sources of fixed carbon for early season growth (Moore 1984). Roots and crowns of perennial species have been implicated as carbohydrate and nitrogen storage tissues for spring growth (McIlvanie 1942; Cook et al. 1958), but the extent that winter-green leaves act in a similar fashion has not been investigated specifically. Decreases in specific leaf mass and protein concentration between fall and spring (Nowak and Caldwell 1984a) provide circumstantial evidence for a storage function, but the carbon and nitrogen lost from these overwintering

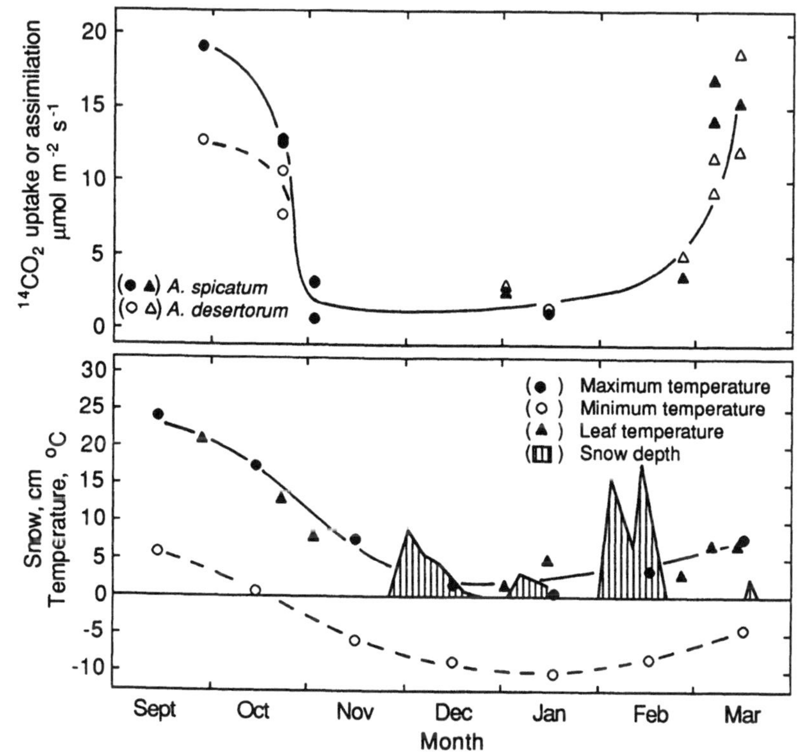

Figure 6.13. Photosynthetic and climatic measurements for the fall, winter, and early spring. At the top of the figure, means and standard errors of photosynthetic rate for *Agropyron desertorum* (open symbols, dashed line) and *A. spicatum* (closed symbols, solid lines) are shown. Infrared gas exchange measurements are indicated by circles, and $^{14}CO_2$ uptake measurements by triangles. All photosynthetic measurements were at saturating irradiance and at leaf temperatures shown by the triangles in the bottom graph. The bottom graph also shows the 30-year, mean monthly maximum and minimum air temperatures and the snow cover over plants during the 1980-1981 winter (vertical cross hatching). Redrawn from Nowak and Caldwell (1984a).

leaves is very small compared to storage in other organs. A detailed, winter-time energetic balance that compares carbon gain from winter-green leaves against respiratory maintenance costs has not been completed for any cold-desert species. However, results from a number of independent studies suggest that winter-green foliage is important for spring growth.

Cold soil temperature (Anderson and McNaughton 1973) and cold prevailing air temperature in the fall (Nowak and Caldwell 1984a) reduce photosynthetic capacity, but bunchgrass species (Figure 6.13) as well as sagebrush (Caldwell 1979) have the capacity to fix carbon during the winter and early spring, even when plants are partially covered with snow. The respiratory maintenance costs of winter-green foliage are unknown, but in roots they decrease greatly during the fall due to decreased ambient temperatures and acclimation (Holthausen and Caldwell 1980). Decreases in carbohydrates (Coyne and Cook 1970; Nowak and Caldwell 1984a) suggest a maintenance cost, but new root and shoot growth accounts for some of the carbohydrate loss. New root growth at cold soil temperatures can be high (Harris and Wilson 1970; Fernandez and Caldwell 1975; Eissenstat and Caldwell 1989), and shoot growth rapidly accelerates in the spring as soil and air temperatures increase (Eissenstat and Caldwell 1987). Thus, winter-green foliage may provide carbohydrates for winter and early spring growth.

Current photosynthetic carbon gain and shoot meristematic activation are probably the most important criteria for plant regrowth after defoliation. Traditionally, carbohydrate reserves had been thought to be the most important energy source for regrowth, but carbohydrate concentrations at the time of defoliation were not related clearly to regrowth for eight Great Basin species (Trlica and Cook 1971). Also regrowth is not correlated with size or changes of carbohydrate pools (Richards and Caldwell 1985). Furthermore, current photosynthetic carbon gain quickly becomes the major source of carbon for regrowth. Leaves of many plants also exhibit compensatory photosynthesis after defoliation, and new regrowth foliage often has relatively high photosynthetic rates (Caldwell et al. 1981; Nowak and Caldwell 1984b). These higher assimilation rates enhance carbon gain after defoliation, but are not related to grazing tolerance. Grazing tolerance is, however, correlated with meristematic activation: active meristems or the ability to activate meristems quickly may control the amount of regrowth that can be produced (Richards 1986).

Productivity: Effects of Abiotic and Biotic Factors

Net annual primary productivity (NAPP) of plant communities in the Intermountain lowlands is among the lowest in the world. Aboveground

NAPP (g m^{-2} yr^{-1}) for the six primary vegetation associations are: 13-57 for creosote-bursage scrub (Szarek 1979; Ludwig 1987), 16-57 for mixed Mojave shrublands (Lane et al. 1984), 30-170 for blackbrush, 50-150 for Great Basin sagebrush, 20-100 for saltbush-greasewood, and 80-250 for sagebrush steppe (West 1983). There are a few values for NAPP belowground, but these greatly exceed aboveground values. For example, belowground NAPP was 3.0-3.5 times greater than aboveground NAPP in a salt-desert shrubland (Caldwell et al. 1977a).

Water availability is the abiotic factor that limits NAPP most (MacMahon and Schimpf 1981; Skujins 1981). Irrigation increases aboveground NAPP in sagebrush steppe vegetation (Pearson 1965). Aboveground NAPP is greater in areas with similar vegetation but greater precipitation (Hull and Klomp 1974). Precipitation and NAPP often are correlated linearly, with r^2 values ranging from 0.7 (Blaisdell 1958; Sneva 1977) to 0.9 (Hutchings and Stewart 1953). In the salt-desert shrubland, midsummer irrigation increases root growth (Hodgkinson et al. 1978), but enhanced shoot growth is delayed until the next growing season (Johnson and Norton 1980).

Nitrogen is thought to be the second most limiting factor for NAPP of semiarid vegetation (James and Jurinak 1978; Skujins 1981). A striking example of nutrient limitation is found in the outcrops of hydrothermally-altered andesite in the western Great Basin where higher-elevation, Sierra Nevada conifers displace Great Basin sagebrush vegetation (Billings 1950). Billings found that production of sagebrush plants grown in the nutrient-poor altered andesite was an order of magnitude less than that of plants grown in unaltered andesite. The absence of sagebrush vegetation from the altered andesite outcrops probably is due to infertility (Billings 1950) rather than competition for water (DeLucia et al. 1988) (see Chapter 5).

Other evidence of nitrogen limitation of NAPP is meager. Annual species generally are limited by nitrogen, and fertilization of exotic, weedy annuals greatly enhances aboveground NAPP (Eckert and Evans 1963; Cline and Rickard 1973), even when co-occurring perennial species do not exhibit a fertilization response (Goodman 1973) or are inhibited (Kay and Evans 1965; Wilson et al. 1966). Even in the absence of weedy annuals, perennial species do not always respond to nitrogen fertilization. The lack of a fertilization response simply may be due to a low rate of nitrogen application (Sneva et al. 1958; Wight and Black 1972; Sneva 1973; James and Jurinak 1978; Williams et al. 1979;

Fairbourn and Rauzi 1982). For other species, the response to fertilization may be delayed until the next growing season (Hodgkinson et al. 1978; James and Jurinak 1978). Finally, cold soil temperature may suppress the response to increased nutrients (Eissenstat and Caldwell 1987). Differences in methods used to estimate biomass also may influence results. Wang (1988) concluded that biomass of sagebrush was increased by nitrogen fertilization, but Carpenter and West (1987) found no effects of nitrogen fertilization. Carpenter and West used the reference unit method to estimate biomass, which would not have detected the increased shoot density that was measured by Wang. However, similar types of measurements also may yield conflicting results (James and Jurinak 1978; Wang 1988).

The variability in aboveground NAPP among species and the difficulty in interpreting results was demonstrated in a field experiment where water and nitrogen availability were manipulated (Table 6.4). The bunchgrass *Sitanion hystrix* showed no significant response to irrigation or fertilization in both years, and no clear patterns of treatment effects were evident from the trends of mean values. In the rhizomatous grass *Agropyron dasystachyum*, significant differences occurred in 1987, but not in 1988. In both years, mean aboveground NAPP on irrigated plots (+W and +N+W) was at least twice that on controls, suggesting a water limitation for *A. dasystachyum*. The effects of nitrogen fertilization on NAPP in *A. dasystachyum* were less clear. In 1987, mean NAPP of the +N plots was greater than that of control plots, but fertilization did not have an additive effect with irrigation. In contrast, the data from 1988 suggested the opposite: current-year fertilization alone did not increase NAPP, but fertilization had an additive effect with irrigation. The data for *A. dasystachyum* also suggested a residual nitrogen effect: plots fertilized in 1987 (+N and +N+W) had greater mean NAPP than plots fertilized in 1988. Aboveground NAPP of forbs showed a significant response to irrigation in both years. Current-year fertilization had an additive effect in 1987, but neither current-year nor residual nitrogen effects were evident in 1988. These results suggest that water limits NAPP of *A. dasystachyum* and forbs, but a nitrogen limitation is not obvious. However, both years had below normal precipitation, particularly 1988. The differences in fertilization effects between years may be due to the extreme drought in 1988, favoring the hypothesis that water is the primary limiting factor for NAPP and nitrogen is a secondary factor.

In contrast to NAPP, physiological processes are limited by water and nitrogen to a similar extent. In addition to the NAPP measurements

Table 6.4. Aboveground net annual primary production for two grass species, *Sitanion hystrix* and *Agropyron dasystachyum*, and all forbs as affected by field manipulations of water and nitrogen availability. In addition to a set of control plots, five treatments were established: irrigated (+W) throughout the growing season for two consecutive years, fertilized once in the spring (+N) and only in the first year (1987), fertilized once in the spring and only in the second year (1988), fertilized once in 1987 but irrigated both years, and fertilized once in 1988 and irrigated only during that year. Treatments were applied to a native, sagebrush steppe community. Details of the field plots as well as irrigation and fertilization treatments are given in Toft et al. 1989. For each species, treatments were compared with a Kruskal-Wallis analysis of variance and values greater than 5% were considered not significant (ns).

Treatments		*S. hystrix*		*A. dasystachyum*		All Forbs	
1987	1988	1987	1988	1987	1988	1987	1988
		Aboveground Net Annual Primary Production (g m^{-2})					
Control	Control	12.5	17.8	32.1	14.3	2.1	0.3
+W	+W	2.7	10.1	80.8	36.5	10.2	2.4
+N	—	6.6	6.8	56.8	25.7	10.9	0.3
—	+N	—	13.7	—	10.7	—	0.5
+N+W	+W only	22.3	32.4	73.8	66.2	23.3	2.4
—	+N+W	—	23.4	—	50.0	—	4.0
Significance		ns	ns	ns	1%	1%	2%

described in the preceding paragraph, two measurements of water-use efficiency were made. If both water and nitrogen limit plant gas exchange and if both conductance and assimilation are affected differentially by water and nitrogen availability, then leaves of plants growing on fertilized plots should have the greatest water-use efficiency, those on irrigated plots the lowest, and those on control plots and plots receiving both irrigation and fertilization in between (Toft et al. 1989). This prediction was consistent with both measurements of water-use efficiency across two grass and two shrub species (Figure 6.14).

The relative difference in the limitation of nitrogen to productivity versus the limitation to physiological processes may be related to how nitrogen cycles in the semiarid Intermountain lowlands as well as to how plants partition their resources. Unlike more mesic ecosystems, the

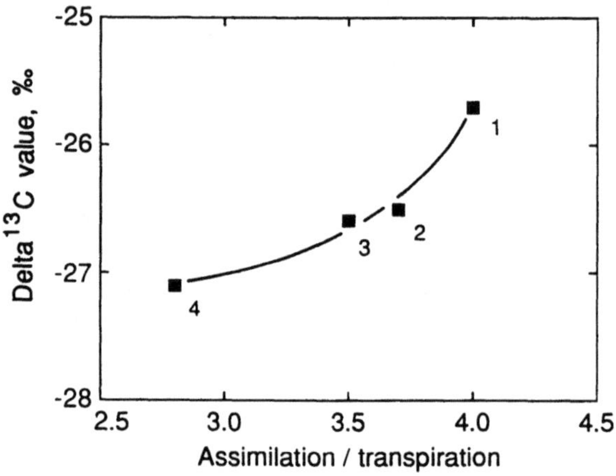

Figure 6.14. Correlation between the ratio of assimilation to transpiration from gas exchange measurements (A/E) and carbon stable isotope ratio (delta ^{13}C value), representing the mean values from 4 species used in the 1987 experiment of Table 6.4 (*Agropyron dasystachyum*, *Sitanion hystrix*, *Ceratoides lanata*, and *Artemisia tridentata*). The numbered treatments refer to +N (1), control (2), +W+N (3) and +W (4) respectively. Data from Toft et al. (1989).

major pathways for nitrogen loss from desert ecosystems are by denitrification and ammonia volatilization (Skujins 1981). A large fraction of nitrogen applied as fertilizers can be lost through denitrification and ammonia volatilization (Klubek et al. 1978; Westerman and Tucker 1978). Thus, much of the applied fertilizer nitrogen may never be assimilated by the vegetation.

How plants allocate the fraction of nitrogen that is assimilated influences the response to fertilization. Fertilization increases tissue nitrogen concentrations of shoots (Sneva et al. 1958; Eckert and Evans 1963; Sneva 1973; Power 1986; Wang 1988), without necessarily changing productivity (James and Jurinak 1978; Fairbourn and Rauzi 1982). Plants in low nitrogen environments conserve nitrogen by translocating it out of senescing plant tissue (Field 1983; Field and Mooney 1983), but a higher nitrogen content of fertilized plants may obviate the need for this intra-plant recycling of nitrogen. Fertilization also increases root nitrogen content (Eckert and Evans 1963; Power 1986) and root production (Eckert and Evans 1963; Eissenstat and Caldwell 1988b).

Biotic factors, such as interspecific competition and grazing, also affect NAPP. Areas dominated by sagebrush tend to have lower aboveground NAPP (Mueggler 1972; Hull and Klomp 1974), and production of individual plants growing with reduced or no competition from sagebrush averages much greater than that of plants growing with intense sagebrush competition (Robertson 1947). Grazing decreases aboveground NAPP, but belowground NAPP of the grazed community is greater than the ungrazed (Pearson 1965). The increase in belowground NAPP with grazing probably is due to root growth of shrubs because two grass species had slight to large decreases in root production after clipping (Richards 1984).

Areas for Future Research

A great deal of information is now available on the distribution patterns of desert plants in the lowlands of the Intermountain Region. Unfortunately, the physiological basis for these distribution patterns is still poorly known for much of the region. More information is sorely needed on cellular water and ionic relations, on cellular aspects of plant stress, on plant tolerance limits, and on the photosynthetic and structural factors that contribute to whole plant productivity and water-use efficiency. It is also imperative that ecophysiological studies are sufficiently pragmatic in nature to result in meaningful ecological conclusions (Osmond et al. 1987a). This requires that the alternatives between maximization of productivity, reproduction, and stress tolerance be addressed in an ecological context.

The water relations of many important taxa of Great Basin plants are not well known, particularly at the cellular level. The role of osmoregulation and cellular adjustments in turgor maintenance as soil water potential declines needs to be investigated. In many areas of the Great Basin, there is a potentially short window of activity between the breaking of winter dormancy and the onset of summer drought, yet we know little about the ability of roots to compete with evaporation in the surface layers of the soil (Fischer and Turner 1978). The resulting effects of declining soil and plant water potentials on root/shoot allocation, canopy dynamics, water-use efficiency, and reproductive activity are known only for a few species at present. We also need more information on the salinity tolerances of Great Basin plants and the role of salinity in vegetation patterns within the basins of the region. From a hydrologic perspective, more thorough analyses of water flux at the ecosystem level will have important implications for groundwater recharge and vegetation water sources along spatial and temporal gradients.

From a carbon balance perspective, we presently have only a moderate understanding of photosynthesis, productivity, and water-use efficiency in Great Basin plants. What are the effects of declining soil moisture or temperature extremes on these parameters, and what is the significance of each parameter for plant performance and survival under field conditions? A majority of plant gas exchange observations to date have been conducted during the "growing season" using young, vigorous shoots or leaves. But what is the role of overwintering foliage or green stems for whole plant carbon balance and growth patterns? Scaling up individual gas exchange measurements to more long-term patterns at the canopy or community level should be an integral part of future gas exchange studies.

Many more studies are needed on the population biology and demography of plants in the Intermountain lowlands, and physiological ecologists need to participate in these studies. The establishment phase is often the most stressful for desert plants, yet studies of stress responses in desert plants rarely have focused on seedling survival or characterization of the microclimates in which successful recruitment occurs. The integration of plant productivity, stress tolerance, and reproductive biology in determining plant fitness is also an area that has been neglected and that requires a combination of physiological and population approaches. For example, many studies have quantified photosynthetic capacity, water-use efficiency, or nitrogen-use efficiency in plants, but rarely have these studies definitively shown that maximization of any one of these parameters results in increased fitness and thus increased survival in the ecological community. To vigorously apply advances in physiological ecology to integrated ecological communities, we need to take more of a population/demographic approach (Billings 1980).

Finally, a large number of applied questions will become increasingly important in the next several decades as humans continue to develop more of the Great Basin for habitation, recreation, and agricultural uses. Human impacts on water tables and increased utilization of marginal lands may have profound impacts on the vegetation of the Intermountain lowlands. When compared to other ecological regions, we know little of the long term successional processes that are occurring in the region. The Great Basin is of particular interest from a climate change perspective due to the potentially significant changes that may occur in the cool-desert steppe regions of the world as atmospheric CO_2 continues to increase. These changes could have profound impacts on the vegetation patterns and range conditions of the region. In a preliminary glasshouse study, Smith et al. (1987) found a short-term CO_2 doubling to result in substantial growth increases in four species of Great Basin grasses, but found *Bromus tectorum* to respond most strongly. Given the current cycle of range degradation caused by *Bromus* throughout much of the Great Basin, it is not clear whether the direct and indirect effects of a CO_2 doubling will have positive or negative impacts on range ecosystems in the region. Obviously, the Great Basin would be an ideal and important location for future long-term studies on the ecosystem-level effects of elevated CO_2 and associated climatic change.

Acknowledgements

Research support was provided by the Department of Energy, Nevada Waste Project Office (to S.D.S.) and the Ecological Studies Program, Idaho National Engineering Laboratory, U.S. Department of Energy Office of Health and Environmental Research, and Nevada Agricultural Experiment Station (to R.S.N.). Research collaboration of J.E. Anderson, J. Nachlinger, C.B. Osmond, N.L. Toft, A.B. Wellington, and research assistance of T. DeRocher, C. Herr, C.L. Nowak, and R. Waschmann is gratefully acknowledged. Excellent technical assistance in manuscript preparation was provided by S. Barnson and in compilation of figures by G. Marrs-Smith. J. McAuliffe kindly provided several of the pictures used in the color plates.

References

Ackerman TL, Romney EM, Wallace A, and Kinnear JE (1980) Phenology of desert shrubs in southern Nye County, Nevada. Great Basin Nat. Mem. 4:4-23.

Aide M and Van Auken OW (1985) Chihuahuan Desert vegetation of limestone and basalt slopes in West Texas. Southwest Nat. 30:533-542.

Anderson JE and Holte KE (1981) Vegetation development over 25 years without grazing on sagebrush-dominated rangeland in southeastern Idaho. J. Range Manage. 34:25-29.

Anderson JE and McNaughton SJ (1973) Effects of low soil temperature on transpiration, photosynthesis, leaf relative water content, and growth among elevationally diverse plant populations. Ecology 54:1220-1233.

Anderson JE, Shumar ML, Toft NL, and Nowak RS (1987) Control of the soil water balance by sagebrush and three perennial grasses in a cold-desert environment. Arid Soil Res. Rehab. 1:229-244.

Bailey HP (1981) Climatic features of deserts. pp. 13-41. In Evans DD and Thames JL (editors), Water in Desert Ecosystems. Dowden, Hutchinson and Ross, Stroudsburg, PA.

Ball JT, Woodrow IE, and Berry JA (1987) A model predicting stomatal conductance and its contribution to the control of photosynthesis under different environmental conditions. pp. 221-224. In Biggins J (editor), Progress in Photosynthesis Research, Vol IV. Martinus Nijhoff, Dordrecht.

Bamberg SA, Kleinkopf GE, Wallace A, and Vollmer A (1975) Comparative photosynthetic production of Mojave Desert shrubs. Ecology 56:732-736.

Barbour MG (1968) Germination requirements of the desert shrub *Larrea divaricata*. Ecology 49:915-923.

Barbour MG (1969) Age and space distribution of the desert shrub *Larrea divaricata*. Ecology 50:679-685.

Barbour MG (1973) Desert dogma re-examined: root/shoot ratios, productivity and plant spacing. Amer. Midl. Nat. 89:41-57.

Barbour MG (1981) Plant-plant interactions. pp. 33-49. In Goodall DW and Perry RA (editors), Arid-land Ecosystems: Structure, Functioning and Management, Vol 2. Cambridge Univ. Press, Cambridge.

Barbour MG, Diaz DV, and Briedenbach RW (1974) Contributions to the biology of *Larrea* species. Ecology 55:1199-1215.

Barker JR and McKell CM (1983) Habitat differences between Basin and Wyoming big sagebrush in contiguous populations. J. Range Manage. 36:450-454.

Barkworth ME and Dewey DR (1985) Genomically based genera in the perennial Triticeae of North America: identification and membership. Amer. J. Bot. 72:767-776.

Bazzaz FA and Carlson RW (1979) Photosynthetic contribution of flowers and seeds to reproductive effort of an annual colonizer. New Phytol. 82:223-232.

Bazzaz FA, Carlson RW, and Harper JL (1979) Contribution to reproductive effort by photosynthesis of flowers and fruits. Nature 279:554-555.

Beatley JC (1966) Ecological status of introduced brome grasses (*Bromus* spp.) in desert vegetation of southern Nevada. Ecology 47:548-554.

Beatley JC (1967) Survival of winter annuals in the northern Mojave Desert. Ecology 48:745-750.

Beatley JC (1970) Perennation in *Astragalus lentiginosus* and *Tridens pulchellus* in relation to rainfall. Madroño 20: 326-332.

Beatley JC (1974a) Effects of rainfall and temperature on the distribution and behavior of *Larrea tridentata*. Ecology 55: 245-261.

Beatley JC (1974b) Phenological events and their environmental triggers in Mojave Desert ecosystems. Ecology 55:856-863.

Beatley JC (1975) Climates and vegetation pattern across the Mojave/Great Basin transition of southern Nevada. Amer. Midl. Nat. 93:53-70.

Bell FC (1979) Precipitation. pp. 373-392. In Goodall DW and Perry RA (editors), Arid-land Ecosystems: Structure, Functioning and Management. Cambridge Univ. Press, Cambridge.

Bell KL, Hiatt HD, and Niles WE (1979) Seasonal changes in biomass allocation in eight winter annuals of the Mojave Desert. J. Ecol. 67:781-787.

Bennert WH and Mooney HA (1979) The water relations of some desert plants in Death Valley, California. Flora 168:405-427.

Berkowitz GA and Kroll KS (1988) Acclimation of photosynthesis in *Zea mays* to low water potentials involves alterations in protoplast volume reduction. Planta 175:374-379.

Berry JA (1975) Adaptation of photosynthetic processes to stress. Science 188:644-650.

Bethenfalvay GJ and Dakessian S (1984) Grazing effects on mycorrhizal colonization and floristic composition of the vegetation on a semiarid range in northern Nevada. J. Range Manage. 37: 312-316.

Billings WD (1945) The plant associations of the Carson Desert region, western Nevada. Butler Univ. Bot. Studies 7:89-123.

Billings WD (1949) The shadscale vegetation zone of Nevada and eastern California in relation to climate and soils. Amer. Midl. Nat. 42:87-109.

Billings WD (1950) Vegetation and plant growth as affected by chemically altered rocks in the western Great Basin. Ecology 31:62-74.

Billings WD (1951) Vegetational zonation in the Great Basin of western North America. pp. 101-122. In Compt. Rend. du Colloque sur les Bases Ecologique de la Regeneration de la Vegetation des Zones Arides. IUBS, Paris.

Billings WD (1980) American deserts and their mountains: an ecological frontier. Ecol. Soc. Amer. Bull. 61:203-209.

Black RA and Mack RN (1986) Mount St. Helens ash: recreating its effects on the steppe environment and ecophysiology. Ecology 67:1289-1302.

Blaisdell JP (1958) Seasonal development and yield of native plants on the upper Snake River plains and their relation to certain climatic factors. USDA Technical Bulletin 1190, Washington, DC.

Bock CE, Bock JH, Kenney WR, and Hawthorne VM (1984) Responses of birds, rodents, and vegetation to livestock exclosure in a semi desert grassland site. J. Range Manage. 37:239-242.

Boyd RS and Brum GD (1983a) Predispersal reproductive attrition in a Mojave Desert population of *Larrea tridentata* (Zygophyllaceae). Amer. Midl. Nat. 110:14-24.

Boyd RS and Brum GD (1983b) Postdispersal reproductive biology of a Mojave Desert population of *Larrea tridentata* (Zygophyllaceae). Amer. Midl. Nat. 110:25-36.

Branson FA, Miller RF, and McQueen IS (1976) Moisture relationships in twelve northern desert shrub communities near Grand Junction, Colorado. Ecology 57:1104-1124.

Brown DE and Minnich RA (1986) Fire and changes in creosote bush scrub of the western Sonoran Desert, California. Amer. Midl. Nat. 116:411-422.

Burk JH (1982) Phenology, germination, and survival of desert ephemerals in Deep Canyon, Riverside County, California. Madroño 29:154-163.

Cable DR (1977) Soil water changes in creosotebush and bur-sage during a dry period in southern Arizona. J. Ariz. Acad. Sci. 12:15-20.

Caldwell MM (1974) Physiology of desert halophytes pp. 355-378. In Reimold RJ and Queen WJ (editors), Ecology of Halophytes. Academic Press, NY.

Caldwell MM (1979) Physiology of sagebrush. pp. 74-85. In The Sagebrush Ecosystem: A Symposium. Utah State Univ., Logan, UT.

Caldwell MM (1985) Cold desert. pp. 198-212. In Chabot BF and Mooney HA (editors), Physiological Ecology of North American Plant Communities. Chapman and Hall, NY.

Caldwell MM and Fernandez OA (1975) Dynamics of Great Basin shrub root systems. pp. 38-51. In Hadley NF (editor), Environmental Physiology of Desert Organisms. Dowden, Hutchinson and Ross, Stroudsburg, PA.

Caldwell MM, Dean TJ, Nowak RS, Dzurek RS, and Richards JH (1983) Bunchgrass architecture, light interception, and water-use efficiency: assessment by fiber optic point quadrats and gas exchange. Oecologia 59:178-184.

Caldwell MM, Meister H-P, Tenhunen JD, and Lange OL (1986) Canopy structure, light microclimate and leaf gas exchange of *Quercus coccifera* L. in a Portugese macchia: measurements in different canopy layers and simulations with a canopy model. Trees 1:25-41.

Caldwell MM, Richards JH, Johnson DA, Nowak RS, and Dzurek RS (1981) Coping with herbivory: photosynthetic capacity and resource allocation in two semiarid *Agropyron* bunchgrasses. Oecologia 50:14-24.

Caldwell MM, White RS, Moore RT, and Camp LB (1977a) Carbon balance, productivity, and water use of cold-winter desert shrub communities dominated by C_3 and C_4 species. Oecologia 29:275-300.

Caldwell MM, Osmond CB, Nott DL (1977b) C_4 pathway photosynthesis at low temperature in cold-tolerant *Atriplex* species. Plant Physiol. 60:157-164.

Callison J, Brotherson JD, and Downs JE (1985) The effects of fire on the blackbrush (*Coleogyne ramosissima*) community of southwestern Utah. J. Range Manage. 38:535-538.

Campbell GS and Harris GA (1977) Water relations and water use patterns for *Artemisia tridentata* Nutt. in wet and dry years. Ecology 58:652-659.

Carpenter AT and West NE (1987) Indifference of mountain big sagebrush growth to supplemental water and nitrogen. J. Range Manage. 40:448-451.

Chabot BF and Billings WD (1972) Origins and ecology of the Sierran alpine flora and vegetation. Ecol. Monogr. 42:163-199.

Charley JL and West NE (1977) Micro-patterns of nitrogen mineralization activity in soils of some shrub-dominated semi-desert ecosystems of Utah. Soil Biol. Biochem. 9:357-365.

Chew RM and Chew AE (1965) The primary productivity of a desert shrub (*Larrea tridentata*) community. Ecol. Monogr. 35:355-375.

Clark DD and Burk JH (1980) Resource allocation patterns of two California-Sonoran Desert ephemerals. Oecologia 46:86-91.

Cline JF and Rickard WH (1973) Herbage yields in relation to soil water and assimilated nitrogen. J. Range Manage. 26:296-298.

Cline, JF, Uresk DW, and Rickard WH (1977) Comparison of soil water used by a sagebrush-bunchgrass and a cheatgrass community. J Range Manage. 30:199-201.

Cluff GJ, Young JA, and Evans RA (1983) Edaphic factors influencing the control of Wyoming big sagebrush and seedling establishment of crested wheatgrass. J. Range Manage. 36: 786-792.

Cody ML (1986) Spacing patterns in Mojave Desert plant communities and near-neighbor analyses. J. Arid Env. 11: 199-217.

Cohen D (1970) The expected efficiency of water utilization in plants under different competition and selection regimes. Israel J. Bot. 19:50-54.

Comstock JP and Ehleringer J (1984) Photosynthetic responses to slowly decreasing leaf water potentials in *Encelia frutescens*. Oecologia 61:241-248.

Comstock J and Ehleringer J (1986) Canopy dynamics and carbon gain in response to soil water availability in *Encelia frutescens* Gray, a drought-deciduous shrub. Oecologia 68:271-278.

Comstock JP and Ehleringer JR (1988) Contrasting photosynthetic behavior in leaves and twigs of *Hymenoclea salsola*, a green-twigged warm desert shrub. Amer. J. Bot. 75:1360-1370.

Comstock JP, Cooper TA, and Ehleringer JR (1988) Seasonal patterns of canopy development and carbon gain in nineteen warm desert shrub species. Oecologia 75:327-335.

Cook CW, Stoddart LA, and Kinsinger FE (1958) Responses of crested wheatgrass to various clipping treatments. Ecol. Monogr. 28: 237-272.

Coyne PI and Cook CW (1970) Seasonal carbohydrate reserve cycles in eight desert range species. J. Range Manage. 23:438-444.

Cronquist A, Holmgren AH, Holmgren NH, and Reveal JL (1972) Intermountain Flora, Vol 1. Hafner, NY.

Davis TD, Sankhla N, Anderson WR, Weber DJ, and Smith BN (1985) High rates of photosynthesis in the desert shrub *Chrysothamnus nauseosus* ssp. *albicaulis*. Great Basin Nat. 45:520-526.

DeLucia EH, Schlesinger WH, and Billings WD (1988) Water relations and the maintenance of Sierran conifers on hydrothermally altered rock. Ecology 69:303-311

DePuit EJ and Caldwell MM (1973) Seasonal patterns of net photosynthesis of *Artemisia tridentata*. Amer. J. Bot. 60:426-435.

DePuit EJ and Caldwell MM (1975a) Gas exchange of three cool semidesert species in relation to temperature and water stress. J. Ecol. 63:835-858.

DePuit EJ and Caldwell MM (1975b) Stem and leaf gas exchange of two arid land shrubs. Amer. J. Bot. 62:954-961.

Detling JK and Klikoff LG (1973) Physiological response to moisture stress as a factor in halophyte distribution. Amer. Midl. Nat. 90:307-318.

Doerr TB, Redente EF, and Reeves FB (1984) Effects of soil disturbance on plant succession and levels of mycorrhizal fungi in a sagebrush-grassland community. J. Range Manage. 37:135-139.

Doescher PS, Miller RF, and Winward AH (1984) Soil chemical patterns under eastern Oregon plant communities dominated by big sagebrush. Soil Sci. Soc. Amer. J. 48:659-663.

Downton WJS, Berry JA, and Seemann JR (1984) Tolerance of photosynthesis to high temperature in desert plants. Plant Physiol. 74:786-790.

Eckert RE Jr and Evans RA (1963) Responses of downy brome and crested wheatgrass to nitrogen and phosphorous in nutrient solution. Weeds 11:170-174.

Eddleman LE and Romo JT (1987) Sodium relations in seeds and seedlings of *Sarcobatus vermiculatus*. Soil Sci. 143:120-123.

Ehleringer J (1982) The influence of water stress and temperature on leaf pubescence development in *Encelia farinosa*. Amer. J. Bot. 69:670-675.

Ehleringer JR (1984) Intraspecific competitive effects on water relations, growth and reproduction in *Encelia farinosa*. Oecologia 63:153-158.

Ehleringer J (1985a) Comparative microclimatology and plant responses in *Encelia* species from contrasting habitats. J. Arid Env. 8:45-56.

Ehleringer J (1985b) Annuals and perennials of warm deserts. pp. 162-180. In Chabot BF and Mooney HA (editors), Physiological Ecology of North American Plant Communities. Chapman and Hall, NY.

Ehleringer JR, Comstock JP, and Cooper TA (1987) Leaf-twig carbon isotope ratio differences in photosynthetic-twig desert shrubs. Oecologia 71:318-320.

Eissenstat DM and Caldwell MM (1987) Characteristics of successful competitors: an evaluation of potential growth rate in two cold desert tussock grasses. Oecologia 71:167-173.

Eissenstat DM and Caldwell MM (1988a) Competitive ability is linked to rates of water extraction: a field study of two aridland tussock grasses. Oecologia 75:1-7.

Eissenstat DM and Caldwell MM (1988b) Seasonal timing of root growth in favorable microsites. Ecology 69:870-873.

Eissenstat DM and Caldwell MM (1989) Invasive root growth into disturbed soil of two tussock grasses that differ in competitive effectiveness. Func. Ecol. 3:345-353.

Evans RA and Young JA (1984) Microsite requirements for downy brome (*Bromus tectorum*) infestation and control on sagebrush rangelands. Weed Sci. 32S:13-17.

Fairbourn ML and Rauzi F (1982) Effect of annual low-nitrogen fertilization of crested wheatgrass. Soil Sci. 134:126-132.

Fautin RW (1946) Biotic communities of the northern desert shrub biome in western Utah. Ecol. Monogr. 16:251-310.

Fenneman NH (1931) Physiography of Western United States. McGraw-Hill, NY.

Fernandez OA and Caldwell MM (1975) Phenology and dynamics of root growth of three cool semi-desert shrubs under field conditions. J. Ecol. 63:703-714.

Field C (1983) Allocating leaf nitrogen for the maximization of carbon gain: leaf age as a control on the allocation program. Oecologia 56:341-347.

Field C and Mooney HA (1983) Leaf age and seasonal effects on light, water, and nitrogen use efficiency in a California shrub. Oecologia 56:348-355.

Fiero B (1986) Geology of the Great Basin. Univ. Nevada Press, Reno, NV.

Fischer RA and Turner NC (1978) Plant productivity in the arid and semiarid zones. Annu. Rev. Plant Physiol. 29:277-317.

Flaschka I, Stockton CW, and Boggess WR (1987) Climatic variation and surface water resources in the Great Basin region. Water Resour. Bull. 23:47-57.

Fonteyn PJ and Mahall BE (1981) An experimental analysis of structure in a desert plant community. J. Ecol. 69:883-896.

Forseth IN, Ehleringer JR, Werk KS, and Cook CS (1984) Field water relations of Sonoran Desert annuals. Ecology 65:1436-1444.

Frank AB (1981) Effect of leaf age and position on photosynthesis and stomatal conductance of forage grasses. Agron. J. 73:70-74.

Frank AB (1983) Plant-water relationships of crested, pubescent, slender, and western wheatgrass. pp. 399-401. In Smith JA and Hays VW (editors), Proc. XIV Intern. Grassland Cong. Westview Press, Boulder, CO.

Frank AB and Barker RE (1976) Rates of photosynthesis and transpiration and diffusive resistance of six grasses grown under controlled conditions. Agron. J. 68:487-490.

Freeman DC, Klikoff LG, and Harper KT (1976) Differential resource utilization by the sexes of dioecious plants. Science 193: 597-599.

Garcia-Moya E and McKell CM (1970) Contribution of shrubs to the nitrogen economy of a desert-wash plant community. Ecology 51:81-88.

Gile LH, Peterson FF, and Grossman RB (1966) Morphological and genetic sequences of carbonate accumulation in desert soils. Soil Sci. 101:347-360.

Gollan T, Turner NC, and Schulze E-D (1985) The responses of stomata and leaf gas exchange to vapor pressure deficits and soil water content. III. In the sclerophyllous woody species *Nerium oleander*. Oecologia 65:356-362.

Goodman PJ (1973) Physiological and ecotypic adaptations of plants to salt desert conditions in Utah. J. Ecol. 61:473-494.

Harris GA (1967) Some competitive relationships between *Agropyron spicatum* and *Bromus tectorum*. Ecol. Monogr. 37:89-111.

Harris, GA and Wilson AM (1970) Competition for moisture among seedlings of annual and perennial grasses as influenced by root elongation at low temperature. Ecology 51:530-534.

Hodgkinson KC, Johnson PS, and Norton BE (1978) Influence of summer rainfall on root and shoot growth of a cold-winter desert shrub, *Atriplex confertifolia*. Oecologia 34:353-362.

Holthausen RS and Caldwell MM (1980) Seasonal dynamics of root system respiration in *Atriplex confertifolia*. Plant Soil 55:307-317.

Hulbert LC (1955) Ecological studies of *Bromus tectorum* and other annual bromegrasses. Ecol. Monogr. 25:181-213.

Hull AC Jr and Klomp GL (1974) Yield of crested wheatgrass under four densities of big sagebrush in southern Idaho. USDA Technical Bulletin 1483.

Hutchings SS and Stewart G (1953) Increasing forage yields and sheep production on Intermountain winter ranges. USDA Circular 925.

Jaeger EC (1957) The North American Deserts. Stanford Univ. Press, Stanford.

James DW and Jurinak JJ (1978) Nitrogen fertilization of dominant plants in the northeastern Great Basin desert. pp. 219-231. In West NE and Skujins JJ (editors), Nitrogen in Desert Ecosystems. Dowden, Hutchinson, and Ross, Stroudsburg, PA.

Jarvis PG and McNaughton KB (1986) Stomatal control of transpiration: scaling up from leaf to region. pp. 1-49. In MacFadyen A and Ford ED (editors), Advances in Ecological Research, Vol 15. Academic Press, London.

Jeffries DL and Klopatek JM (1987) Effects of grazing on the vegetation of the blackbrush association. J. Range Manage. 40:390-392.

Johnson PS and Norton BE (1980) The effects of subsurface irrigation on current and subsequent year's growth in shadscale. J. Range Manage. 33:331-336.

Jordan PW and Nobel PS (1979) Infrequent establishment of seedlings of *Agave deserti* (Agavaceae) in the northwestern Sonoran Desert. Amer. J. Bot. 66:1079-1084.

Kay BL and Evans RA (1965) Effects of fertilization on a mixed stand of cheatgrass and intermediate wheatgrass. J. Range Manage. 18:7-11.

Klikoff LG (1966) Competitive response to moisture stress of a winter annual of the Sonoran Desert. Amer. Midl. Nat. 75: 383-391.

Klubek B, Eberhardt PJ, and Skujins J (1978) Ammonia volatilization from Great Basin desert soils. pp. 1-49. In West NE and Skujins JJ (editors), Nitrogen in Desert Ecosystems. Dowden, Hutchinson, and Ross, Stroudsburg, PA.

Lane LJ, Romney EM, and Hakonson TE (1984) Water balance calculations and net production of perennial vegetation in the northern Mojave Desert. J. Range Manage. 37:12-18.

Levitt J (1980) Responses of Plants to Environmental Stress, Vol. II. Water, Radiation, Salt, and Other Stresses. Academic Press, NY.

Ludwig JA (1987) Primary productivity in arid lands: myths and realities. J. Arid Environ. 13:1-7.

Lunt OR, Letey J, and Clark SB (1973) Oxygen requirements for root growth in three species of desert shrubs. Ecology 54: 1356-1362.

Mack RN (1981) The invasion of *Bromus tectorum* L. into Western North America: an ecological chronicle. Agro-Ecosystems 7:145-165.

Mack RN and Pyke DA (1983) The demography of *Bromus tectorum*: variation in time and space. J. Ecol. 71:69-93.

MacMahon JA (1979) North American deserts: their floral and faunal components. pp. 21-82. In Goodall DW and Perry RA (editors), Arid-land Ecosystems: Structure, Functioning and Management, Vol 1. Cambridge Univ. Press, Cambridge.

MacMahon JA (1987) Warm deserts. pp. 231-264. In Barbour MG and Billings WD (editors), North American Terrestrial Vegetation. Cambridge Univ. Press, Cambridge.

MacMahon JA and Schimpf DJ (1981) Water as a factor in the biology of North American desert plants. pp. 114-117. In Evans DD and Thames JL (editors), Water in Desert Ecosystems. Dowden, Hutchinson, and Ross, Stroudsburg, PA.

McAuliffe JR (1988) Markovian dynamics of simple and complex desert plant communities. Amer. Nat. 131:459-490.

McIlvanie SK (1942) Carbohydrate and nitrogen trends in bluebunch wheatgrass, *Agropyron spicatum*, with special reference to grazing influences. Plant Physiol. 17:540-557.

Miller RF (1988) Comparison of water use by *Artemisia tridentata* spp. *wyomingensis* and *Chrysothamnus viscidiflorus* spp. *viscidiflorus*. J. Range Manage. 41:58-62.

Miller RF and Schultz LM (1987) Development and longevity of ephemeral and perennial leaves on *Artemisia tridentata* Nutt. ssp. *wyomingensis*. Great Basin Nat. 47:227-230.

Miller RF, Branson FA, McQueen IS, and Snyder CT (1982) Water relations in soils as related to plant communities in Ruby Valley, Nevada. J. Range Manage. 35:462-468.

Monson RK and Smith SD (1982) Seasonal water potential components of Sonoran Desert plants. Ecology 63:113-123.

Monson RK and Szarek SR (1981) Life cycle characteristics of *Machaeranthera gracilis* (Compositae) in desert habitats. Oecologia 49:50-55.

Mooney HA, Björkman O, and Collatz GJ (1977a) Photosynthetic acclimation to temperature in the desert shrub, *Larrea divaricata*. I. Carbon dioxide exchange characteristics of intact leaves. Plant Physiol. 61:406-410.

Mooney HA, Ehleringer J, and Björkman O (1977b) The energy balance of leaves of the evergreen desert shrub *Atriplex hymenelytra*. Oecologia 29:301-310.

Mooney HA, West NE, and Brayton R (1966) Field measurements of the metabolic responses of bristlecone pine and big sagebrush in the White Mountains of California. Bot. Gaz 127:105-113.

Moore PD (1984) Why be an evergreen? Nature 312:703.

Moore RT, White RS, and Caldwell MM (1972) Transpiration of *Atriplex confertifolia* and *Eurotia lanata* in relation to soil, plant, and atmospheric moisture stresses. Can. J. Bot. 50:2411-2418.

Morrow LA and Stahlman PW (1984) The history and distribution of downy brome (*Bromus tectorum*) in North America. Weed Sci. 32S:2-6.

Mozafar A and Goodin JR (1970) Vesiculated hairs, a mechanism for salt tolerance in *Atriplex halimus* L. Plant Physiol. 45:62-65.

Mueggler WF (1972) Influence of competition on the response of bluebunch wheatgrass to clipping. J. Range Manage. 25:88-92.

Mulroy TW and Rundel PW (1977) Annual plants: adaptations to desert environments. BioScience 27:109-114.

Neilson RP (1987) Biotic regionalization and climatic controls in western North America. Vegetatio 70:135-147.

Nobel PS and Franco AC (1986) Annual root growth and intraspecific competition for a desert bunchgrass. J. Ecol. 74:1119-1126.

Nowak RS and Caldwell MM (1984a) Photosynthetic activity and survival of foliage during winter for two bunchgrass species in a cold-winter steppe environment. Photosynthetica 18: 192-200.

Nowak RS and Caldwell (1984b) A test of compensatory photosynthesis in the field: implications for herbivory tolerance. Oecologia 61:311-318.

Nowak RS and Caldwell MM (1986) Photosynthetic characteristics of crested wheatgrass and bluebunch wheatgrass. J. Range Manage. 39:443-450.

Nowak RS, Anderson JE, and Toft NL (1988) Gas exchange of *Agropyron desertorum*: diurnal patterns and responses to water vapor gradient and temperature. Oecologia 77:289-295.

Noy-Meir I (1973) Desert ecosystems: environment and producers. Annu. Rev. Ecol. System. 4:25-51.

Noy-Meir I (1979/80) Structure and function of desert ecosystems. Israel J. Bot. 28:1-19.

O'Leary JF and Minnich RA (1981) Postfire recovery of creosote bush scrub in the western Colorado Desert. Madroño 28:61-66.

Osmond CB (1979) Ion uptake, transport and excretion. pp. 607-625. In Goodall DW, and Perry RA (editors), Arid-land Ecosystems: Structure, Functioning and Management, Vol 1. Cambridge Univ. Press, Cambridge.

Osmond CB, Austin MP, Berry JA, Billings WD, Boyer JS, Dacey JWH, Nobel PS, Smith SD, and Winner WE (1987a) Stress physiology and the distribution of plants. BioScience 37:38-48.

Osmond CB, Smith SD, Ben G-Y, and Sharkey TD (1987b) Stem photosynthesis in a desert ephemeral, *Eriogonum inflatum*: characterization of leaf and stem CO_2 fixation and H_2O vapor exchange under controlled conditions. Oecologia 72:542-549.

Pearson LC (1965) Primary production in grazed and ungrazed desert communities of eastern Idaho. Ecology 46:278-285.

Phillips DL and MacMahon JA (1981) Competition and spacing patterns in desert shrubs. J. Ecol. 69:97-115.

Power JF (1986) Nitrogen cycling in seven cool-season perennial grass species. Agron. J. 78:681-687.

Power JF and Alessi J (1970) Effects of nitrogen source and phosphorous on crested wheatgrass growth and water use. J. Range Manage. 23:175-178.

Rasmussen LL and Brotherson JD (1986) Response of winterfat (*Ceratoides lanata*) communities to release from grazing pressure. Great Basin Nat. 46:148-156.

Rice B and Westoby M (1978) Vegetation responses of some Great Basin shrub communities protected against jackrabbits or domestic stock. J. Range Manage. 31:28-34.

Richards JH (1984) Root growth response to defoliation in two *Agropyron* bunchgrasses: field observations with an improved root periscope. Oecologia 64:21-25.

Richards JH (1986) Plant response to grazing: the role of photosynthetic capacity and stored carbon reserves. pp. 428-430. In Joss PJ, Lynch PW, and Williams OB (editors), Rangelands: A Resource Under Siege. Proceedings Second Rangeland Congress, Australian Acad. Sci., Canberra.

Richards JH and Caldwell MM (1985) Soluble carbohydrates, concurrent photosynthesis and efficiency in regrowth following defoliation: a field study with *Agropyron* species. J. Appl. Ecol. 22:907-920.

Rickard WH (1967) Seasonal soil moisture patterns in adjacent greasewood and sagebrush stands. Ecology 48:1034-1038.

Robberecht R, Mahall BE, and Nobel PS (1983) Experimental removal of intraspecific competitors - effects on water relations and productivity of a desert bunchgrass, *Hilaria rigida*. Oecologia 60:21-24.

Robertson JH (1947) Responses of range grasses to different intensities of competition with sagebrush (*Artemisia tridentata* Nutt.). Ecology 28:1-16.

Romo JT and Haferkamp MR (1988) Comparative water relations in *Artemisia tridentata* Nutt. subsp. *wyomingensis* and *Kochia prostrata* (L) Schred. J. Arid Env. 15:53-64.

Sage RF and Sharkey TD (1987) The effect of temperature on the occurrence of O_2 and CO_2 insensitive photosynthesis in field grown plants. Plant Physiol. 84:658664.

Schlesinger WH (1985) The formation of caliche in soils of the Mojave Desert, California. Geochim. Cosmochim. Acta 49:57-66.

Schulze E-D (1986) Carbon dioxide and water vapor exchange in response to drought in the atmosphere and in the soil. Annu. Rev. Plant Physiol. 37:247-274.

Shantz HL (1927) Drought resistance and soil moisture. Ecology 8:145-157.

Sheps LO (1973) Survival of *Larrea tridentata* seedlings in Death Valley National Monument, California. Israel J. Bot. 22:8-17.

Shown LM, Lusby GC, and Branson FA (1972) Soil-moisture effects of conversion of sagebrush cover to bunchgrass cover. Water Res. Bull. 8:1265-1272.

Shreve F (1942) The desert vegetation of North America. Bot. Rev. 8:195-246.

Shreve F and Mallery TD (1933) The relation of caliche to desert plants. Soil Sci. 35:99-113.

Shumar ML and Anderson JE (1986) Water relations of two susbspecies of big sagebrush on sand dunes in southeastern Idaho. Northwest Sci. 60:179-185.

Sinclair R and Thomas DA (1970) Optical properties of leaves of some species in arid South Australia. Aust J. Bot. 18:261-273.

Skujins J (1981) Nitrogen cycling in arid ecosystems. pp. 477-491. In Clark FE and Rosswall T (editors), Terrestrial Nitrogen Cycles: Processes, Strategies and Management Impacts. Swedish National Science Research Council, Stockholm.

Smith SD and Ludwig JA (1978) The distribution and phytosociology of *Yucca elata* in southern New Mexico. Amer. Midl. Nat. 100: 202-212.

Smith SD and Nobel PS (1986) Deserts. pp. 13-62. In Baker NR and Long SP (editors), Photosynthesis in Contrasting Environments. Elsevier, Amsterdam.

Smith SD and Osmond CB (1987) Stem photosynthesis in a desert ephemeral, *Eriogonum inflatum*: morphology, stomatal conductance and water-use efficiency in field populations. Oecologia 72:533-541.

Smith SD, Hartsock TL, and Nobel PS (1983) Ecophysiology of *Yucca brevifolia*, an arborescent monocot of the Mojave Desert. Oecologia 60:10-17.

Smith SD, Strain BR, and Sharkey TD (1987) Effects of CO_2 enrichment on four Great Basin grasses. Func. Ecol. 1:139-143.

Smith WK and Nobel PS (1978) Influence of irradiation, soil water potential, and leaf temperature on leaf morphology of a desert broadleaf, *Encelia farinosa* Gray (Compositae). Amer. J. Bot. 65:429-432.

Sneva FA (1973) Crested wheatgrass response to nitrogen and clipping. J. Range Manage. 26:47-50.

Sneva FA (1977) Correlations of precipitation and temperature with spring regrowth and mature crested wheatgrass yields. J. Range Manage. 30:270-275.

Sneva FA, Hyder DN, and Cooper CS (1958) The influence of ammonium nitrate on the growth and yield of crested wheatgrass on the Oregon high desert. Agron. J. 50:40-44.

Strojan CL, Turner FB, and Castetter R (1979) Litter fall from shrubs in the northern Mojave Desert. Ecology 60:891-900.

Sturges DL (1977) Soil water withdrawal and root characteristics of big sagebrush. Amer. Midl. Nat. 98:257-273.

Stutz HC (1978) Explosive evolution of perennial *Atriplex* in western America. Great Basin Nat. Mem. 2:161-168.

Szarek SR (1979) Primary production in four North American deserts: indices of efficiency. J. Arid. Environ. 2:187-209.

Thill DC, Schirman RD, and Appleby AP (1979) Influence of soil moisture, temperature, and compaction on the germination of downy brome (*Bromus tectorum*). Weed Sci. 27:625-630.

Thomson WW, Berry WL, and Liu LL (1970) Localization and secretion of salt by the salt glands of *Tamarix aphylla*. Proc. Nat. Acad. Sci. USA 63:310-317.

Thorgeirsson H (1985) Temporal and spatial partitioning of the soil water resource between two *Agropyron* bunchgrasses and *Artemisia tridentata*. Thesis, Utah State Univ., Logan, UT.

Thornthwaite CW (1948) An approach toward a rational classification of climate. Geogr. Rev. 38:55-94.

Toft NL and Pearcy RW (1982) Gas exchange characteristics and temperature relations of two desert annuals: a comparison of a winter-active and a summer-active species. Oecologia 55:170-177.

Toft NL, Anderson JE, and Nowak RS (1989) Water use efficiency and carbon isotope composition of plants in a cold desert environment. Oecologia (in press).

Trlica MJ Jr and Cook CW (1971) Defoliation effects on carbohydrate reserves of desert species. J. Range Manage. 24:418-425.

Turner FB and Randall DC (1987) The phenology of desert shrubs in southern Nevada. J. Arid Env. 13:119-128.

Vasek FC (1980) Creosote bush: long-lived clones in the Mojave Desert. Amer. J. Bot. 67:246-255.

Wallace A, Romney EM, and Hale VQ (1973) Sodium relations in desert plants: 1. Cation contents of some plant species from the Mojave and Great Basin deserts. Soil Sci. 115:284-287.

Wallace A, Bamberg SA, and Cha JW (1974a) Quantitative studies of roots of perennial plants in the Mojave Desert. Ecology 55: 1160-1162.

Wallace A, Romney EM, Cha JW, and Alexander GV (1974b) Sodium relations in desert plants: 3. Cation-anion relationships in three species which accumulate high levels of cations in leaves. Soil Sci. 118:397-401.

Wallace A, Romney EM, and Cha JW (1980a) Depth distribution of roots of some perennial plants in the Nevada Test Site area of the northern Mojave Desert. Great Basin Nat. Mem. 4:201-207.

Wallace A, Romney EM, and Kinnear JE (1980b) Frequency distribution of three perennial plant species to nearest neighbor of the same species in the northern Mojave Desert. Great Basin Nat. Mem. 4:89-93.

Walter H, Harnickell E, and Mueller-Dombois D (1975) Climate-Diagram Maps of the Individual Continents and the Ecological Climate Regions of the Earth. Springer-Verlag, Heidelberg.

Wang J (1988) Response of *Artemisia tridentata* ssp. *wyomingensis* and *Stipa thurberiana* to nitrogen amendments. Thesis, Oregon State Univ., Corvallis.

Welch BL and Jacobson TLC (1988) Root growth of *Artemisia tridentata*. J. Range Manage. 41:332-334.

Went FW (1948) Ecology of desert plants. I. Observations on germination in the Joshua Tree National Monument, California. Ecology 29:242-253.

Went FW (1949) Ecology of desert plants. II. The effect of rain and temperature on germination and growth. Ecology 30:1-13.

Went FW (1979) Germination and seedling behavior of desert plants. pp. 477-489. In Goodall DW and Perry RA (editors), Arid-land Ecosystems: Structure, Functioning and Management, Vol. 1. Cambridge Univ. Press, Cambridge.

Wentworth TR (1981) Vegetation on limestone and granite in the Mule Mountains, Arizona. Ecology 62:469-482.

Wentworth TR and Davidson EA (1987) Foliar mineral elements in native plants on contrasting rock types: multivariate patterns and nutrient balance regulation. Soil Sci. 144: 190-201.

Werk KS and Ehleringer JR (1983) Photosynthesis by flowers in *Encelia farinosa* and *Encelia californica* (Asteraceae). Oecologia 57:311-315.

Werk KS and Ehleringer J (1985) Photosynthetic characteristics of *Lactuca serriola* L. Plant Cell Environ. 8:345-350.

West NE (1983) Ecosystems of the World, Vol 5. Temperate Deserts and Semideserts. Elsevier, NY.

West NE (1987) Intermountain deserts, shrub steppes, and woodlands. pp. 209-230. In Barbour MG, Billings WD (editors), North American Terrestrial Vegetation. Cambridge Univ. Press, Cambridge.

West NE and Gasto J (1978) Phenology of the aerial portions of shadscale and winterfat in Curlew Valley, Utah. J. Range Manage. 31:43-45.

West NE, Tausch RJ, Rea KH, and Tueller PT (1978) Taxonomic determination, distribution, and ecological indicator values of sagebrush within the pinyon-juniper woodlands of the Great Basin. J. Range Manage. 31:87-92.

West NE, Provenza FD, Johnson PA, and Owens MK (1984) Vegetation change after 13 years of livestock grazing exclusion on sagebrush semidesert in west central Utah. J. Range Manage. 37:262-264.

Westerman RL and Tucker TC (1978) Denitrification in desert soils. pp. 75-106. In West NE and Skujins JJ (editors), Nitrogen in Desert Ecosystems. Dowden, Hutchinson, and Ross, Stroudsburg, PA.

Wiebe HH, Al-Saadi HA, and Kimball SL (1974) Photosynthesis in the anomalous secondary wood of *Atriplex confertifolia* stems. Amer. J. Bot. 61:444-448.

Wight JR and Black AL (1972) Energy fixation and precipitation-use efficiency in a fertilized rangeland ecosystem of the northern Great Plains. J. Range Manage. 25:376-380.

Wight JR and Skiles JW (1987) SPUR: simulation of production and utilization of rangelands. Documentation and user guide. USDA Agricultural Research Service Publication 63.

Williams RJ, Boersma K, and Van Ryswyk AL (1979) The effects of nitrogen fertilization on water use by crested wheatgrass. J. Range Manage. 32:98-100.

Wilson AM, Harris GA, and Gates DH (1966) Fertilization of mixed cheatgrass-bluebunch wheatgrass stands. J. Range Manage. 19: 134-137.

Woodell SRJ, Mooney HA, and Hill AJ (1969) The behaviour of *Larrea divaricata* (creosote bush) in response to rainfall in California. J. Ecol. 57:37-44.

Wraith JM, Johnson DA, Hanks RJ, and Sisson DV (1987) Soil and plant water relations in a crested wheatgrass pasture: response to spring grazing by cattle. Oecologia 73:573-578.

Yeaton RI and Cody ML (1976) Competition and spacing in plant communities: the northern Mojave Desert. J. Ecol. 64:689-696.

Young JA and Evans RA (1973) Downy brome: intruder in the plant succession of big sagebrush communities in the Great Basin. J. Range Manage. 26:410-415.

Young JA and Evans RA (1978) Population dynamics after wildfires in sagebrush rangelands. J. Range Manage. 31:283-289.

Young JA and Evans RA (1986) Erosion and deposition of fine sediments from playas. J. Arid Env. 10:103-115.

Young JA, Evans RA, and Tueller PT (1975) Great Basin plant communities - pristine and grazed. pp. 187-212. In Elston R (editor), Holocene Climate in the Great Basin. Nevada Archeological Survey, Reno, NV.

Plate 13.1. (upper) Blackbrush-Joshua tree (*Coleogyne ramosissima - Yucca brevifolia*) vegetation type in southern Nevada.

Plate 13.2. (lower) Shadscale (*Atriplex confertifolia*) vegetation type in western Nevada.

Plate 14.1. (upper) Sagebrush-steppe vegetation in southeastern Idaho, dominated by *Artemisia tridentata*, with *Chrysothamnus viscidiflorus* an important subdominant species.

Plate 14.2. (lower) *Elymus cinereus* in southeastern Idaho, a bunchgrass that occurs in almost pure monocultures in small playas throughout the cold desert.

Plate 15. Interface of a shrub coppice dune (upper left) and adjacent interspace in a stand of shadscale (*Atriplex confertifolia*). Herbaceous species emergence is restricted to the coppice dune and interspace cracks. Microphytes (algae, bacteria, lichen, moss) cover the interspace polygons. Photograph by K. Ewing.

Plate 16. Adjacent experimental plots of mountain big sagebrush (*Artemisia tridentata* ssp *vaseyana*) mixed with either bluebunch wheatgrass (upper, *Agropyron spicatum* [Syn: *Pseudoroegneria spicata*]) or crested wheatgrass (lower, *Agropyron desertorum*). The plots had been established 3 years earlier with small transplants of similar size. The large difference in sagebrush canopy volumes between the plots was the result of differences in belowground interference between the species of wheatgrass. This pattern was repeated in several plots.

Chapter 7

Basin Hydrology and Plant Root Systems

James P. Dobrowolski, Martyn M. Caldwell, and James H. Richards

Introduction

The Great Basin of western North America represents more than one-fifteenth the area of the United States and can be defined and circumscribed on the basis of drainage, that is, no outlet except to the atmosphere, and general aridity (Houghton 1976). The Great Basin is a physiographic unit, covering about 70% (373,000 km^2) of the Basin and Range Province outlined by Fenneman (1931). Although the hydrologic boundaries are well defined (Figure 7.1), less consensus exists regarding the physiographic boundaries. The Basin extends from the Sierra Nevada-Cascade ranges in the west to the Wasatch-Rockies with tributary valleys into Wyoming on the east. The north-southeast boundaries are generally delimited by the Columbia lava plateau and Colorado River plateaus, respectively. Southward, the Basin extends into the Sonoran Province (Blackwelder 1948).

The Basin and Range Province includes portions of the Mojave, Sonoran and Chihuahuan deserts. This chapter will focus on the Great Basin section, where internal or partial drainage characteristics confer many unique features to the region. The hydrologic properties of Great Basin lowlands, from the special features of soil surface crusts to the basin drainage and flow patterns, will be described. Root systems of lowland Great Basin plants exhibit unique properties. The structure and function of these root systems will be described from experiments on selected species. These descriptions of root systems range from characterization of the form and quantity of roots to the function of roots in immediate competition with roots of neighboring plants.

Great Basin Hydrology

The Physiographic Setting

The Great Basin is not composed of a single large basin as its name implies, but a collection of over 100 basins interspersed with more than 150 mountain ranges (Young and Evans 1986). Many of these basins exhibit internal drainage, though some are drained partially either directly or via the eight major river systems (Bear, Weber, Jordan, Sevier, Humboldt, Carson, Truckee, and Walker) into the five principal sinks or lakes (Eakin et al. 1976). These sinks include the Humboldt-Carson Sink, Pyramid Lake, and Walker Lake in Nevada, and the Great Salt Lake and Sevier Lake in Utah. The Great Basin is characterized by generally parallel, north-to-northeast trending mountain ranges separated by alluvial basins. Plate extension and intensive block faulting from the Miocene to the present have produced the current structure. Displacement and erosion in the mountains have provided alluvium of Quaternary age, which has been deposited at the canyon mouths (Snyder 1962). Coalescence of these alluvial fans tends to form aggradational piedmont plains or bajadas. These piedmont plains typically lead to fine-textured flats in closed basins (Mabbutt 1979).

Most of the basins in the Great Basin contain flat and generally unvegetated lower areas that periodically flood, desiccate, and accumulate sediment. Generally these areas are referred to as playas (Neal 1975a). Playas are a very important component of closed basin hydrology. They transform from dry to wet or the reverse in response to changing conditions. Duffy and Al-Hassan (1988) observed the dry playa in Pilot Valley, Nevada to become moist as a result of a recharge-produced water table rise or due to surface flooding. An ephemeral lake may form on a moist playa in response to a single thunderstorm. They also observed the appearance of a lake during the winter months due to surface flooding when evaporation could not keep up with water table rise.

During the wetter periods of the Quaternary, the saline lakes and playas of the Great Basin went through successive wet and dry cycles. Snyder and Langbein (1962) found evidence for a Pleistocene lake of considerable area in the closed basin of Spring Valley, Nevada. Using climatic inferences from the current dry playa and extrapolations from closed basins with existing lakes, they found lake levels to be sensitive to even small climatic changes. Morrison (1968) referred to these lakes as

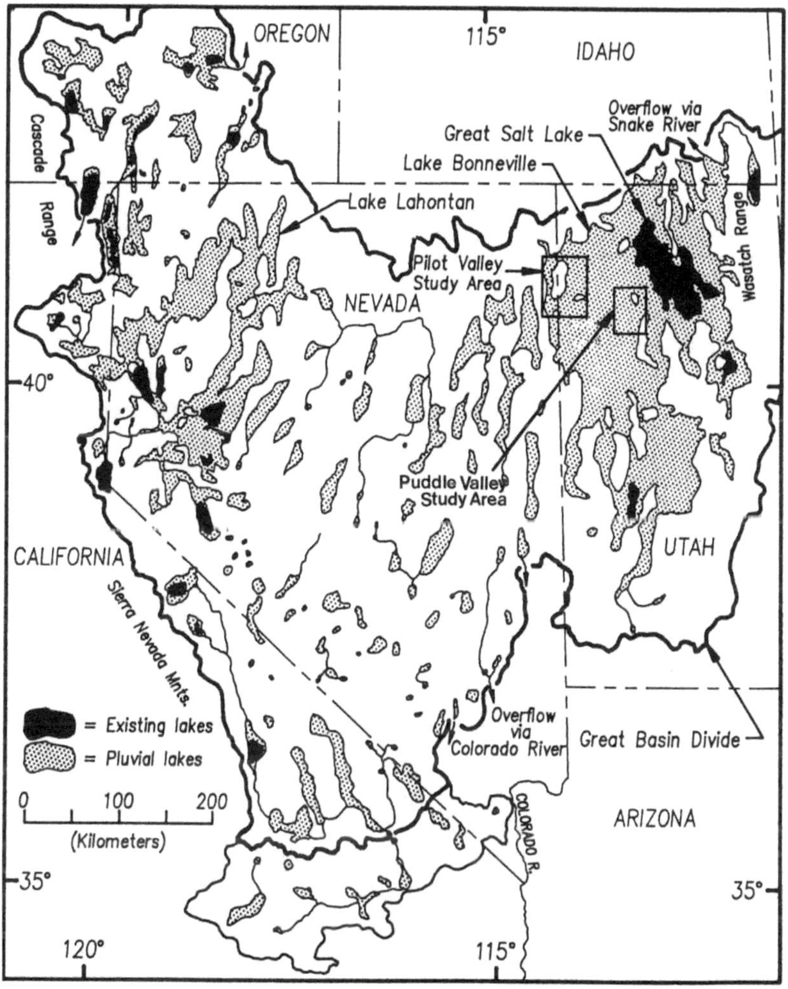

Figure 7.1. Map identifying the drainage divide for the Great Basin Physiographic Province. The present day distribution of pluvial lakes and their maximum expansion during the late Pleistocene is indicated. Pilot Valley, NV and Puddle Valley, UT study areas are shown. From Morrison (1965) and Duffy and Al-Hassan (1988).

pluvial lakes, and considers pluvial-lake successions to be one of the best records of Pleistocene climatic change. Motts (1970) distinguished between lakes, playa lakes, and playas by the average proportion of time the basin is submerged. Climatic changes influence the salinity of these pluvial lakes. Salinity has been observed to vary inversely with lake volume (Eardley et al. 1957).

Table 7.1. Classification of ephemeral lakes and playas in the Great Basin Physiographic Province based on hydrologic regime (Adapted from Mabbutt 1979; Motts 1965; Snyder 1962.)

Topographically closed basin		Topographically open basin
No groundwater outlet	External groundwater outlet	
Moist surface	Moist in wet season	Dry surface
Saline groundwater	Saline or nonsaline, depending on climate, topography and subsurface drainage	Principally nonsaline, depending on climate and degree of topographical closure
Salt-crust or wet mud flats	Silty-clay surface subject to deflation	Claypan
Primarily large in size (i.e.,Great Salt Lake)		Primarily small in size

Increasing importance of surface inflow →

← Increasing importance of water table rise

Whether a basin is hydrologically closed depends upon the character and location of the basin outlet. Surface and subsurface flows that have no outlet except to the atmosphere constitute hydrologic closure (Snyder 1962). The existence of hydrologic closure is maintained when incoming water is less than or equal to evaporation over extended time periods. Hydrologic closure ceases to exist when excess water remains for long periods of time and results in external outlets (Duffy and Al-Hassan 1988). Whether a playa is fed from surface inflow or from groundwater determines the distinction between dry and moist playas, which has important consequences for Great Basin sediment and salinity budgets, and the type of plant community that develops (Table 7.1).

Precipitation Processes

Physiographic characteristics (i.e., aspect, slope, and elevation) and mountain range orientation are known to exert an important influence on the delivery and cycling of water in the Great Basin. The great variability in distribution and timing of precipitation is a function of the relative importance of the two oceanic sources (Pacific Ocean and Gulf of California), complicated by triggering mechanisms such as convection, orography, and cyclonic activity (Houghton 1969). Snyder (1962) describes the mountains surrounding a playa as catchment areas for primarily orographic precipitation, and explains that the mountains have little capacity for natural water storage. The basins, in contrast, have a greater water storage capacity than the amount of direct annual precipitation. Thus, high elevation precipitation provides the principal source of water in the Great Basin and low elevation evaporation from playas and closed basin lakes is the principal source of water loss (Duffy and Al-Hassan 1988).

Orographic lifting of moist air masses, which results in cloud formation, has its greatest influence during the winter. This process produces the majority of precipitation as mountain snowpack which melts and moves through surface and groundwater systems during the spring and summer (Eakin et al. 1976). Duffy and Al-Hassan discovered a direct relationship between precipitation and elevation using a data set from Chang (1969) for west facing slopes of the Wasatch and Bear River Ranges east of the Great Salt Lake, Utah. Peck and Brown (1962) suggested a similar relationship for leeward slopes of the Wasatch Range. To describe adequately the orographic precipitation of a hydrologically closed basin, Duffy and Al-Hassan added the spatial distribution of the elevation-precipitation relationship. By taking advantage of natural length scales of the topography, they constructed a generalized scaling relationship for elevation versus distance. Figure 7.2 shows elevation-distance profiles from five east facing slopes of the Pilot Valley study area (see Figure 7.1) and the fitted orographic function $f(X/\sigma)$ incorporating elevation-precipitation and distance. The value X/σ is the horizontal distance from the topographic divide to the "hingeline" or hydrologic boundary between upslope net recharge to the subsurface system and the playa, where net discharge predominates. Local variability in precipitation along the Wasatch Range in Utah is produced by storm direction, slope, and aspect (Peck and Brown 1962).

Hydrologic Circulation

Basin scale circulation in the Great Basin can be exemplified by describing the distribution and movement of water throughout a small, topographically closed and undrained valley. Small drainage basins have

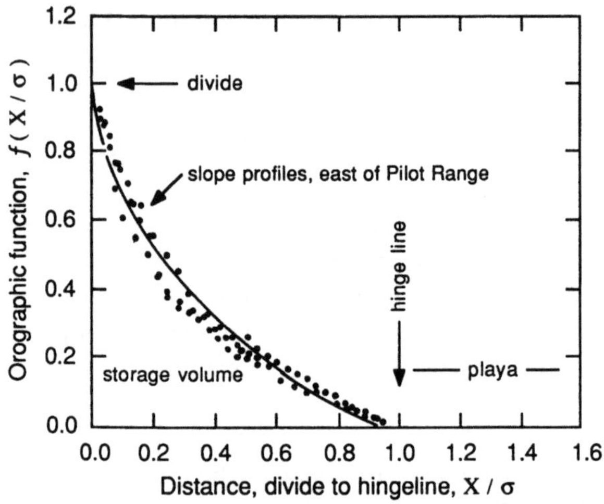

Figure 7.2. Fitted orographic function, $f(X/\sigma)$, and normalized elevation-distance profiles from five east facing slopes of Pilot Valley, Nevada (see Figure 7.1). The change in elevation between the topographic divide and the playa, and the horizontal distance from the divide to the hingeline (X/σ) were used to normalize the slopes. Redrawn from Duffy and Al-Hassan (1988).

been shown by Toth (1962) to comprise the most important units in the hydrologic regime, particularly when attempting to understand groundwater movement. An idealized fluvial system of a topographically and hydrologically closed basin proposed by Schumm (1977) can be represented by three zones in the downstream direction (Figure 7.3). Zone 1 is composed of one or more drainage basins; this is the source for water, sediment, and potential energy to the system. Water resulting from snowmelt or rainfall is captured at different mountain elevations and moves downslope via surface runoff, subsurface flow in unsaturated soil (i.e., interflow), streamflow, or as saturated groundwater flow along a path towards the valley. The streams in this Zone may be ephemeral, intermittent, or perennial. Zone 2 is considered the predominant transfer zone for water and sediment. Fluvial morphology in this area is often characterized by linear channels without tributaries. These channels tend to lose water as they cross the alluvial fans towards the playa. Zone 3 is the area of deposition or discharge, which is occupied by the playa in a closed basin. Water reaching the playa may persist and form a shallow lake until lost by evaporation. Within the Great Basin, flooding and alluvial deposition may occur regularly in this zone (Schumm 1977).

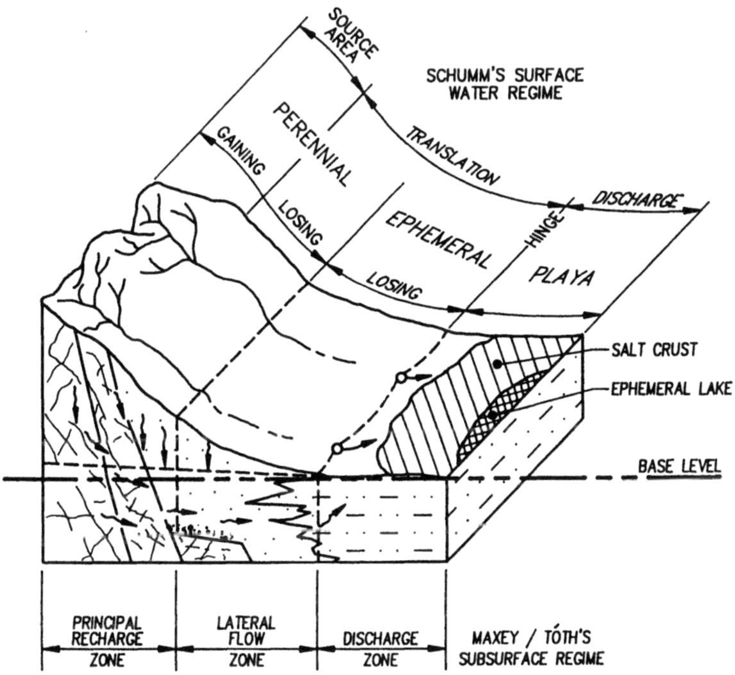

Figure 7.3. Abstraction of the hydrology of a topographically closed and undrained basin. Surface water flow and sediment transport regime are ascribed to Schumm (1977). The subsurface regime was originally proposed by Toth (1962) and Maxey (1967). Redrawn from Duffy and Al-Hassan (1988).

Recharge to groundwater occurs in Zone 1, where the movement of water is downward and hydraulic head decreases with depth. The groundwater flows laterally out of the mountain aquifers and enters the alluvial fans (Zone 2) at the valley margin (Snyder 1962). Streams become ephemeral at the valley margin and lose a considerable flow to the groundwater reservoir in the upper portions of the alluvial fan. The playa (Zone 3) is the area of discharge where water movement is upward and hydraulic head increases with depth. The discharge area is the site of greatest human development due to water availability, and typically has been the most intensively studied (Maxey 1967).

Hydrologic conditions in these basins are controlled by physiography; the mountains recharge groundwater and provide the necessary potential energy to generate discharge in the valleys. Valley isolation by surrounding mountains may preclude movement of water from one valley to another. In other instances valley isolation is not entire, and valleys may have surface or groundwater outlets. If a groundwater outlet exists, the playa remains dry for most of the year. Partially-drained valleys, such as Puddle (Sink) Valley, Utah (see Figure 7.1) have water tables that may be tens of meters below the playa surface. The water table slopes toward the outlet which in this case is the Great Salt Lake Desert (Price and Bolke 1970). Accumulation of sediments in dry playas is determined by the flooding frequency. These areas are characteristically covered with silt and clay settled from surface waters. Vegetation in these closed, but drained, valleys is composed primarily of xerophytic species such as sagebrush (*Artemisia tridentata*) with isolated pockets of salt-tolerant plants such as shadscale (*Atriplex confertifolia*) and greasewood (*Sarcobatus vermiculatus*) in areas of locally poor drainage (Neal 1975b).

When no effective groundwater outlet exists, the basin is closed, water accumulates in the alluvium, and movement is upward and produces a moist playa. In these basins, the material below the playa is saturated and the water table is near the soil surface. Evaporation from basins with moist playas results in the accumulation of soluble salts and production of a white crust during the dry season. The ratio of precipitated salts to sediment increases in proportion to the dominance of saline groundwater discharge in basin hydrology. The playa surface may be extremely flat and hard without a perennial salt crust, undulating from the accumulation of salts on the soil surface, or perennially salt encrusted with saline sediments and pure salts left from the evaporation of water (Eardley et al. 1957, Mabbutt 1979). Pilot Valley, Nevada (see Figure 7.1) exemplifies a closed and undrained valley and contains an increasing ratio of salt precipitate to sediment from a hingeline to the central portion of the playa (Lines 1979). A thin, ephemeral crust exists at the valley margin that dissolves as water accumulates. The ephemeral salt crust may be transported eventually via dissolution and surface runoff towards the central playa, where the brine accumulates as standing water and produces a thick perennial crust (Duffy and Al-Hassan 1988). Where this perennial salt crust exists, vegetation is nearly absent, with the exception of occasional plants of pickleweed (*Allenrolfia occidentalis*) or other very salt-tolerant species such as Torrey seepweed (*Suaeda torreyana*). Less salt-tolerant phreatophytic species such as greasewood circumscribe the perennial crust. Radiating away from the playa, xerophytes replace phreatophytes where the water table is below the rooting depth.

Surface Control of Infiltration

Soils of these playas and alluvial basin margins are typically fine-textured Aridisols, easily eroded by surface water and wind. The A-horizons have been influenced by loessial deposition of silt and very fine sand transported from Pleistocene lake bottoms (Young and Evans 1986; Eckert et al. 1987). The arid to semiarid climate restricts vegetation development to perennial nonsprouting shrubs and sparse grass cover. Recent rehabilitation efforts on vast areas denuded by catastrophic fire have met with limited success. Rehabilitation difficulties and concomitant accelerated soil loss may be due, in part, to the propensity of the native soil to form soil crusts and other structural impediments to infiltration and percolation.

On dry playas and alluvial basin margins, small scale spatial variability exists in soil hydrologic properties. Frequently perennial shrubs are found in a contagious distribution across these areas, individually surrounded by apparently bare interspaces. In reality these interspaces are often covered by nonvascular microphytes composed of mosses, lichens, liverworts, algae, fungi, and bacteria (Friedmann and Galun 1974). Interspace surfaces are characterized frequently by a network of shrinkage cracks producing a polygonal surface pattern (Hugie and Passey 1964). These polygons are the crowns of coarse prisms that constitute the primary structural units of the A horizon (Eckert et al. 1986). Cracks between these prisms are considered preferential pathways for infiltration, percolation and root growth. Herbaceous species often are restricted to the area of accumulation of litter and soil beneath shrubs, referred to as the coppice dune (Blackburn et al. 1975), and when in the interspaces, then exclusively in soil cracks (Schlatterer 1968) (Plate 15). Different soil-surface morphological types on these sites can be defined by surface soil structure, the size and shape of the polygons, and the cracks separating them (Figure 7.4). Eckert et al. (1978, 1986) presented a detailed description of these soil morphologic types and associated microtopographic positions.

Type I and Type II microtopographic positions seldom are crusted, and have lower bulk density, greater organic matter, greater aggregate stability attributes and, thereby, better infiltration characteristics than the lower positions. Type III and Type IV positions contain noninterstitial macroscopic vesicles in the surface 1 to 8 cm, which form a resistant, physically and chemically stabilized crust. Infiltration is slowed in the presence of soil horizons containing vesicular porosity (foam soils)

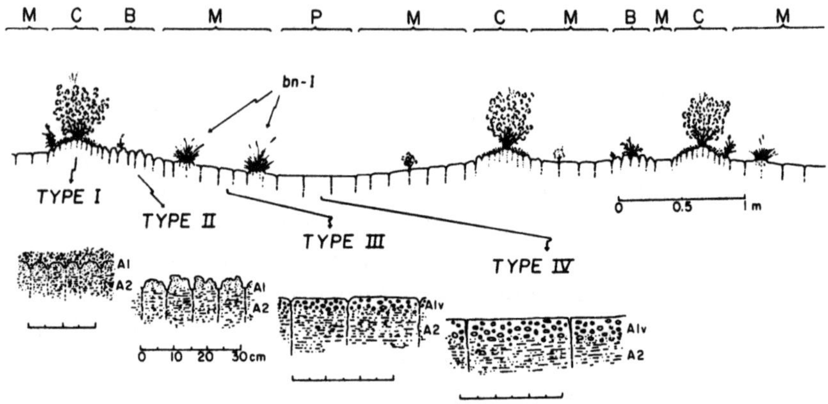

Figure 7.4. Schematic cross-section of microtopographic positions and associated surface-soil morphologic types with sagebrush (*Artemisia tridentata*) plant communities. Microtopographic positions include: C, coppice, B, coppice bench, M, intercoppice microplains, P, playette. Vertical scale is exaggerated somewhat; intercoppice microplains and playettes can be much wider than illustrated; several coppices can be linked together. Vertical lines under the soil surface define sides of crust polygons (A1v) that continue downward as sides of prisms in the compoundly weak prismatic and moderately platy A2 horizon. Type I is litter-covered. Circles indicate vesicular porosity in crust (A1v). Only Types III and IV are significantly crusted. Redrawn from Eckert et al. (1978).

(Jones 1969; Blackburn 1975; Blackburn et al. 1975). Vesicular horizons (A1v, Eckert et al. 1987) have been described in arid and semiarid regions throughout the world (Volk and Geyger 1970). Putative vesicle formation mechanisms include the formation of spherical pores from platy soil by repeated wetting and drying and vesicle enlargement by gas expansion from soil heating (Springer 1958; Miller 1971), efflux of gasses entrapped during the drying process (Evenari et al. 1974), and gas entrapment with repeated freezing and thawing (Brown 1969). During snowmelt periods or intense thunderstorms water will not pond on Type I or Type II surfaces. However, runoff and standing water have been observed for short periods on microplains (Type III) and for several days on playettes (Type IV). Miller (1971) proposed that vesicle growth decreases the surface-saturated hydraulic conductivity by reducing the cross-sectional area of water-conducting micropores. He and others have noted that even under saturated conditions vesicles remained gas-filled. Horizontal platelets found beneath the vesicular

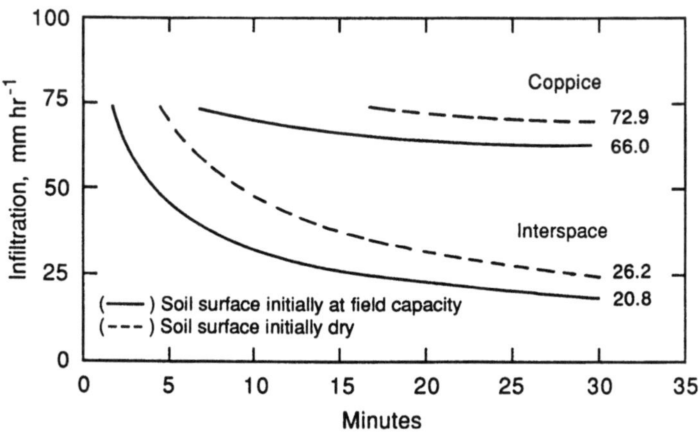

Figure 7.5. Mean infiltration rates for coppice dune and interspace areas in a sagebrush (*Artemisia tridentata*) plant community. Values to the right of curves indicate the mean infiltration rate after 30 minutes. Redrawn from Blackburn et al. (1975).

horizon also offer a tortuous path to water flow, thus providing the potential for transient surface soil saturation (Eckert et al. 1987).

Widening of cracks during successive wetting and drying events may encourage macropore flow, thereby compensating for the restriction to infiltration within the polygon. Blackburn et al. (1975) found infiltration rates and mechanical soil strength in sagebrush steppe to vary 5-fold along a 2-m transect from Type I through Type III positions (Figure 7.5). Increased soil strength or hardness resulting from the formation of vesicular porosity reduced seedling germination and emergence of crested wheatgrass (*Agropyron desertorum)* and bottlebrush squirreltail *(Sitanion hystrix* [Syn: *Elymus elymoides*]) seedlings and significantly reduced overall seedling emergence when compared to coppice soils in three Nevada watersheds (Figure 7.6) (Wood 1976; Wood et al. 1978). Although these results have implications for rehabilitation success in arid and semi-arid areas, the effect of vesicular porosity on the emergence of native or exotic shrub seedlings has to be determined yet.

Microphytes, which may form an almost continuous biological crust within shrub interspaces, are typically found to stabilize vesicular horizons throughout the Great Basin (Anderson et al. 1982). In their own right, microphytic crusts have been shown by some to promote water infiltration when compared to bare, physically and chemically crusted, or

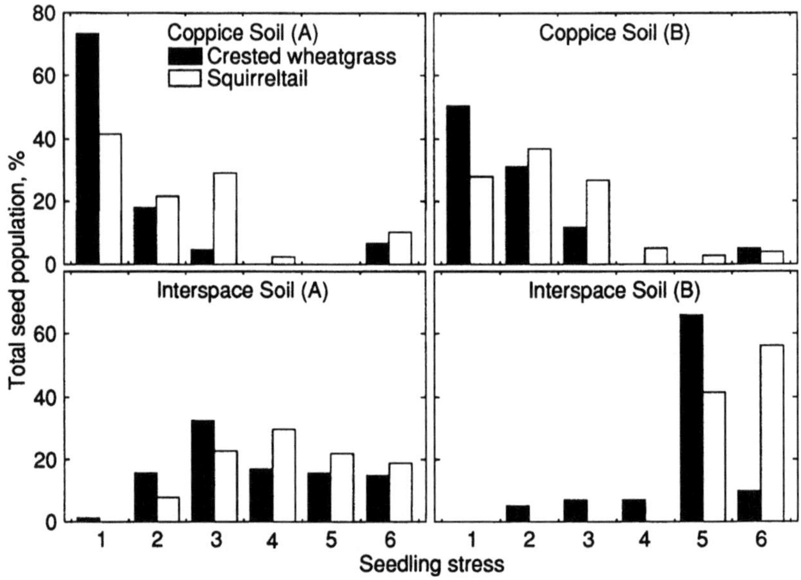

Figure 7.6. Percentage of crested wheatgrass (*Agropyron desertorum*) and bottlebrush squirreltail (*Sitanion hystrix* [Syn: *Elymus elymoides*]) seedlings exhibiting various degrees of seedling stress when planted in coppice dune and interspace soils and watered every 3(A) and 6(B) days. Redrawn from Wood et al. (1978).

"scalded" sites (Fletcher and Martin 1948; Gifford 1972; Loope and Gifford 1972; Blackburn 1975; Hacker 1986) (Figure 7.7). Hydrophobicity or restriction of gas exchange associated with certain microphytes has led others to infer a negative influence of microphytic crusts on infiltration (Walker 1979; Rogers 1977; Danin 1978; Stanley 1983). Infiltration rate into a sandy soil at the Koonamore Vegetation Reserve in South Australia was 7 cm h^{-1} in an area with lichen crust. Without the lichen crust on similar soils, infiltration rate was almost 14 cm h^{-1} (Rogers 1977).

Raindrop impact on bare soil reduces infiltration and promotes splash erosion by causing soil aggregates to slake, sealing the surface with splashed particles. Microphytic crusts were considered more desirable than exposed mineral soil on poor condition grassland by Osborn (1952), under the assumption that the crust attenuated the raindrop energy required for particle dislodgement. Recently, Tchoupopnou (1989) has shown that splash erosion was less on moss-dominated soil

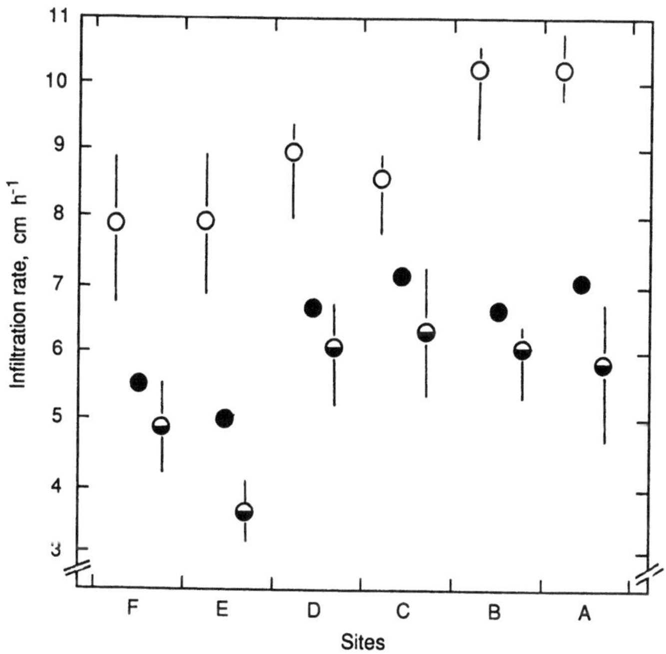

Figure 7.7. Infiltration rates for each microphytic crust condition, with cryptogamic cover increasing left to right. Closed symbols indicate mean infiltration rates over the 28 minute rainfall simulation event. Half-closed symbols depict mean infiltration rates during the final 5 minutes. Open symbols show mean infiltration rates during the first 5 minute period. Study areas are designated by capital letters: A, Relict areas, microphytic crust undisturbed; B, Grazed areas within pinyon (*Pinus edulis*)-juniper (*Juniperus* spp.) plant community, microphytic crust still intact; C, Grazed areas within pinyon-juniper, microphytic crust in an intermediate stage of breakdown; D, Microphytic crust severely disturbed by animal trampling or waterway; E, Microphytic crust mechanically disturbed by 2-way chaining (a large anchor chain pulled behind 2 tractors, first in one direction, then in the opposite direction) of pinyon-juniper leaving debris in place; F, Microphytic crust mechanically disturbed in the fall by chaining pinyon-juniper and windrowing the debris. Redrawn from Loope and Gifford (1972).

monoliths than on lichen-dominated soil monoliths from central Utah shrub interspaces. Moss- and lichen-dominated monoliths produced significantly less splash erosion than either algae-dominated or bare surfaces. The structure and function of microphytic crusts in arid and semiarid ecosystems is reviewed by West (1990).

Controls of Soil Surface Evaporation

Although primary emphasis in this chapter is placed upon the control of infiltration of rain falling directly on playas and piedmont fans of the Great Basin, the influence of crusts on the subsequent return flow toward the surface upon drying may cause significant but subtle changes in evaporation. A physical-chemical soil crust affects the upward movement of liquid-phase water by at least two mechanisms. Initial restriction of infiltration by the crust reduces the moisture content of the subtending soil. Simply, lowered moisture content below the crust results in less upward movement of water in response to increased evaporative demand compared to the uncrusted condition (Cary and Evans 1974).

A more direct way in which a soil crust may control the upward movement of water results from changes in the pore size distribution with depth. Soil moisture flux is restricted across horizon boundaries with differing pore sizes in the downward direction (Moore 1939; Grismer 1986). In the upward direction, although the matric potential gradient remains continuous, differing water holding capacities at a given matric potential will cause an abrupt change in the moisture content gradient (Marshall and Holmes 1979). Increasing the upward flux rate as a result of an increased evaporative demand requires a greater matric potential gradient, produced by a decrease in soil moisture content. As water content decreases, conductivity also decreases forcing a greater change in the matric potential gradient. Eventually, the flow volume of liquid-phase water is extremely small (Cary and Evans 1974). Interspace vesicular horizons may behave similarly and represent an abrupt change in pore size distribution. Field observations of interspace moisture flux in sagebrush steppe provide evidence for restriction of liquid-water movement upward through the vesicular layer after thorough wetting of the surface horizons by spring snowmelt.

Interference with soil gas exchange by a surface crust has been shown to occur when the crust is extremely wet or completely impermeable (Miller and Gifford 1974). Changes in the pore size distribution influence the displacement of gas by water during infiltration. As pore sizes change as a result of crust formation, more gas is encapsulated. A greater volume of encapsulated gas reduces the available porosity for water transmission and increases the resistance to flow (Constantz et al. 1988). When the encapsulated gas escapes ahead of the wetting front, a "gas counterflow" is established, which reduces infiltration (Grismer 1986).

Since the viscous flow of air is proportional to the square of the pore radius, vapor-phase water flux is reduced when the density of the soil is increased. Cracking of the soil crust may encourage the convective transfer of water vapor and compensate for the increased soil density (Cary and Evans 1974).

Microphytic crusts are believed to seal the soil surface, thereby reducing evaporation (Brotherson and Rushforth 1983). In contrast, Harper and Marble (1988) established that heavily crusted sites in Utah lost significantly more water from the upper 7.5 cm of soil than did intermixed, scalped plots. The dark cyanobacteria and lichen crusts tended to absorb more solar radiation and the soil surface attained higher temperatures than did light-colored, scalped plots, which resulted in greater evaporation.

Surface Runoff from Great Basin Lowlands

The type and extent of vegetation are of primary importance in determining the amount and timing of runoff. The perennial xerophytes comprising the majority of Great Basin playa and basin-margin plant communities consist of woody shrubs and small trees, which survive by adaptations to restrict transpiration and withstand considerable osmotic stress, and perennial grasses and forbs, which exhibit summer dormancy. Characteristics such as narrow leaves and short stature reduce the effectiveness of these plants to attenuate raindrop energy. Sparse canopies offer little resistance to runoff, rill, and interrill erosion. Additionally, surface soils contain very small quantities of litter and have a propensity to crust under the beating action of falling rains.

Surface runoff supplies energy and mass for the operation of fluvial processes in the Great Basin (Graf 1988). The determination of runoff relationships on arid and semi-arid rangelands is complicated by the variability of precipitation and infrequency of runoff events. Runoff as a percent of precipitation was computed from data by Dortignac (1956) for the semi-arid Rio Grande Basin of Colorado and New Mexico. These data show a threshold level of 25 cm or more precipitation must occur on an annual basis before there is a significant percentage of rainfall transported by surface runoff (Branson et al. 1981). Flooding occurs in most parts of the Great Basin. This results from melting snow in the spring, frontal-type rainfall at lower elevations in winter (when soils are frozen or saturated), and summer convective storms (Burkham 1988). Infrequent but localized convective rainfall on sparsely vegetated surfaces leads to flash runoff heavily laden with sediment (Mabbutt 1979). Torrential stream flooding may occur immediately following rainfall, but is short-lived due to high evaporative losses.

Typically, runoff processes in arid and semiarid areas are dominated by Hortonian (infiltration controlled) runoff; interflow and groundwater flow are relatively unimportant (Yair and Lavee 1985). In the Great Basin however, groundwater flows are very important in closed basin circulation. Hortonian runoff is particularly important during intense rainfall events and rapid snowmelt.

The allocation of water to different components of the hydrologic cycle can be influenced greatly by vegetation. Precipitation falling on a basin may be retained in storage, e.g.,on plant surfaces, in plant tissue, in soil and rock, and in surface water systems; it ultimately departs the basin as vapor or liquid. Above the soil surface, foliage and stems intercept and detain a portion of the precipitation, which consequently may be evaporated or redistributed in space (West and Gifford 1976). At the soil surface, living vegetation and litter attenuate raindrop impact energies, which promotes infiltration and restricts surface runoff (Thompson and James 1985). Decomposition of plant litter and roots provides additional organic matter that binds soil particles and promotes water absorption and storage by improved soil structure (Oades 1984).

Water may be lost from basins through evaporation from free water surfaces, the mineral soil surface, or microphytic crusts. However, vascular plant transpiration represents a major pathway of water loss to the atmosphere. The partitioning of water loss to evaporation versus transpiration and the total amount of water lost to the atmosphere in different sites is dependent on many factors of soil, local climate, and the vegetation itself. Great Basin plant root characteristics such as rooting depth and root length density influence rates and patterns of soil moisture extraction (Caldwell 1985). Root growth patterns and phenology influence the competitive ability of plants for soil water (Harris 1977, Richards 1984, Caldwell and Richards 1986). Root system morphology of Great Basin plants and several aspects of root system production, physiology, function in water extraction and movement, and competitive interactions below ground are discussed subsequently.

Plant Root Systems of Great Basin Lowlands

Root System Morphology

Root characteristics of many species important in the Great Basin were described by Weaver (1919) during his investigations in the steppes of southern Idaho and the Columbia Plateau. He observed that, with few exceptions, the form of the root systems of nongraminaceous plants in those areas could be classified as generalized where the primary root (taproot) and laterals develop to approximately equal extent (Cannon 1911). Most dicots and gymnosperms of Great Basin lowlands also can be categorized as having generalized root systems. This contrasts with the Sonoran and other warm deserts that have many species with specialized root systems where either the taproot or lateral roots dominate (Cannon 1911).

Weaver (1919) made extensive excavations of roots of perennial grasses important in the Great Basin, e.g.,bluebunch wheatgrass (*Agropyron spicatum* [Syn: *Pseudoroegneria spicata*]) Idaho fescue, (*Festuca idahoensis*), Sandberg's bluegrass (*Poa sandbergii*), and Junegrass (*Koeleria cristata*), and found that their root systems exhibited two distinct depth-distribution patterns. The large tussock grass, bluebunch wheatgrass, had extensive, deep roots that penetrated at least 1.5 m, whereas the other three species had their entire root systems restricted to the upper 0.5 m of soil. In agreement with Weaver (1919), recent studies of the roots of the large perennial grasses, including both native and introduced species important in the Great Basin, e.g.,bluebunch wheatgrass, crested wheatgrass, streambank wheatgrass (*A. riparium* [Syn: *Elymus lanceolatus*]), Great Basin wildrye (*Elymus cinereus* [Syn: *Leymus cinereus*]), needle-and-thread grass (*Stipa comata*), and Indian ricegrass (*Oryzopsis hymenoides*), have found that the maximum rooting depth of these species is usually <2 m (Klepper et al. 1985; Anderson et al. 1987; Richards and Caldwell 1987; Wraith et al. 1987; Abbott 1989; Reynolds and Fraley 1989). These studies also show that the large perennial grasses nearly always have roots penetrating to 1.5 m. Similarly, recent investigations of the roots of smaller statured grasses, e.g.,bottlebrush squirreltail, indicate that their roots are restricted to the upper 1 m of soil and often to even shallower depths (Abbott 1989; Reynolds and Fraley 1989). The extent of exploitation of the soil profile by the exotic annual, cheatgrass (*Bromus tectorum*), is dependent on the size the plants achieve. In competitive situations their roots may penetrate only 15 cm (Hulbert 1955). Isolated plants and pure stands of this annual grass, however, exploit the soil to at least 1 m and can have roots as deep as 1.5 to 1.7 m (Hulbert 1955; Harris 1967; Harris and Wilson 1970; Cline et al. 1977).

In contrast to the grasses, Weaver (1919) found that all of the perennial forbs he examined [21 species total, including species of balsamroot (*Balsamorhiza*), mule's-ears (*Wyethia*), lupine (*Lupinus*), milk-vetch (*Astragalus*), buckwheat (*Eriogonum*), etc.] in southern Idaho and the Columbia Plateau, adjacent to the northern edge of the Great Basin, had deeply penetrating roots that reached depths over 1.5 m and commonly extended as deep as 2.4 to 3.0 m. Similar results, but with shallower maximum depths, were reported for four perennial forbs by Klepper et al. (1985), who conducted their studies at a drier site than that of Weaver. Even many annual forbs, both introduced and native, e.g.,bursage (*Ambrosia acanthicarpa*), bushy birdbeak (*Cordylanthus ramosus*) and Russian thistle (*Salsola kali*), often have roots penetrating at least 1.5 m (Klepper et al. 1985, Reynolds and Fraley 1989). Smaller statured annual forbs probably are rooted less deeply, e.g., prickly lettuce (*Lactuca serriola*) (Klepper et al. 1985).

Roots of the dominant shrubs of the Great Basin are branched highly in the upper soil layers, and fine root production in those layers is high resulting in root densities that, although lower than those achieved by grasses, are nevertheless quite high (Figure 7.8). The vertical distribution of the roots of big sagebrush in numerous habitats is skewed toward the surface 1 to 1.5 m of soil, with more than an order of magnitude decrease in root density in the deeper horizons (Branson et al. 1976; Sturges 1977; Sturges and Trlica 1978; Wallace et al. 1980; Richards and Caldwell 1987). Although only a few roots may extend to the maximum rooting depth, these roots may be very important for several aspects of plant function, including survival, escape from competition with the perennial grasses whose roots are less deep (Figure 7.8), and supply of water during the summer when transpirational demand is high (Sturges 1977). Uptake efficiency of these roots is improved by the phenomenon of hydraulic lift (Richards and Caldwell 1987; Caldwell and Richards 1989).

The maximum rooting depth of big sagebrush is usually 2 to 3 m (Tabler 1964; Sturges 1977; Sturges and Trlica 1978; Tisdale and Hironaka 1981; Richards and Caldwell 1987), although roots as deep as 9 m (Cannon 1960) or restricted to <1 m depth (Fernandez and Caldwell 1975) have

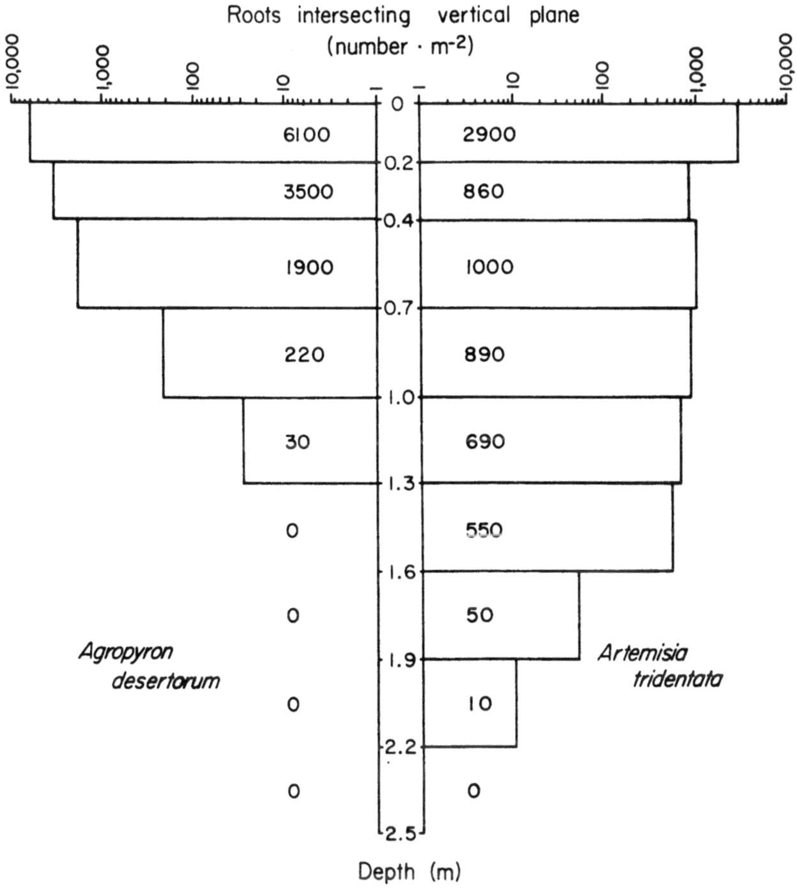

Figure 7.8. Relative root density of mountain big sagebrush (*Artemisia tridentata* ssp. *vaseyana*) and crested wheatgrass (*Agropyron desertorum*) with depth based on profile wall mapping at a site near Logan, UT. Shrub and grass main root axes (> ~0.3 mm diameter) intersecting a vertical plane were counted in 50 cm wide quadrats at each depth interval and averaged (n=6 for depths <1.0 m and n=3 for deeper depths). Sagebrush data from Richards and Caldwell (1987); grass data from Richards et al. unpublished. Total root length density, including all live fine roots,between plants of sagebrush and crested wheatgrass in these stands is ~12 cm cm^{-3} averaged over the 0 to 50 cm soil layer (Caldwell et al. unpublished).

been reported for this widespread, variable species. The several subspecies of big sagebrush appear to have different requirements and capacities for developing deep root systems (Shumar and Anderson 1986;

Barker and McKell 1983). At a site near Logan, Utah, evidence of roots of mountain big sagebrush (*Artemisia tridentata* ssp. *vaseyana*) were found at depths >3 m in excavations in 1986 when the deepest live roots extended only to 2.2 m (Richards and Caldwell 1987) (Figure 7.8). This excavation was done at the end of a series of years when precipitation was far above average, and observations at the site suggest that anaerobic conditions probably occurred temporarily in the deep soil layers. A nearby (~300 m) stream course that according to long-term local residents had been dry since the 1930s, ran from 1983-1986. Big sagebrush roots are known to be highly susceptible to anoxia (Lunt et al. 1973; Ganskopp 1986) and may have been killed at deeper depths during the period of high soil moisture. Presumably, the sagebrush roots would reexplore the soil at depths >2.2 m during a period of dry years, as subsurface flow in sand and gravel lenses appears to assure recharge of those deep horizons at the site in northern Utah. These observations suggest that the maximum rooting depth of big sagebrush, and other Great Basin species susceptible to soil anoxia, is dynamic and responds to changing conditions of soil aeration, water table depth and water availability. Reexploration of deep soil layers where roots have been killed by anoxia (or severe drying in years of limited recharge) occurs in addition to the continual reexploration within the rooted zone that has been documented for Great Basin shrubs (Fernandez and Caldwell 1975; Caldwell and Camp 1974).

Most other important shrubs of the Great Basin also appear to have roots throughout the profile extending to considerable depths. This contrasts with the relative paucity of roots in the surface soil (<10 cm depth) and shallow maximum depth of roots of shrubs in warm deserts (Cannon 1911; Barbour 1973; Wallace et al. 1980). Important Great Basin shrubs such as Utah serviceberry (*Amelanchier utahensis*), Mormon tea (*Ephedra viridis*), one-seed juniper (*Juniperus monosperma*), Gambel oak (*Quercus gambelii*), and greasewood all have roots capable of penetrating as deep as 4 to 6 m, and occasionally deeper (Cannon 1960). Other deep-rooted (>2 m) Great Basin species include: fourwing saltbush (*Atriplex canescens*), spiny hopsage (*Atriplex spinosa*), bitterbrush (*Purshia tridentata*), and rabbitbrush (*Chrysothamnus nauseosus*) (Branson et al. 1976; Fairchild and Brotherson 1980; Klepper et al. 1985; Reynolds and Fraley 1989). Although many Great Basin shrubs are deep rooted, some smaller species [e.g., shadscale, winterfat (*Ceratoides lanata*), green rabbitbrush (*Chrysothamnus viscidifloris*)] are not rooted as deeply as the larger shrubs listed above (Fernandez and Caldwell 1975; Wallace et al. 1980; Klepper et al. 1985; Reynolds and

Fraley 1989). Also, in topographic or regional locations where recharge is limited to the shallow soil layers or where indurated, anoxic or saline horizons occur at shallow depths, even the deep-rooted species will have limited rooting depth (Cannon 1911; Barbour 1973; Fernandez and Caldwell 1975; Richards 1986).

As with root depth, the lateral extent of roots is highly dependent on site conditions. Plant density is also important in determining the horizontal distribution of roots (Atkinson et al. 1976). Roots of big sagebrush extend up to 1.5 m laterally from the plant, whereas roots of the perennial grasses and forbs that have been studied appear to be limited to <1 m lateral spread (Tabler 1964; Sturges 1977; Sturges and Trlica 1978; Abbott 1989; Reynolds and Fraley 1989). At the site in northern Utah mentioned above, the interspaces between neighboring mountain big sagebrush and two tussock wheatgrass species were filled thoroughly with roots of both species and there was no detectable decrease in root density of either grass or shrub roots at a distance away from the respective plant (Caldwell and Richards 1986). Individual roots of both the shrub and the grasses were observed to pass directly under neighboring plants. Such intimate intertmingling of roots of neighboring plants results in intense belowground competition for resources, even in the plant interspaces where there may be no canopy cover.

Vast areas of the Great Basin are dominated by deep-rooted shrubs, mixed with grasses and forbs on well drained soils. If a general pattern is present, it is that the shrubs and perennial forbs have roots accessing the deepest soil horizons, escaping competition to some extent with the dense, and moderately deep roots of the large perennial grasses. Small perennial grasses, annual grasses, and forbs are restricted to shallower soil layers.

Root System Turnover, Production, and Costs

Great Basin shrubs invest a large proportion of their biomass in root systems — particularly in the fine, diffuse root mass. If one forcefully lifts a sagebrush shrub by its stem, the apparent root system attached to the main stem is the tap root and a few major branches. But, this is less than 5% of the belowground biomass. A careful excavation and recovery of the entire root system reveals that the bulk of the root mass of sagebrush is a fine, diffuse network of roots.

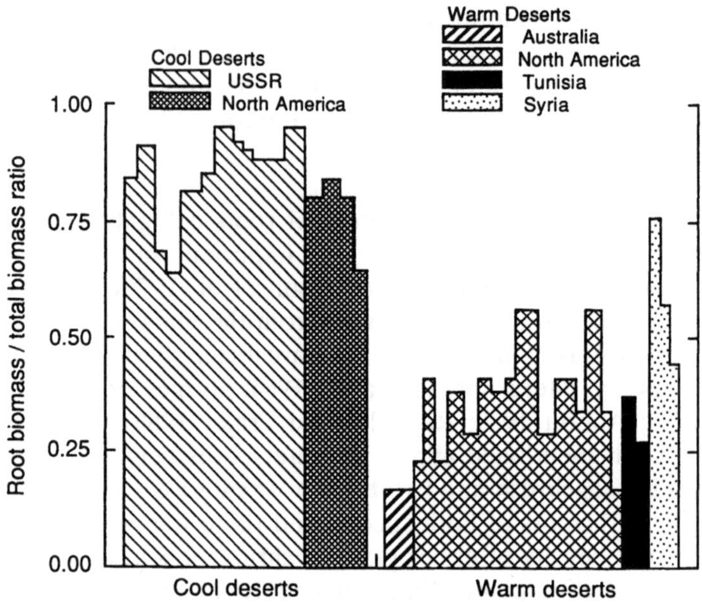

Figure 7.9. The ratio of root to total-plant biomass in cool- and warm-desert communities compiled largely from unpublished reports of the International Biological Programme for various continents. The Asian data were taken from Rodin and Basilevich (1967).

Intuitively, one expects a large proportion of the plant biomass to be roots in water-limited environments. Yet, the root/total-plant biomass ratios in most desert environments often do not exceed 50%, and usually are much less (Barbour 1973). Although very few species have been examined intensively, the proportion of biomass below ground in Great Basin shrubs is quite high — sometimes approaching 90% (Caldwell and Fernandez 1975; Caldwell et al. 1977). For perspective, the absolute mass of root system carbon of a shadscale stand is about the same as that of a mesic deciduous forest, although the amount of carbon in the shadscale stand shoot system is only 2% of that in the aboveground biomass of the forest (Caldwell 1985). This high proportion of root mass in Great Basin plants seems a bit curious in view of the lower proportion of root mass in warm desert plant systems. One might expect the opposite trend in that the ratio of precipitation to potential evaporation is usually much lower in true deserts than in the Great Basin. Certainly, too few species have been examined for robust generalization, but a synoptic view of root mass proportions for perennial plants in warm and cool deserts does suggest a pattern. The ratios of root/total biomass in cool, shrub-steppe regions of the Great Basin and the Soviet Union appear to be greater than those reported in warm deserts of North America, Australia and the Middle East (Figure 7.9). Why this should be is open to speculation. In the Great Basin, soil moisture recharge usually occurs once during the year, in the winter to early spring. During the long, dry summers, water is extracted progressively from shallow to deeper soil layers and maintenance of a fine network of roots over a depth of a meter or two may be of advantage in this sequential extraction. In warm desert environments where soil moisture extraction and recharge tend to be interspersed during the year, a less extensive network of roots located mainly in the shallower layers may be adequate to harvest water from short-term recharge events. Furthermore, the maintenance respiration costs of root systems in the warm soils of deserts should be much greater than in the relatively cool soils of the Great Basin. Thus, maintenance of a large root biomass may be energetically too costly in warm deserts.

The energetic costs of the root biomass in Great Basin plants are high owing to the large proportion of the plant mass belowground. The two major components of these costs are (1) the annual renewal of the diffuse root system, i.e., replacing the fraction of roots that die during the course of the year, and (2) the maintenance respiration of the entire root system.

The annual renewal of shrub root systems can represent 75% of the total-plant biomass production. Data on root system turnover and carbon balance of Great Basin plants are scant, but examples exist for two perennial shrub species, shadscale and winterfat (Caldwell et al. 1977). In stable stands of these shrubs, new root growth represents replacement of roots rather than exploration of new regions of the soil (but see pp. 260-262). For nearly monospecific stands of these two shrubs, 78 and 76% of the annual biomass produced was belowground for winterfat and shadscale, respectively. Most of this belowground production was fine root system; tap root growth was only 1 to 2% of the annual investment in root production.

When plants are pruned or grazed, they tend to curtail root production so that a suitable root/shoot biomass relationship is reestablished (Leopold and Kriedemann 1975). Reduced root mass and depth of rooting have been reported for rangeland species when grazed heavily (Hanson and Stoddart 1940; Schuster 1964). Although a reduction of

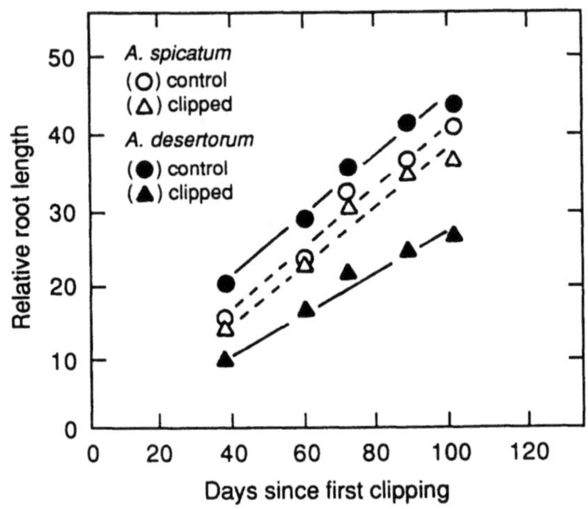

Figure 7.10. Relative increase in root length of bluebunch (*Agropyron spicatum* [Syn: *Pseudoroegneria spicata*]) and crested wheatgrasses (*Agropyron desertorum*) in the first 100 days following two successive severe defoliations. Data for nondefoliated control plants are also shown. Day 0 was the first defoliation and day 13 was the second. The lines are weighted least squares fits. Redrawn from Richards (1984).

root mass may well reduce root system costs for the plant, it also can compromise its ability to acquire soil resources and to compete with neighboring plants. Whether the diminished root growth following grazing is simply a passive phenomenon resulting from foliage loss and, thus, reduced carbon gain by the plant or a sensitive adjustment that takes place before the plant is depleted of energy, appears to differ among Great Basin species. Richards (1984) compared the root growth response of bluebunch wheatgrass and crested wheatgrass following two severe defoliations (85% foliage removal) in a field experiment. Crested wheatgrass generally is considered to be much more tolerant of grazing than bluebunch wheatgrass (Caldwell et al. 1981; Richards and Caldwell 1985). Root growth was observed with a root periscope system beginning some 40 days after the first defoliation. Although both species significantly reduced root growth and respiration within 24 to 48 h after the shock of a severe reduction in canopy leaf area (Richards and Caldwell 1985; Thorgeirsson 1988), the species differed considerably in their long-term response. Over a three-month period, crested wheatgrass curtailed root growth considerably following defoliation while bluebunch wheatgrass continued to exhibit root growth at about

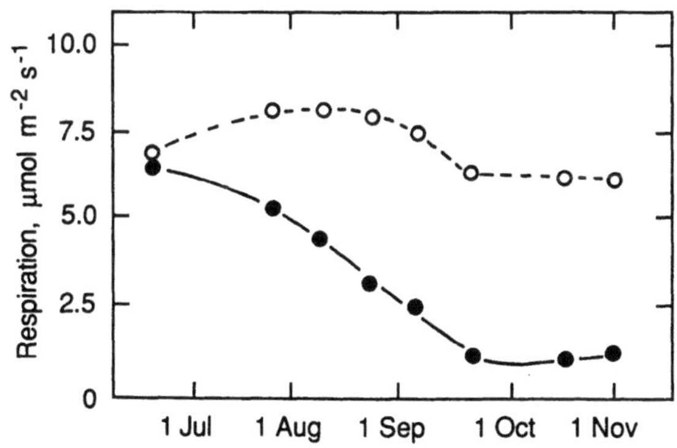

Figure 7.11. Potential root system respiration for a stand of shadscale (*Atriplex confertifolia*) expressed on a unit ground area basis. The solid line represents values derived by accounting for the seasonal metabolic adjustments of root respiration and the temperature dependencies of respiration at different depths in the profile. The dotted line represents values which account for the temperature dependencies of respiration, but does not account for the seasonal metabolic curtailment in respiratory activity in the late summer and autumn. Redrawn from Holthausen and Caldwell (1980).

the same rate as control plants that had not experienced the defoliations (Figure 7.10). This sensitive adjustment of root growth following foliage removal is considered to be of advantage for crested wheatgrass in diverting energy to rapid foliage growth (Richards 1984). This, in turn, contributes to a more effective recovery from defoliation and subsequent root growth for crested wheatgrass. In the winter following these defoliation experiments, bluebunch wheatgrass tussocks that had been defoliated the previous spring suffered a greater decline in root length compared to control plants than did defoliated crested wheatgrass plants compared with their control plant counterparts (Richards 1984). Thus, the energy savings resulting from root growth suppression, appear to be of great advantage in the recovery from defoliation and ultimately in future root production.

In addition to the costs of renewing a portion of a perennial root system each year, root maintenance respiration stands as a major cost of large root systems. Since Great Basin soils are cool, or cold, much of the year, maintenance respiration costs are much lower than they might be in the soils of warm deserts, as was mentioned earlier. Apart from this, there is evidence that roots may adjust their basal maintenance costs under field conditions to effect further energy savings. Roots of shadscale were found to lower their respiration activity considerably during the late summer and autumn independent of soil temperature (Holthausen and Caldwell 1980). Individual roots of this shrub were excavated in the field and their respiration rates were determined at constant assay temperatures. The roots used in these assays were not actively growing and thus the values were thought to reflect primarily maintenance respiration, rather than growth respiration. There was some apparent shift in respiratory capacity with depth in the profile that was roughly parallel to the shifting growth activity with depth. However, the most pronounced shift in respiration capacity was a seasonal shift resulting in extremely low rates in the early fall. An estimated savings in respiratory energy expenditure associated with this seasonal shift of respiratory capacity is represented in Figure 7.11.

Although respiration and annual root system renewal represent the most prominent costs of root systems in the Great Basin, other carbon costs associated with phenomena such as root tissue sloughing and carbon exudation may not be trivial (Caldwell 1979). Furthermore, biotic associations such as mycorrhizae or belowground herbivory also may contribute to these energy expenditures. As yet, however, the magnitude of these other costs are poorly known, especially for Great Basin plants.

Root Phenology

Root growth is initiated by established plants of both native and introduced perennial wheatgrasses in the fall when rain or snow moistens the soil. At the site in the northern Great Basin near Logan, Utah, crested wheatgrass initiated new adventitious roots very quickly following the dry season as precipitation occurred in October and had produced over 3.5 cm of new adventitious root per tiller by the time the native species, bluebunch wheatgrass, initiated (<0.25 cm/tiller) new adventitious root production (Thorgeirsson 1985). This advantage in new root initiation was maintained by crested wheatgrass into the following spring. New root initiation peaked in early April for crested wheatgrass, but not until late April or early May for bluebunch wheatgrass. Earlier root initiation by crested wheatgrass in the autumn and early spring correlates with greater foliage area per tiller, but not number of leaves or biomass per tiller, during those cool periods (Nowak and Caldwell 1984b; Eissenstat and Caldwell 1987). Root growth of these plants may be enhanced during brief periods of warmer temperatures in the cool seasons partly because some photosynthate can be produced by overwintering basal green leaves (Nowak and Caldwell 1984b).

Winter root growth into disturbed soil patches at the same site in the northern Great Basin was greater for crested wheatgrass than for bluebunch wheatgrass, but the benefits of this root growth were not detected immediately by uptake of radioactive phosphorus placed in the disturbed soils (Eissenstat and Caldwell 1989). These authors discussed the difficulty of extrapolating results showing greater root growth at low soil temperatures (Harris and Wilson 1970; Eissenstat and Caldwell 1987) to an immediate advantage in resource acquisition under field conditions. Nevertheless, crested wheatgrass appears to have greater ability than the native bluebunch wheatgrass to grow both new adventitious root axes and lateral roots from preexisting axes during the cool conditions of autumn, winter and early spring. Destructive harvests of plants on the same field site indicated that total root biomass and nitrogen acquisition of crested wheatgrass increased one month earlier in the spring than it did for bluebunch wheatgrass (Caldwell et al. 1981). Early root growth may be important not only for acquisition of some resources, such as nitrogen, that are only available temporarily, but also may result in the placement of roots in locations or horizons of the soil where resources remain available longer as resource supply becomes limiting later in the season.

Under cold conditions an advantage similar to that of crested wheatgrass is likely to be important for the success of introduced annual grasses like cheatgrass and medusahead (*Taeniatherum asperum*), seedlings of which are capable of root growth at soil temperatures as low as 2 to 3° C (Harris 1967, 1977; Harris and Wilson 1970). In the field studies by Harris (1967, 1977), much of the root growth of seedlings of these species occurred during late autumn and winter, placing roots deep in the profile before substantial canopy development and water use occurred in May. Here again, the benefit of root growth under cool conditions may not result in immediate resource gain, but this early growth places roots where they can acquire resources, especially water, when resource availability becomes limiting later in the season.

Root growth duration is greater than that of aboveground growth in perennial Great Basin plants. Root growth continued for several months into, or through, the dry summer period after shoot vegetative

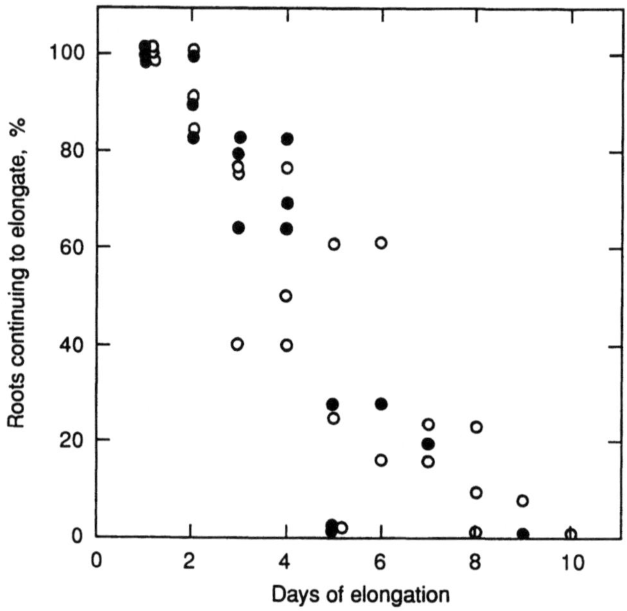

Figure 7.12. Duration of individual lateral root elongation for crested wheatgrass (*Agropyron desertorum*, solid symbols) and bluebunch wheatgrass (*A. spicatum* [Syn: *Pseudoroegneria spicata*], open symbols). Roots were observed through three 60 x 200 cm glass panels facing monocultures of each species growing at a field site in the northern Great Basin near Logan, Utah. Roots from all depths are included. Redrawn from Thorgeirsson (1985).

and reproductive development were complete in bluebunch and crested wheatgrass, shadscale, winterfat, and sagebrush (Fernandez and Caldwell 1975; Thorgeirsson 1985). The period of elongation for individual lateral roots of all five species listed above is limited to < three-weeks (Fernandez and Caldwell 1975; Thorgeirsson 1985) (Figure 7.12). Main axes continue growth much longer, but quickly pass through any particular soil layer. Thus, new laterals must be initiated continually for root growth to continue in any zone. This continual reexploration of the soil would contribute to water uptake in drying soils (Gardner 1960; Cowan 1965; Caldwell 1976). Main root axes of both of these growth forms are functional, at least as conducting elements, for a minimum of two years, and in the case of the shrubs, probably for substantially longer. Root hairs were persistent for over one year in shadscale and remained even on suberized portions of the roots in the three shrubs studied by Fernandez and Caldwell (1975).

Water Uptake and Transport

Winter and early spring precipitation typically recharge the upper soil layers in the Great Basin. Soil moisture is depleted in a nearly monotonic pattern both seasonally and with increasing depth in the profile. The depth of recharge limits the rooting depth of the perennial species and those plants can extract water from the complete profile (Campbell and Harris 1977; Cline et al. 1977; Sturges 1977, 1979, 1980; Wraith et al. 1987; Anderson et al. 1987). In some topographic locations, however, lateral flow from off site or increased infiltration due to snow or runoff can recharge soils to depths greater than would be expected due to average recharge alone (Sturges 1979, Klepper et al. 1985, Reynolds and Fraley 1989). In these locations, herbaceous perennial or annual vegetation may not extract all of the moisture that would be utilized by shrub-dominated vegetation (Cline et al. 1977; Sturges 1979).

Fernandez and Caldwell (1975) examined root growth rates of shadscale, winterfat, and sagebrush throughout the growing season at a site in Curlew Valley, Utah. They found that growth rates of individual rootlets followed a wave-like pattern, such that as the upper soil layers dried and root growth declined, root growth in the deeper soil layers increased. Thus, there was nearly constant activity throughout the growing season, but the zone of most pronounced root growth occurred at different depths in the soil profile. Patterns of water extraction by these and other deep-rooted shrubs parallel this pattern of root growth (Campbell and Harris 1977; Caldwell et al. 1977; Black and Mack 1986; Anderson et al. 1987; Abbott 1989; Romo and Haferkamp 1989). Similar patterns of root growth were documented for two perennial grasses by Thorgeirsson (1985) and can be inferred from water extraction patterns for a number of other grass species of the Great Basin (Wraith et al. 1987; Anderson et al. 1987; Abbott 1989).

Although the root systems of the dominant perennial plants of the Great Basin commonly extend to great depths, the bulk of their roots are found in the upper meter or two of soil. The timing of root activity and associated resource use in these shallow soil layers varies among species and seems to be an important determinant of competitive ability in the predictable spring growing season of the Great Basin (Thorgeirsson 1985; Eissenstat and Caldwell 1987). For example, both sagebrush and crested wheatgrass extract water earlier, especially from depths >60 cm, than does the less competitive bluebunch wheatgrass (Thorgeirsson 1985).

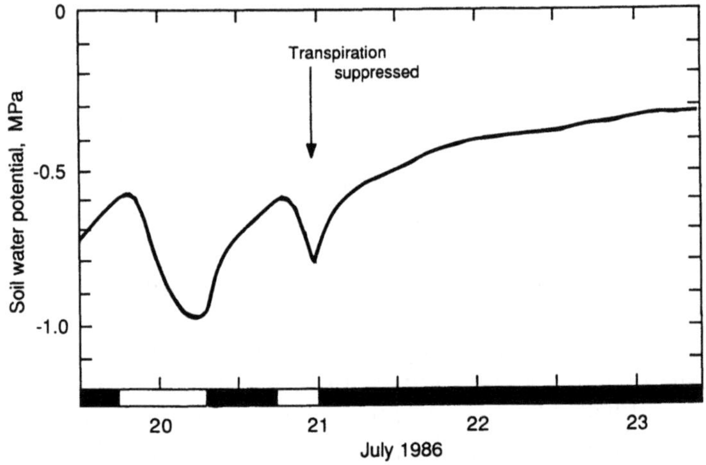

Figure 7.13. Course of soil water potential at 80 cm depth determined from hourly measurements with soil thermocouple psychrometers beneath mountain big sagebrush (*Artemisia tridentata* ssp. *vaseyana*) plants shows the recharge of soil water overnight (short dark bars on abscissa) due to hydraulic lift. When the sagebrush canopies were covered to suppress transpiration, and plant water potential rose to predawn values, recharge of the shallow soil began immediately and continued progressively as long as the plant remained covered. Parallel patterns of soil water potential were observed at 35 and 50 cm depths. Data are from Richards et al. unpublished; see also Richards and Caldwell (1987).

Following depletion of the majority of the water in the shallow soil, the deep-rooted species, such as sagebrush, are able to maintain photosynthetic activity and transpiration, albeit at reduced levels, throughout much of the dry portion of the summer and early autumn (Moore et al. 1972; Depuit and Caldwell 1973; Caldwell et al. 1977; Caldwell et al. 1981; Black and Mack 1986; Anderson et al. 1987). During these periods, transpirational demands are high, yet root length density is low in the deep, moist soil horizons (Klepper et al. 1985; Caldwell and Richards 1986; Richards and Caldwell 1987) (Figure 7.8) and, thus, water uptake capacity is limited. Also, water absorbed by deep roots must be transported through an ever increasing depth of dry soil. At night and other times of low transpiration, when plant water potential rises, the water potential in the roots in the shallow dry soil layers then will be greater than the water potential in the shallow soil (Campbell and Harris 1977; Cline et al. 1977; Caldwell et al. 1981; Shumar and Anderson 1986; Romo and Haferkamp 1989).

Under these conditions, which may persist for several months each summer in the Great Basin, water will tend to move from the roots to the dry soil. If water is lost temporarily into the shallow, dry soil layers at night, it will be resorbed the next day and utilized for transpiration. The phenomenon of water absorption by deep roots, transport in the root system, and leakage from the roots into drier shallow soil, followed by resorption the next day is called hydraulic lift and has been documented in the field for roots of sagebrush (Richards and Caldwell 1987; Caldwell and Richards 1989) (Figure 7.13). Some evidence for hydraulic lift also has been observed for the perennial wheatgrasses (Williams et al. unpublished), for sand sagebrush (*Artemisia filifolia*), and creosote bush (*Larrea tridentata*) (Richards unpublished). Water lost into the shallow soil horizons may not be used exclusively by the plant that absorbed and transported it. Competitors with intermingled roots may also access some of the water deposited in the shallow layers, but the quantitative significance of this phenomenon remains to be investigated (Caldwell and Richards 1989).

Although water leakage from roots would appear to be costly for plants, there are several advantages that aridland species may gain from this phenomenon. Hydraulic lift improves the effectiveness of water uptake by sparse deep roots by allowing them to function day and night rather than only during the day (Caldwell and Richards 1989). In field experiments where hydraulic lift was prevented experimentally by nighttime illumination of sagebrush plants, the maximum rate and total daily transpiration the following day was reduced by an average of 34% (Figure 7.14). Study of the water relations of sagebrush indicates that it maintains a surprisingly constant root hydraulic conductance as the soil dries (Black and Mack 1986). Without hydraulic lift this would be unlikely to occur, because of the exponentially declining root density at increasing depth (Richards and Caldwell 1987, Caldwell and Richards 1989) (Figure 7.8).

A second potentially important consequence of the leakage of water from roots is that it maintains a moist local environment for the roots and associated rhizosphere organisms (Richards and Caldwell 1987; Caldwell et al. 1989). This may prolong root life and the activity of mycorrhizae, nitrogen-fixing organisms and other rhizosphere microbes in dry soil. Water lost into the rhizosphere and soil around roots should allow both continued mineralization and nutrient uptake from shallow, relatively nutrient-rich, but dry, soil layers (Nambiar 1976, 1977). Nightly rehydration and daily dehydration of the bulk of roots in the

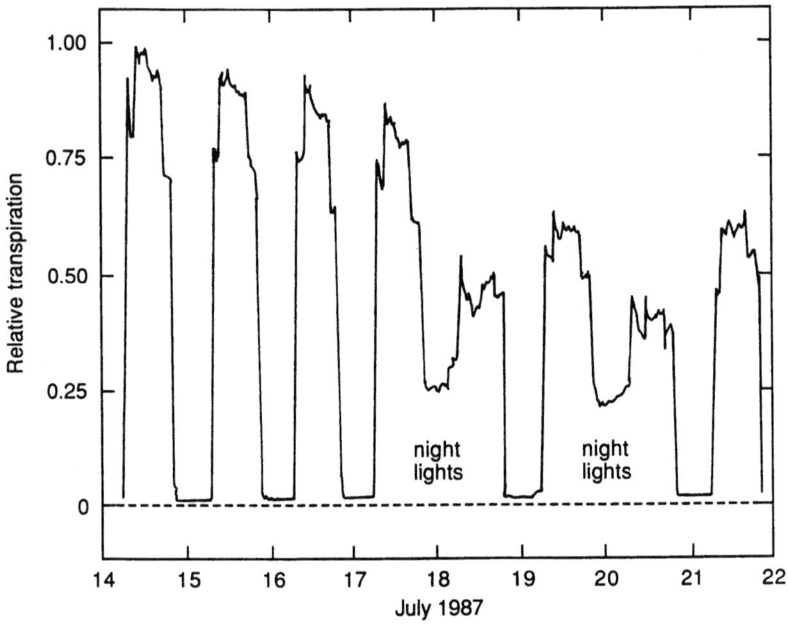

Figure 7.14. Transpiration of a mountain big sagebrush (*Artemisia tridentata* ssp *vaseyana*) shrub in a climatized whole-plant cuvette that had the same conditions of high evaporative demand and high light during each day (maximum air temperature = 34° C, maximum vapor pressure deficit = 47 mb, 1220 μmole m^{-2} s^{-1} photosynthetic photon flux density). Conditions in the cuvette were cool (12° C) at night, even when lamps illuminated the plant in the cuvette. Soil water potential measurements showed that soil water recharge by hydraulic lift did not occur on the nights of illumination. Redrawn from Caldwell and Richards (1989).

shallow, drying soil layers also may be important for transport of root signals of soil water status, such as abscisic acid, to the shoot.

The root systems of many Great Basin species appear well suited for the performance of hydraulic lift. For hydraulic lift to occur, active roots in deep moist soil layers must be connected anatomically to active fine roots in shallower dry soil layers. If fine roots in the shallow layers have been killed or shed because of inherent phenological development, water stress, or high temperatures, only well suberized main root axes will remain and little water loss is likely to occur. The water-stress-tolerant shrubs and grasses of the Great Basin are capable of maintaining live lateral roots and root hairs in the shallow soil (<1.5 m) down to water potentials as low as -3.5 to -4.0 MPa (Fernandez and Caldwell 1975; Thorgeirsson 1985). In addition to these plant characteristics, the soil profile must be dried sequentially from the surface downward so that roots pass through shallow drier soil layers. Deep soil recharge by winter-spring precipitation followed by summer drought, typical of the Great Basin, meets this criteria and would be expected to promote the existence of hydraulic lift. Summer precipitation, typical of warm deserts and arid grasslands, usually does not recharge deep soil layers as is necessary for vertical hydraulic lift. Also, summer precipitation events that increase the water potential of the shallow soil layers would prevent hydraulic lift. Based on these considerations, hydraulic lift may be of widespread importance for productivity, and water relations of many Great Basin species, particularly the water-stress-tolerant, deep-rooted species. Hydraulic lift also may have important implications for nutrient cycling and hydrology in the basins in the Great Basin region (Richards and Caldwell 1987; Caldwell et al. 1989).

Susceptibility of Roots to Anoxia

Roots of Great Basin shrubs are remarkable for their ability to grow in dry soil. Roots of shadscale have been reported to continue extension at soil water potentials near -7.0 MPa (Fernandez and Caldwell 1975). At night these water potentials may be ameliorated somewhat by hydraulic lift; however, growth in such dry soil is still impressive. Much emphasis has been placed on the ability of aridland species to survive and acquire water from dry soils. On the other hand, the growth and function of Great Basin and other desert species in flooded or continuously wet soil has received much less attention.

Roots of big sagebrush of the Great Basin and creosote bush of the Mojave Desert have been reported to have very high oxygen requirements, at least when compared to crop species (Lunt et al. 1973). Their susceptibility to flooding was attributed to this high oxygen requirement. Ganskopp (1986) reported that big sagebrush and green rabbitbrush quickly succumbed to experimental flooding and elevation of water tables to 10-cm depth. Greasewood was more tolerant of this treatment, but after 40 days it had also succumbed.

Yet, several Great Basin shrub species inhabit valley floor locations and are occasionally faced with prolonged periods of flooding or high water table. Fine-textured soils in these locations contribute to poor aeration.

Groeneveld and Crowley (1988) investigated the survival and root characteristics under flooding for three valley floor species, Torrey saltbush (*Atriplex torreyi*), rabbitbrush (*Chrysothamnus nauseosus* ssp. *viridulus*), and greasewood, of the western Great Basin and northern Mojave. Although greasewood is very tolerant of shallow water table conditions, it could not survive a 6-month inundation. This was similar to the observations of Ganskopp (1986). In contrast, Torrey saltbush and rabbitbrush were able to survive long-term flooding and this was attributed to production of new sets of flooding-tolerant roots. These roots tended to be thick (1 to 3 mm diameter) primary unsuberized roots with well developed aerenchyma. In the field, root systems in flooded soils were restricted to the upper 30 cm of the soil and had less lateral extension than the root systems of well aerated sites. In flooded-container studies, Groeneveld and Crowley (1988) found that newly produced roots of all three species possessed aerenchyma. The superior tolerance of saltbush and rabbitbrush to flooding in the field study when compared to greasewood is surprising in that greasewood commonly is considered to be a phreatophyte and tolerant of high water tables, whereas saltbush and rabbitbrush are thought to have more xeric-adapted ancestry.

Large-scale shrub die-off in the Great Basin has been reported in the last few years; in Utah alone some half million hectares have been estimated to be affected. The species most involved are shadscale, sagebrush (both basin big sagebrush, *Artemisia tridentata* ssp. *tridentata*, and Wyoming big sagebrush, *A. tridentata* ssp. *wyomingensis*), budsage (*A. spinescens*), black sagebrush (*A. nova*), winterfat, greasewood, horsebrush (*Tetradymia spinosa*), spiny hopsage (*Grayia spinosa*), bitterbrush, and narrowleaf low rabbitbrush (Nelson et al. 1989). This die-off has been linked to the record-high precipitation in the period 1978 to 1986.

Decline and die-off in temperate and tropical forest vegetation is receiving much attention (Mueller-Dombois 1988). Assigning causes for die-off is difficult and, in the end, it is largely a matter of surmise and inference. This applies to the Great Basin shrub die-off phenomenon. Nelson et al. (1989) discuss several hypotheses regarding causes of shrub decline and die-off including overgrazing, drought, winter injury, insects and plant pathogens. Although the symptoms exhibited by the shrubs do not permit diagnosis of a single cause of the syndrome, Nelson et al. reasoned that anoxia could be a major contributing factor during this period of particularly high precipitation. Elevated water tables and prolonged soil waterlogging in the lower portions of basins and other depressions may have led to periods of excessive oxygen deprivation to which the roots could not accommodate either physiologically or by formation of more tolerant roots. Although the symptoms exhibited by the affected shrubs were consistent with this hypothesis, they could not easily be distinguished from symptoms caused by salinity or certain diseases. Furthermore, the high precipitation would likely cause changes in the normal soil salinity patterns so that shrubs might have been exposed to salinity not normally experienced in certain sites. Anoxia and/or salinity also may predispose the shrubs to pathogen attack that they might otherwise resist. Attempting to identify a single causative factor to explain all instances of shrub die-off in the Great Basin would be unduly difficult (Nelson et al. 1989). Several factors might interact to cause the symptoms and the primary factors responsible may well differ among sites and species, but circumstantial evidence suggests anoxia to be implicated heavily in the Great Basin die-off of the past few years.

Competition Among Plants Belowground

As in most ecosystems, belowground competition among Great Basin plants is pronounced (Caldwell 1987, Eissenstat and Caldwell 1988). Nevertheless, an understanding of the processes and the mechanisms of belowground competition is just beginning to unfold. Experiments that compare competitive effectiveness between species are useful in examining the process of competition as well as the characteristics of species that contribute to competitiveness. A series of experiments has compared competitive abilities of crested wheatgrass and bluebunch wheatgrass. Bluebunch wheatgrass is a widespread native of the Great Basin and crested wheatgrass is an exotic that has been sown widely in rangelands throughout the region because it is more grazing tolerant and relatively easy to establish where as little as 200 mm annual precipitation occurs.

Although these two tussock grass species are native to different continents, they are remarkably similar in many morphological and physiological traits including phenological patterns of the shoot system, seasonal course of shoot water potential, light and temperature dependencies of photosynthesis at different times of year, biomass allocation patterns and water-use and nitrogen-use efficiencies of photosynthesis (Caldwell et al. 1981; Caldwell et al. 1983; Nowak and Caldwell 1984a and b, 1986). Yet, they differ greatly in their ability to compete with each other and with sagebrush in the field and in experimental garden plots. For example, Wyoming big sagebrush indicator plants were transplanted into nearly monospecific neighborhoods of the two bunchgrass species in a rangeland setting in northern Utah (Eissenstat and Caldwell 1988). The sagebrush indicator plants had lower survival, growth, reproduction and late-season water potentials in the neighborhoods dominated by crested wheatgrass than in stands dominated by bluebunch wheatgrass. Since there was not much canopy overlap of neighboring plants, most of this interference was belowground. Similar responses were found in experimental field plots with mountain big sagebrush when interplanted with the two species of tussock grass (Richards et al. unpublished). An example of the difference in sagebrush canopy size when planted with neighboring tussock grass of these two species is shown in Plate 16. This difference in competitive ability also was reflected below ground in the amount of root length in these experimental plots (Richards et al. unpublished; Caldwell and Richards 1986). Sagebrush roots became more prevalent than grass roots both with depth and with advancing age of the experimental plots. However, in those plots where crested wheatgrass was planted, grass roots maintained their prominence more effectively in mixtures with sagebrush than they did in plots of bluebunch wheatgrass mixed with sagebrush.

Inferences from water uptake patterns. Circumstantial evidence suggests that competition for water might explain at least part of the below ground interference evidenced in the field studies described above. Monocultures of these grass species differed in the rate of water extraction from soils both in a rangeland setting (Eissenstat and Caldwell 1988) and in the experimental field plots described above (Thorgeirsson 1985). Crested wheatgrass stands extracted water at a more rapid rate than bluebunch wheatgrass in both situations (Figure 7.15). Furthermore, in the study of Thorgeirsson, the rate and extent to which monocultures of crested wheatgrass extracted moisture was about the same as that at which mixtures of sagebrush and crested wheatgrass or mixtures of sagebrush and bluebunch wheatgrass removed water. The density of all these plantings was the same. This suggests that sagebrush in mixtures with bluebunch wheatgrass was able to exploit the moisture not removed rapidly by bluebunch wheatgrass, whereas sagebrush in mixtures with crested wheatgrass did not accelerate moisture removal beyond that effected by monocultures of this grass.

Although these results strongly suggest competition for water and a difference in the ability of the two tussock grass species to compete with sagebrush for soil moisture, the conclusions still must be based on inference because such experiments do not demonstrate directly an

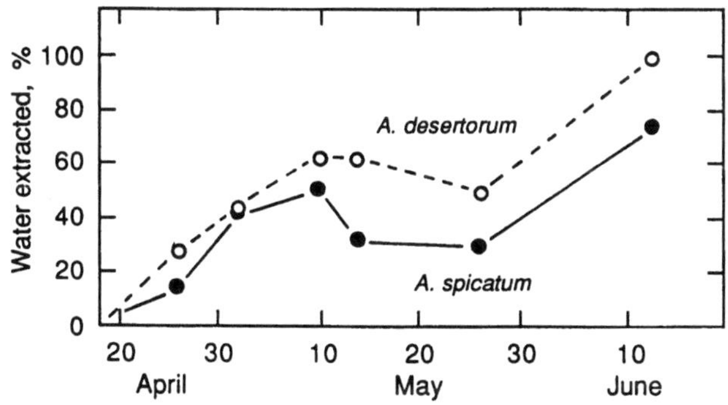

Figure 7.15. Soil water depletion by nearly monoculture stands of bluebunch (*Agropyron spicatum* [Syn: *Pseudoroegneria spicata*]) and crested wheatgrass (*Agropyron desertorum*) in a rangeland setting. These values are expressed as a percentage of total water extracted by crested wheatgrass during the time period 18 April to 13 June from 1 to 67 cm depth. The area between the two lines indicates average difference in amount of water extracted between 6 pairs of plots of the two grass species. Redrawn from Eissenstat and Caldwell (1988).

interspecific competition for the same moisture resource. In the absence of established perennial plants, the timing of root growth and water use has been implicated in the success of seedlings of cheatgrass and medusahead when competing with seedlings of both the native and introduced perennial grasses, bluebunch and crested wheatgrass (see pp. 269-270).

Competition for mineral nutrients: case studies with phosphorus. Use of isotopes as tracers allows more direct testing of resource competition among neighboring plants than is possible with the experiments such as those described above. Although plants likely compete for several soil resources in Great Basin environments, phosphate has been of particular interest for several reasons. Exchangeable phosphate is often very low in Great Basin soils, especially if the soils are calcareous (Caldwell et al. 1985). Unlike nitrate and many other more diffusible soil nutrient ions, phosphate is almost immobile in soils and is accessible only when it is within a few millimeters of a root or its associated mycorrhizae (Nye and Tinker 1977). Furthermore, there are two radioisotopes of phosphorus and when both are employed in experiments, much greater resolution of competitive interactions can be obtained than if only a single isotope tracer is used (Caldwell and Eissenstat 1987).

Dual-isotope experiments have shown that (1) there is an immediate competition for phosphate, (2) species differ greatly in their ability to compete for phosphate, (3) the physiological activity of the root system plays an important role in competition, and (4) the balance of competition can shift rapidly if a neighboring plant is put at a disadvantage such as by partial defoliation. These experiments were conducted in the field experimental plots mentioned earlier (Caldwell et al. 1985; Caldwell et al. 1987). Sagebrush was used as an indicator plant to test the competitive ability of neighboring tussock grass species in plots where sagebrush was interplanted with either bluebunch wheatgrass or crested wheatgrass, or with both wheatgrasses as neighbors. The indicator plants were given access to phosphorus radioisotopes by placing labelled orthophosphoric acid in the soil on opposite sides of the sagebrush. The labelled P was placed into the soil halfway between the indicator plant and a neighboring tussock grass using ^{32}P on one side of the sagebrush and ^{33}P on the other side. During the course of the experiment, the ratio of the two P isotopes was determined in tissues of the indicator plant.

There were striking differences in the ability of the two wheatgrass species to compete for phosphate. When the sagebrush plants were located between neighbors of the two grass species, the isotope ratios in the shrubs indicated that the ability of the shrubs to acquire P from the soil interspace shared with bluebunch wheatgrass was 6-fold greater than their ability to acquire P from the soil interspace shared with crested wheatgrass (Caldwell et al. 1985). Yet, the shrubs invested the same amount of biomass in roots and supported the same mycorrhizal infection rate in the interspaces shared with both species of grass. Actively absorbing mycorrhizal roots of the grass species were also present in these soil interspaces.

The immediate competition for phosphate also was indicated in the experiments in which the indicator shrubs were in plots with a single species of grass and one grass neighbor was defoliated partially at the time of labeling (Caldwell et al. 1987). Initially, the isotope ratios in the sagebrush indicated that the shrubs were acquiring equal amounts of P isotope from interspaces shared with intact and partially defoliated neighbors. However, within two weeks there was an opportunistic shift by the shrubs such that P isotope uptake from interspaces shared with the defoliated grasses increased to as much as six times that of uptake from interspaces shared with unperturbed neighbors (Figure 7.16). The rapid and sizeable shift in P uptake exhibited by the shrubs indicated an immediate competition for P. Because of the virtual immobility of P in

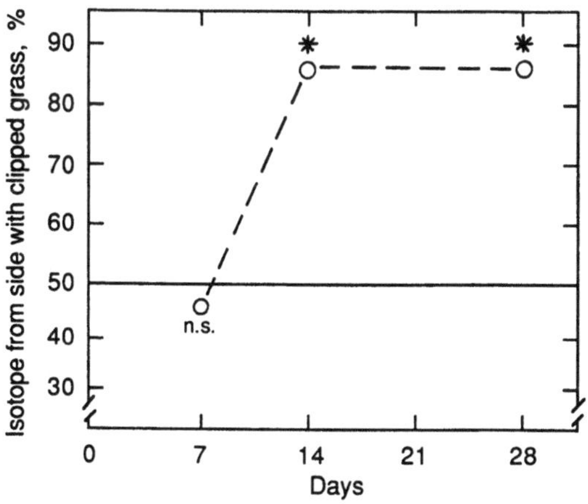

Figure 7.16. The proportion of P isotope acquired by mountain big sagebrush (*Artemisia tridentata* ssp. *vaseyana*) indicator plants from the soil interspace shared with the partially defoliated bluebunch wheatgrass (*Agropyron spicatum* [Syn: *Pseudoroegneria spicata*]) neighbor at different times following defoliation and P isotope labeling. Each point represents the isotope proportion in the preceding time interval. Statistical significance (<0.01) is indicated by an asterisk or as nonsignificant (ns). These data are from experiments of sagebrush planted with bluebunch wheatgrass. Similar results were obtained from analogous experiments with crested wheatgrass (*Agropyron desertorum*). Redrawn from Caldwell et al. (1987).

these soils, competition for P must be very localized and roots of neighboring plants (or their associated mycorrhizae) must be within millimeters for this immediate competition to occur. Microscale mapping showed roots of neighboring plants to be in this range of proximity.

The competition for P appears to be influenced substantially by the physiological activity of the roots rather than simply by general root morphology and distribution in the soil. Processes such as root phosphate absorption, continued fine-root maintenance and extension, and support of mycorrhizae all depend on a continuing energy supply to the roots and on plant demand for P. Partial defoliation can influence the energy status of the root system and also the demand for P. Apparently, P uptake of the grasses was curtailed soon following clipping of the grasses. Since the general distribution and physiological activity of the shrub roots would not be expected to change over this short time interval following clipping of the grasses, the increased P uptake by the shrubs was the result of less competitive pressure by the grass roots.

These experiments not only demonstrate the immediacy of resource competition, but also indicate how quickly the balance of competition might change when selective grazing takes place. Although these experiments involved only P as a resource, it would not be unreasonable to predict that competition for other soil resources operates in a similar manner.

Summary

The Great Basin is a collection of over 100 hydrologically closed and partially drained basins. High elevation precipitation captured by surrounding mountains as snow, provides the major source of water in the Great Basin. This water is transported downslope via surface runoff, interflow, streamflow or along a saturated groundwater flow path. Low-elevation evaporation from playas and closed basin lakes is the primary source of water loss. Valleys in the Great Basin are characterized by the degree of hydrologic closure. A basin is closed when no effective groundwater or surface water outlet exists. Surface or groundwater inflows distinguish a dry from a moist playa, which has significant consequences for sediment and salinity budgets, and for the type of plant community that develops.

Surface soils on playa and basin margins are encrusted often with nonvascular microphytes, salt, or vesicular porosity. These crusts often restrict water infiltration but also may restrict evaporative water loss. The type and extent of vegetation in the Great Basin are of primary importance in altering the timing and amount of runoff. Flooding, which results from rain on frozen or saturated soil, snowmelt runoff, and summer thunderstorms, occurs in most parts of the Great Basin.

Great Basin plants have moderately deep root systems, usually on the order of 3 m or less, with some species extending their roots to as much as 6 m. Although these roots systems may not be as deep as notable examples of roots in warm deserts, Great Basin plants have much greater root/shoot ratios than do species in warm deserts. Much of the root mass of Great Basin plants is in the form of fine, diffuse roots. Root system production represents a large proportion of total primary production in the Great Basin. Soil moisture extraction during the year tends to be rather monotonic starting with the major recharge of the soils in the early spring and progressing through the dry summer periods. Fall precipitation then starts the slow general recharge of the soil during the cold months of the year.

There is often intense competition for both soil moisture and nutrients in the Great Basin. Rapid root growth in the cold soils of early spring appears to be particularly important for the competitive advantage of plants. Other root system characteristics, including physiological traits also appear to be quite important in the intense competition for soil nutrients.

A significant water efflux from roots during the night in the upper, relatively dry soil layers has been demonstrated in the Great Basin, although this probably occurs in other arid areas as well. This phenomenon has been shown to improve substantially the effectiveness of water uptake by the sparse, deep root system when soil moisture is available only in the deep soil layers. This phenomenon has numerous implications for the hydrology of basins and for nutrient cycling during dry summer periods.

Although roots of Great Basin plants appear to be well adapted to the large annual fluctuations in soil temperature and the dry soil conditions of the summer months, they are very susceptible to prolonged wetness. Anoxia may be one of the primary reasons for restricted rooting depth in some areas and, during very long wet periods, for the mortality of some shrub species in the Great Basin.

Acknowledgements

Some of the information and concepts in this review resulted from research supported by the National Science Foundation (BSR 8705492) and the Utah Agricultural Experiment Station.

References

Abbott ML (1989) Root profiles of six cold desert plant species determined by radionuclide uptake. M.S. Thesis, Colorado State Univ., Ft. Collins, CO.

Anderson DC, Harper KT, and Holmgren RC (1982) Factors influencing the development of cryptogamic soil crusts in Utah Deserts. J. Range Manage. 35:180-185.

Anderson JE, Shumar ML, Toft NL, and Nowak RS (1987) Control of the soil water balance by sagebrush and three perennial grasses in the cold-desert environment. Arid Soil Res. Rehab. 1:229-224.

Atkinson D, Naylor D, and Coldrick GA (1976) The effect of tree spacing on the apple root system. Hortic. Res. 16:89-105.

Barbour MG (1973) Desert dogma reexamined: Root/shoot productivity and plant spacing. Amer. Mid. Nat. 89:41-57.

Barker JR and McKell CM (1983) Habitat differences between basin and Wyoming big sagebrush in contiguous populations. J. Range Manage. 36:450-454.

Black RA and Mack RN (1986) Mount St. Helens ash: Recreating its effects on the steppe environment and ecophysiology. Ecology 67:1289-1302.

Blackburn WH (1975) Factors influencing infiltration rates and sediment production of semi-arid rangelands in Nevada. Water Resources Res. 11:929-937.

Blackburn WH, Eckert RE Jr, Wood MK, and Peterson FF (1975) Influence of vesicular horizons on watershed management. Proceedings Watershed Management Symposium, ASCE Irrigation and Drainage Divison, Logan, UT.

Blackwelder E (1948) The Great Basin with emphasis on glacial and postglacial times. Bull. Univ. Utah 38:1-16.

Branson FA, Miller RF, and McQueen IS (1976) Moisture relationships in twelve northern desert shrub communities near Grand Junction, Colorado. Ecology 57:1104-1124.

Branson FA, Gifford GF, Renard KG, and Hadley RF (1981) Rangeland hydrology. Kendall/Hunt Publ, Dubuque, NM.

Brotherson JD and Rushforth SR (1983) Influence of cryptogamic crusts on moisture relationships of a soil in Navajo National Monument, Arizona. Great Basin Nat. 43:73-79.

Brown J (1969) Soils of the Okpilak River Region, Alaska. pp. 91-128. In Pewe TL (editor) The Periglacial Environment. McGill-Queens Univ. Press, Montreal.

Burkham DE (1988) Methods for delineating flood-prone areas in the Great Basin of Nevada and adjacent states. US Geological Survey Water Supply Paper 2316.

Caldwell MM (1976) Root extension and water absorption. pp. 63-85. In Lange OL, Kappen L, and Schulze ED (editors) Water and Plant Life. Ecological Studies Vol. 19, Springer-Verlag, Heidelberg.

Caldwell MM (1979) Root structure: the considerable cost of belowground function. pp. 408-432. In Solbrig OT (editor) Topics in Plant Population Biology. Columbia Univ. Press, NY.

Caldwell MM (1985) Cold desert. pp. 198-212. In Chabot B and Mooney H (editors) Physiological Ecology of North American Plant Communities. Chapman and Hall, London.

Caldwell MM (1987) Competition between root systems in natural communities. pp. 167-185. In Gregory PJ, Lake JV, and Rose DA (editors) Root Development and Function. Cambridge Univ. Press.

Caldwell MM and Camp LB (1974) Belowground productivity of two cool desert communities. Oecologia 17:123-30.

Caldwell MM and Eissenstat DM (1987) Coping with variability: examples of tracer use in root function studies. pp. 95-106. In Tenhunen JD, Catarino F, Lange OL, and Oechel WC (editors) Plant Response to Stress -- Functional Analysis in Mediterranean Ecosystems. Springer-Verlag, Heidelberg.

Caldwell MM and Fernandez OA (1975) Dynamics of Great Basin shrub root systems. pp. 38-51. In Hadley N (editor) Environmental Physiology of Desert Organisms. Dowden Hutchinson and Ross, Stroudsburg, PA.

Caldwell MM and Richards JH (1986) Competing root systems: morphology and models of absorption. pp. 251-273. In Givnish T (editor) On the Economy of Plant Form and Function. Cambridge Univ. Press, Cambridge.

Caldwell MM and Richards JH (1989) Hydraulic lift: Water efflux from upper roots improves effectiveness of water uptake by deep roots. Oecologia 79:1-5.

Caldwell MM, White RS, Moore RT, and Camp LB (1977) Carbon balance, productivity and water use of cold-water desert shrub communities dominated by C_3 and C_4 species. Oecologia 29:275-300.

Caldwell MM, Richards JH, Johnson DA, Nowak RS, and Dzurec RS (1981) Coping with herbivory: photosynthetic capacity and resource allocation in two semiarid *Agropyron* bunchgrasses. Oecologia 50:14-24.

Caldwell MM, Dean TJ, Nowak RS, Dzurec RS, and Richards JH (1983) Bunchgrass architecture, light interception, and water-use efficiency: assessment by fiber optic point quadrats and gas exchange. Oecologia 59:178-184.

Caldwell MM, Eissenstat DM, Richards JH, and Allen MF (1985) Competition for phosphorus: differential uptake from dual-isotope-labeled interspaces between shrub and grass. Science 228:384-386.

Caldwell MM, Richards JH, Manwaring JH, and Eissenstat DM (1987) Rapid shifts in phosphate acquisition show direct competition between neighboring plants. Nature 327:615-616.

Caldwell MM, Richards JH, and Beyschlag W (1989) Hydraulic lift: Ecological implications of water efflux from roots. In Atkinson D (editor) Plant Root Systems: Their Effect on Ecosystem Composition and Structure. Blackwell, Oxford (in press).

Campbell GS and Harris GA (1977) Water relations and water use patterns for *Artemisia tridentata* nutt. in wet and dry years. Ecology 58:652-659.

Cannon HL (1960) The development of botanical methods of prospecting for uranium on the Colorado Plateau, US Geological Survey Bulletin 1085-A, Washington, DC.

Cannon WA (1911) The Root Habits of Desert Plants. Carnegie Institute Washington Publication 131, Washington, DC.

Cary JW and Evans DD (1974) The influence of soil crusts on heat and water storage. pp. 45-55. In Cary JW and Evans DD (editors) Soil Crusts. Univ. Arizona Agr. Exp. Sta. Tech. Bull. 214, Tucson, AZ.

Chang TS (1969) A study of precipitation characteristics for Utah. M.S. Thesis, Utah State Univ., Logan, UT.

Cline JF, Uresk DW, and Rickard WH (1977). Comparison of soil water used by a sagebrush-bunchgrass and a cheatgrass community. J. Range Manage. 30:199-201.

Constantz J, Herkelrath WN, and Murphy F (1988) Air encapsulation during infiltration. Soil Sci. Soc. Amer. J. 52:10-16.

Cowan IR (1965) Transport of water in the soil-plant-atmosphere system. J. Appl. Ecol. 2:221-229.

Danin A (1978) Plant species diversity and plant succession in a sandy area in the Northern Negev. Flora 167:409-422.

DePuit EJ and Caldwell MM (1973) Seasonal pattern of net photosynthesis of *Artemisia tridentata*. Amer. J. Bot. 60:426-435.

Dortignac EJ (1956) Watershed resources and problems of the Upper Rio Grande Basin. US Forest Service Rocky Mountain Forest Range Station Miscellaneous Publication, Ft. Collins, CO.

Duffy CJ and Al-Hassan S (1988) Groundwater circulation in a closed desert basin: topographic scaling and climatic forcing. Water Resources Res. 24:1675-1678 .

Eakin TE, Price D, and Harrill JR (1976) Summary appraisals of the nation's ground-water resources-Great Basin Region. US Geological Survey Professional Paper 813-G, Washington, DC.

Eardley AJ, Gvosdetsky V, and Marsell RE (1957) Hydrology of Lake Bonneville and sediments and soils of its basin. Geol Soc. Amer. Bull. 68:1141-1202.

Eckert RE Jr, Peterson FF, and Belton JT (1986) Relation between ecological range condition and proportion of soil surface types. J. Range Manage. 39:409-414.

Eckert RE Jr., Wood MK, Blackburn WH, Peterson FF, Stephens JL, and Meurisse MS (1978) Effects of surface soil morphology on improvement and management of some arid and semi-arid rangelands. pp. 299-302. In Hyder DN (editor) Proceedings First Intnl. Rangeland Congress, Soc. Range Manage., Denver, CO.

Eckert RE Jr, Peterson FF, and Emmerich FL (1987) A study of factors influencing secondary succession in the sagebrush (*Artemisia* spp. L.) type. pp. 149-168. In Frazier GW and Evans RA (editors) Proceedings of Symposium on Seed and Seedbed of Rangeland Plants, USDA, Tucson, AZ.

Eissenstat DM and Caldwell MM (1987) Characteristics of successful competitors: an evaluation of potential growth rate in two cold desert tussock grasses. Oecologia 71:167-173.

Eissenstat DM and Caldwell MM (1988) Competitive ability is linked to rates of water extraction. A field study of two aridland tussock grasses. Oecologia 75:1-7.

Eissenstat DM and Caldwell MM (1989) Invasive root growth into disturbed soil of two tussock grasses that differ in competitive effectiveness. Functional Ecol. 3:345-353.

Evenari M, Yaalon DH, and Gutterman Y (1974) Note on soils with vesicular structure in deserts. Z. Geomorph. NF 16:162-172.

Fairchild JA and Brotherson JD (1980) Microhabitat relationships of six major shrubs in Navajo National Monument, Arizona. J. Range Manage. 33:150-156.

Fenneman NM (1931) Physiography of the Western United States. McGraw-Hill, NY.

Fernandez OA and Caldwell MM (1975) Phenology and dynamics of root growth of three cool semi-desert shrubs under field conditions. J. Ecol. 63:703-714.

Fletcher JE and Martin WP (1948) Some effects of algae and molds in the raincrust of desert soils. Ecology 29:95-100.

Friedmann EI and Galun M (1974) Desert algae, lichens, and fungi. pp. 165-212. In Brown GW (editor) Desert Biology Vol. 2. Academic Press, NY.

Ganskopp DC (1986) Tolerances of sagebrush, rabbitbrush and greasewood to elevated water tables. J. Range Manage 39:334-337.

Gardner WR (1960) Dynamic aspects of water availability to plants. Soil Sci 89:63-73.

Gifford GF (1972) Infiltration rate and sediment production on a plowed big sagebrush site. J. Range Manage 25:53-55.

Graf WL (1988) Fluvial processes in dryland rivers. Springer-Verlag, Heidelberg.

Grismer ME (1986) Pore size distributions and infiltration. Soil Sci. 141:249-260.

Groeneveld DP and Crowley DE (1988) Root system response to flooding in three desert shrub species. Funct. Ecol. 2:491-497.

Hacker RB (1986) Effects of grazing on chemical and physical properties of an earthy sand in the Western Australian mulga zone. Austr. Range J. 8:11-17.

Hanson WR and Stoddart LA (1940) Effects of grazing upon bunch wheat grass. J. Amer Soc. Agron 32:278-289.

Harper KT and Marble JT (1988) Effects of cryptogamic plants on water relations or Utah desert soils. In Tueller PT (editor) Vegetation Science Applications for Rangeland Analysis and Management. Kluwer, Dordrecht.

Harris GA (1967) Some competitive relationships between *Agropyron spicatum* and *Bromus tectorum*. Ecol. Monogr 37:89-111.

Harris GA (1977) Root phenology as a factor of competition among grass seedlings. J. Range Manage. 30:172-177.

Harris GA and Wilson AM (1970) Competition for moisture among seedlings of annual and perennial grasses as influenced by root elongation at low temperature. Ecology 51:530-534.

Holthausen RS and Caldwell MM (1980) Seasonal dynamics of root system respiration in *Atriplex confertifolia*. Plant Soil 55:307-317.

Houghton SG (1976) A Trace of Desert Waters: The Great Basin Story. Arthur H. Clark, Glendale.

Houghton JG (1969) Characteristics of rainfall in the Great Basin. Univ. Nevada Desert Res. Inst. Rep., Reno, NV.

Hugie VK and Passey HB (1964) Soil surface patterns of some semiarid soils in northern Utah, southern Idaho, and northeastern Nevada. Soil Sci. Soc. Amer. Proc. 28:786-792.

Hulbert LC (1955) Ecological studies of *Bromus tectorum* and other annual bromegrasses. Ecol. Monogr 25:181-213.

Jones RM (1969) Scald reclamation studies in the Hay District Part IV-Scale soils: Their properties and changes with reclamation. J. Soil Cons. Serv. NSW 25:104-120.

Klepper EL, Gano KA, and Cadwell LL (1985) Rooting depth and distribution of deep-rooted plants in the 200 area control zone of the Hanford Site, PNL-5247, Nat. Tech. Info. Serv., Springfield.

Larcher W (1975) Physiological Plant Ecology. Springer-Verlag, Heidelberg.

Leopold AC and Kriedemann PE (1975) Plant Growth and Development. Second Ed., McGraw-Hill, NY.

Lines, GC (1979) Hydrology and surface morphology of the Bonneville Salt Flats and Pilot Valley playa, Utah. US Geological Survey Water Supply Paper 2057, Washington, DC.

Loope WL and Gifford GF (1972) Influence of a soil microfloral crust on select properties of soils under pinyon-juniper in southeastern Utah. J. Soil Water Conserv. 27:164-167.

Lunt OR, Letey J, and Clark SB (1973) Oxygen requirements for root growth in three species of desert shrubs. Ecology 54:1356-1362.

Mabbutt J (1979) Desert Landforms. MIT Press, Cambridge, MA.

Marshall TJ and Holmes JW (1979) Soil Physics. Cambridge Univ. Press, Cambridge.

Maxey GB (1967) Hydrogeology of desert basins. Symp. Groundwater Devel. in Arid Basins, Utah State Univ., Logan, UT.

Miller DE (1971) Formation of vesicular structure in soil. Soil Sci. Soc. Amer. Proc. 35:635-737.

Miller DE and Gifford RO (1974) Modification of soil crust for plant growth. pp. 45-55. In Cary JW and Evans DD (editors) Soil Crusts. Univ. Arizona Agr. Exp. Sta. Tech. Bull. 214, Tucson, AZ.

Moore RE (1939) Water conduction from shallow water tables. Hilgardia 12:383-426.

Moore RT, White RS, and Caldwell MM (1972) Transpiration of *Atriplex confertifolia* and *Eurotia lanata* in relation to soil, plant, and atmospheric moisture stresses. Can. J. Bot. 50:2411-2418.

Morrison RB (1965) Quaternary geology of the Great Basin. pp. 265-285. In HE Wright and DG Frey (editors) The Quaternary of the United States, Princeton Univ. Press, Princeton, NJ.

Morrison RB (1968) Pluvial lakes. pp. 873-883. In Fairbridge RW (editor) The Encyclopedia of Geomorphology. Van Nostrand-Reinhold, NY.

Motts WS (1965) Hydrologic types of playas and closed valleys and relation of hydrology to playa geology, geology, mineralogy, and hydrology of U.S. playas. pp. 73-104. In Air Force Cambridge Res. Lab Environ. Res. Paper 96.

Motts WS (1970) Geology and hydrology of selected playas in western United States. Air Force Cambridge Res. Lab, Office at Aerospace Res., US Air Force, Bedford.

Mueller-Dombois D (1988) Forest decline and dieback - A global ecological problem. Trends Ecol. Evol. 3:310-312.

Nambiar EKS (1976) Uptake of Zn^{65} from dry soil by plants. Plant Soil 44:267-271.

Nambiar EKS (1977) The effects of water content of the topsoil on micronutrient availability and uptake in a siliceous sandy soil. Plant Soil 46:175-183.

Neal JT (1975a) Playas and dried lakes, occurrence and development. Benchmark Papers in Geology, Halsted Press, NY.

Neal JT (1975b) Playas and Dried Lakes. Dowden, Hutchison and Ross, Stroudsburg, PA.

Nelson DL, Harper KT, Boyer KC, Weber DJ, Haws BA, and Marble JR (1989) Wildland shrub dieoffs in Utah: an approach to understanding the cause. pp. 119-135. In Proceedings--Symposium on Shrub Ecophysiology and Biotechnology. USDA For. Serv. Gen. Tech. Rep. INT-256.

Nowak RS and Caldwell MM (1984a) A test of compensatory photosynthesis in the field: implications for herbivory tolerance. Oecologia 61:311-318.

Nowak RS and Caldwell MM (1984b) Photosynthetic activity and survival of foliage during winter for two bunchgrass species in a cold-winter steppe environment. Photosynthetica 18:192-200.

Nowak RS and Caldwell MM (1986) Photosynthetic characteristics of crested wheatgrass and bluebunch wheatgrass. J. Range Manage. 39:443-450.

Nye PH and Tinker PB (1977) Solute Movement in the Soil-Root System. Univ. of California Press, Berkeley.

Oades JM (1984) Soil organic matter and structural stability: mechanisms and implications for management. Plant Soil 76:319-337.

Osborn B (1952) Range conditions influence water uptake. J. Soil Water Conserv. 7:128-132.

Peck E and Brown M (1962) An approach to the development of isohyetal maps for mountainous areas. J. Geophys. Res. 67:681-694.

Price D and Bolke EL (1970) Hydrologic reconnaissance of the Sink Valley area, Tooele and Box Elder Counties, Utah. Utah Dept. Nat. Res. Tech. Pub. Number 26, Salt Lake City, UT.

Reynolds TD and Fraley L (1989) Root profiles of some native and exotic plant species in southeastern Idaho. Env. Exp. Bot. 29:241-248.

Richards JH (1984) Root growth response to defoliation in two *Agropyron* bunchgrasses: field observations with an improved root periscope. Oecologia 64:21-25.

Richards JH (1986) Root form and depth distribution in several biomes. pp. 82-97. In Carlisle D, Berry WL, Kaplan IR, and Watterson JR (editors) Mineral Exploration: Biological Systems and Organic Matter. Rubey Vol. V. Prentice-Hall, Englewood Cliffs, NJ.

Richards JH and Caldwell MM (1985) Soluble carbohydrates, concurrent photosynthesis and efficiency in regrowth following defoliation: a field study with *Agropyron* species. J. Appl. Ecol. 22:907-920.

Richards JH and Caldwell MM (1987) Hydraulic lift: Substantial nocturnal water transport between soil layers by *Artemisia tridentata* roots. Oecologia 73:486-489.

Roden LE and Bazilevich NI (1967) Production and Mineral Cycling in Terrestrial Vegetation. Oliver and Boyd, Edinburgh.

Rogers RW (1977) Lichens of hot and semi-arid lands. pp. 211-252. In Steward MRD (editor) Lichen Ecology. Academic Press, NY.

Romo JT and Haferkamp MR (1989) Water relations of *Artemisia tridentata* ssp. *wyomingensis* and *Sarcobatus vermiculatus* in the steppe of southeastern Oregon. Amer. Midl. Nat. 121:155-164.

Satterlund DR (1972) Wildland Watershed Management. John Wiley, NY.

Schlatterer EF (1968) Establishment and survival of three native grasses under natural and artificial conditions. M.S. Thesis, Univ. of Idaho, Moscow.

Schumm SA (1977) The Fluvial System. Wiley, NY.

Schuster JL (1964) Root development of native plants under three grazing intensities. Ecology 39:63-70.

Shumar ML and Anderson JE (1986) Water relations of two subspecies of big sagebrush on sand dunes in Southeastern Idaho. Northwest Sci. 60:179-185.

Snyder CT (1962). A hydrologic classification of valleys in the Great Basin, western United States. Bull. Internat. Assoc. Sci. Hydrol. 7:53-59.

Snyder CT and Langbein WT (1962) The pleistocene lake in Spring Valley, Nevada and its climatic implications. J. Geophys. Res. 67:2385-2394.

Springer ME (1958) Desert pavement and vesicular layer of some soils of the desert of the Lahontan Basin, Nevada. Soil Sci. Soc. Amer. Proc. 22:63-66.

Stanley RJ (1983) Soils and vegetation: An assessment of current status. pp. 8-18. In Messer J and Mosely G (editors) What Future for Australia's Arid Land? Austr. Conserv. Found, Canberra.

Sturges DL (1977) Soil water withdrawal and root characteristics of big sagebrush. Amer. Midl. Nat. 98:257-274.

Sturges DL (1979) Hydrologic relations of sagebrush lands. pp. 86-100. In Gifford GF and Busby FE (editors) The Sagebrush Ecosystem: A Symposium. Utah State Univ. Press, Logan, UT.

Sturges DL (1980) Soil water withdrawal and root distribution under grubbed, sprayed, and undisturbed big sagebrush vegetation. Great Basin Nat. 40:157-164.

Sturges DL and Trlica MJ (1978) Root weights and carbohydrate reserves of big sagebrush. Ecology 59:1282-1285.

Tabler RD (1964) The root system of *Artemisia tridentata* at 9,500 feet in Wyoming. Ecology 45:633-636.

Tchoupopnou E (1989) Experimental studies of rain splash erosion from soil microphytic crusts on Utah rangelands. M.S. Thesis, Utah State Univ., Logan, UT.

Thompson AL and James LG (1985) Water droplet impact and its effect on infiltration. Trans. ASAE 28:1506-1510.

Thorgeirsson H (1985) Temporal and spatial partitioning of the soil water resource between two *Agropyron* bunchgrass and *Artemisia tridentata*. M.S. Thesis, Utah State Univ., Logan, UT.

Thorgeirsson H (1988) The modeling and measurement of respiratory carbon use and net carbon gain of two *Agropyron* bunchgrasses. PhD Dissertation, Utah State Univ., Logan, UT.

Tisdale EW and Hironaka M (1981) The sagebrush-grass region: A review of the ecological literature. Bull. 33 Forest Wildlife and Range Exp. Sta., Univ. of Idaho, Moscow.

Toth J (1962) A theory of groundwater motion in small drainage basins in Central Alberta, Canada. J. Geophys. Res. 67:4375-4387.

Volk OH and Geyger E (1970) Schaumboden als Ursache der Vegetationslosigkeit in ariden Gebieten. Z. Geomorph. NF 14:79-95.

Walker BH (1979) Game ranching in Africa. pp. 55-81. In Walker BH (editor) Management of Arid Ecosystems. Elsevier, Amsterdam.

Wallace A, Romney EM, and Cha JW (1980) Depth distribution of roots of some perennial plants in the Nevada test site area of the Northern Mojave Desert. Great Basin Nat. Mem. 4:201-207.

Weaver JE (1919) The Ecological Relations of Roots, Carnegie Institute Washington Publication 286, Washington, DC.

West NE and Gifford GF (1976) Rainfall interception by cool-desert shrubs. J. Range Manage. 29:171-172.

West NE (1990) Structure and function of soil microphytic crusts in wildland ecosystems of arid to semi-arid regions. Adv. Ecology (in press).

Wood MK (1976) The effects of vesicular horizons on emergence of grasses and shrubs. M.S. Thesis, Univ. Nevada, Reno, NV.

Wood MK, Blackburn WH, Eckert RE Jr, and Peterson FF (1978) Interrelations of the physical properties of coppice dune and vesicular dune interspace soils with grass seedling emergence. J. Range Manage. 31:189-192.

Wraith JF, Johnson DA, Hanks RJ, and Sisson DV (1987) Soil and plant water relations in a crested wheatgrass pasture: response to spring grazing by cattle. Oecologia 73:573-578.

Yair A and Lavee H (1974) Areal contribution to runoff on scree slopes in an extremely arid environment - a simulated rainstorm experiment. Z. Geomorph. Suppl. 21:106-121.

Young JA and Evans RA (1986) Erosion and deposition of fine sediments from playas. J. Arid Environ. 10:103-115.

Chapter 8

Isotopic Assessment of Vegetation Changes in Grassland and Woodland Systems

Larry L. Tieszen and Steven Archer

Introduction

Dwight Billings is directly responsible for much of our understanding of plant ecophysiology, and he was one of the first to elucidate and interpret plant physiological processes in relation to ecosystem structure and function. Recent advances in stable isotope chemistry offer opportunities to identify, clarify, and trace further the linkages between plant and ecosystem-level processes, both past and present. It is within this framework of a new integrative technology, which facilitates the approaches developed by Billings, that we present this contribution. To date, there have been few studies of Great Basin vegetation processes using the techniques discussed here, and these have been described in Chapters 5 and 6, so most of our examples have been drawn from adjacent ecosystems.

An understanding of factors affecting the composition and productivity of communities through time is a fundamental interest of plant ecologists. However, in many cases we are not sure if vegetation has changed, let alone how or why. It is reasonable to expect that if we can identify and understand causes of past changes, we will be better able to anticipate future events and design management strategies to ensure sustained productivity and minimize the likelihood of undesirable, anthropogenically-induced shifts in vegetation structure. For example, one controversial problem of many arid and semi-arid landscapes in North and South America, Australia, Africa, and India in recent history is the presumed replacement of productive grasslands and savannas with high densities of unpalatable woody perennials. If herbaceous productivity of present-day grasslands and savannas is to be maintained in the face of increasing human population pressures and anticipated global climatic changes, which may favor woody plants (Emanuel et al. 1985a, b), an understanding of the dynamics of grasses and woody plants is essential. The use of stable isotopes, by contributing to our ability to assess vegetation history directly, will aid in anticipating the rate, magnitude, and direction of future vegetation change in such ecosystems.

The existence of stable isotopic forms of a number of elements that are of biogeochemical interest has been known for some time. For example, it was 50 years ago that we realized land plants were depleted in ^{13}C relative to atmospheric CO_2 and carbonates (Nier and Gulbransen 1939) and that some plants were less depleted than others. This initial geochemical discovery was given more biological meaning 20 years ago when Bender (1968) suggested a systematic, predictable relationship between the photosynthetic carbon reduction pathways (C_3 and C_4 systems) and delta ^{13}C values. This discovery, concurrent with rapid advances in our understanding of C_3 and C_4 photosynthetic mechanisms and the realization that these labels are transmitted to other trophic levels, has now made available a powerful research tool, which has contributed to our fundamental understanding of ecology (Tieszen and Boutton 1989).

Carbon stable isotope analysis has been useful especially in quantifying energy flow through higher trophic levels in terrestrial (Ambrose and DeNiro 1986; Tieszen and Imbamba 1980) and aquatic (Schell and Ziemann 1988; Fry and Sherr 1984; Peterson and Fry 1987) systems, in archeology and paleonutrition (DeNiro 1987; Schwartz et al. 1985), and most recently in analyses of water-use efficiency (Hubick et al. 1986; Farquhar et al. 1988; Johnson et al. 1990).

Several recent studies have documented changes in the isotopic labels of soil organic matter (SOM) and have inferred shifts in community composition and distribution (Dzurec et al. 1985) or paleoclimate (Krishnamurthy and DeNiro 1982). These approaches have been refined to provide excellent tools to assess SOM turnover in situ (Schwartz et al. 1986; Volkoff and Cerri 1987; Balesdent et al. 1987, 1988).

Other naturally occurring isotopes are rapidly becoming similarly useful for ecological studies, largely because we both understand the behavior of these isotopes better than we did 10 years ago and because of the increased availability of analytical techniques and services. Isotopic forms of H, N, O, and S are especially relevant and ecological applications are reviewed by Rundel, Ehleringer and Nagy (1989). In this paper we wish to describe the use of natural abundances of stable isotopes to quantify the transfer of carbon from primary producers to other trophic levels, including grazing and detrital food chains. We will concentrate on grassland-woodland systems with examples from North America and appropriate systems elsewhere.

Uses of Stable Isotopes of C in Ecological Research

Most of the carbon in natural systems is in isotopic form carbon-12 (^{12}C) with only 1.11% as carbon-13 (^{13}C). Fractionation of these two forms can result in samples of organic carbon that are enriched or depleted in ^{13}C. In all cases it is important to remember that mass balance is maintained. As one compartment becomes enriched in ^{13}C, another must become depleted. This enrichment or depletion results in different ratios of $^{13}C/^{12}C$, which can be measured following complete combustion to CO_2 and cryogenic purification of the gas before admission to an isotope ratio mass spectrometer.

Biologists usually describe the isotopic relationship of a sample by its delta ^{13}C value. This is calculated by comparing the ratio of $^{13}CO_2$ to $^{12}CO_2$ (mass 45 to mass 44) in a sample to a standard, usually the international PeeDee belemnite (PDB) standard. All results are expressed in the units, "per mil" ($^{o}/oo$) according to the following:

$$\text{delta } ^{13}C\ (^{o}/oo) = \frac{^{13}C/^{12}C\ (\text{sample}) - ^{13}C/^{12}C\ (\text{standard})}{^{13}C/^{12}C\ (\text{standard})} \times 1000$$

Compared to the PDB standard, bulk atmospheric air is depleted in ^{13}C (delta $^{13}C_{air}$ = ca. -7.7$^{o}/oo$), C_4 plants often are depleted slightly more (around -12.5$^{o}/oo$), and C_3 plants are depleted strongly (around -26.5$^{o}/oo$). In this treatment we will attempt to describe consistently samples that are more depleted in ^{13}C as being more negative than the standard or other referenced material.

Isotopic ratios of naturally occurring elements are useful for ecological research because they can be used as quantitative or qualitative tracers, or because they become fractionated during the course of specific ecosystem processes, which then become measurable or quantifiable. Delta ^{13}C values provide a diagnostic label that can be applied to:

1. Distinguish carbon sources and quantitatively estimate their contributions.
2. Trace carbon sources between trophic levels and ecosystems. This allows us to make quantitative estimates of dietary components and energy transfer.
3. Integrate physiological processes over long time periods, thereby providing quantitative, time-integrated estimates of energy sources, water-use efficiency, and organic matter turnover.
4. Provide a paleoecological, historical, as well as contemporary, record of carbon inputs to organic components resistant to diagenetic processes.

As we continue to learn how environmental factors control processes leading to ecosystem-level values of delta ^{13}C and delta ^{15}N, we will have an increasingly powerful tool for making inferences about paleoecological processes and quantitative changes in paleocommunities or paleoclimatic regimes.

Biogeochemistry of Carbon and Operational Assumptions

Fractionation

The primary end members for carbon isotopes in most terrestrial systems are C_3 and C_4 plants. It is clear that the principal factor controlling ^{13}C-depletion from the ambient value (ca. 7.7 $^{o}/oo$, Mook 1986) is the photosynthetic pathway (Smith and Epstein 1971). This results in bimodal distributions of delta ^{13}C values that group around generally accepted mean values between -26 and -28 $^{o}/oo$ for C_3 plants and -12 and -14 $^{o}/oo$ for C_4 plants. O'Leary (1988) calculates mean values of -27.1 ± 1.2 $^{o}/oo$ and -13.1 ± 1.2 $^{o}/oo$ for C_3 and C_4 plants, respectively. However, the range of values within each group is substantial, nearly overlapping in extreme cases (C_3 = -22 to -38 $^{o}/oo$; C_4 = -9 to -21 $^{o}/oo$) (O'Leary 1988; Medina et al. 1986).

Label Fidelity: Environmental Variation

Before utilizing natural abundances of stable carbon isotopes as ecosystem tracers, it is necessary to establish the extent of genetic or environmentally-induced variability in delta ^{13}C values among the end members, those species or components of a system that represent the most negative and least negative potential sources of isotopic signal. The plant value will be a direct and nearly linear function of the ambient air delta ^{13}C value of CO_2 fixed in photosynthesis. Although normally around -7.7 $^{o}/oo$, respiratory release of CO_2 from C_3 plant material will result in locally more negative values that can be significant in closed-canopies or situations where rapid decomposition occurs concurrently with photosynthesis. The magnitude of this delta ^{13}C profile in closed-canopy forests has to be established yet, although Medina et al. (1986) estimated depletion as great as 8 $^{o}/oo$ (air = -16 $^{o}/oo$) to account for plant values of -36 $^{o}/oo$. Van der Merwe and Medina (unpublished) have shown recently that this does result in very negative plant values. However, it is unlikely that local depletion phenomena would cause substantial problems interpreting plant delta ^{13}C values in most grassland, savanna, or open woodland systems.

The genetic and environmental determinants of fractionation operate most strongly on the C_3 end members, which appear to be more affected by relative enzymatic versus stomatal control of photosynthesis. Environmental components may cause variable discrimination because of their influences on concentrations of CO_2 inside the leaf (C_i) relative to the concentration in air (C_a) through their effects on either the carboxylating enzyme and/or stomatal conductances. Several important environmental determinants and their effects on delta ^{13}C values are summarized in Table 8.1.

Limitations on photosynthesis that might occur when the carboxylating enzyme becomes limiting (low nutrient status) or when levels of ATP and reduced NADP generated by the light reactions are low can result in an increase in C_i/C_a and more negative delta ^{13}C values. This is clearly shown in a natural cline by Ehleringer et al. (1986) and in a field experiment by Mulkey (1986). Although the low light effect is independent of any reassimilation of depleted CO_2, in nature the two often will occur together.

As water stress develops stomata close and, with light energy still available, C_i/C_a is reduced, resulting in more negative delta ^{13}C values. This

Table 8.1. Measured and predicted physiological responses to environmental factors that affect carbon isotope discrimination in C_3 plants.

Environmental Parameter	Coupling Plant Process	Effect on C_i/C_a	Expected Range of Delta ^{13}C value ($^o/oo$) Range	Direction	Most Likely Ecological Context
Recycled, respired CO_2	-	small ±	8	-	Very dense and closed tree canopies
Declining irradiance	B[1]	+	5 to 6[3]	-	Dense and closed tree canopies
Increasing water stress	S[2]	-	3 to 6[4]	+	Open environments, arid to semi-arid
Increasing osmotic stress	S	-	5 to 10[5]	+	Localized osmotically impacted areas
Low nutrient content	B	+	3 to 5	-	N or P limited natural areas or depleted soils
Low temperature	B	+	≤ 3	-	High polar latitudes
Reduced CO_2 partial pressure (increased altitude)	S	-	3 to 7[6]	+	Steep elevation gradients
Chilling injury	B	+	?	-	
Photoinhibition	B	+	?	-	

[1] B, Biochemical conductance. [2] S, Stomatal conductance. [3] Ehleringer, et al. 1986 [4] Johnson, et al. 1990
[5] Guy, et al. 1980; Neales, et al. 1983; Downton, et al. 1985 [6] Körner, et al. 1988.

results in less water loss for each unit of photosynthate produced (i.e., a higher water-use efficiency, WUE=g carbon produced/g H_2O lost; Farquhar and Richards 1984; Hubick et al. 1986). In agricultural species such as wheat, these physiological processes are associated with delta ^{13}C values 3 to 4 $^o/oo$ less negative under field conditions.

Assessments of the genetic and environmental components of these responses in native grasses have been made. The range of delta ^{13}C values among thirty cultivars of wheatgrass (*Agropyron desertorum*) grown under high to low water levels was 6 $^o/oo$, and the variation was related closely to yield (Johnson et al. 1990) and water-use efficiency (WUE). Since water availability often limits plant growth, we might expect there to be substantial variation among wet, mesic, and xeric habitats. Our unpublished grassland data from China support this relationship as has now been confirmed by Ehleringer and Cooper (1988). They sampled all species along a decreasing soil moisture gradient in an American desert and clearly identified a habitat effect of 2 to 4 $^o/oo$, with plants from the driest habitats at delta ^{13}C = -24.4 $^o/oo$ and those from the wettest at delta ^{13}C = -26.6 $^o/oo$. These effects on delta ^{13}C values are likely to be of great ecological significance in the Southwest. Increasing osmotic stress exerts an effect identical to that of water stress and could alter delta ^{13}C values by 10 $^o/oo$ (Guy et al. 1980) in extreme cases. Those C_3 species that show an osmotic adjustment should have a broad tolerance and may show more negative delta ^{13}C values than those with no osmotic adjustment.

As discussed earlier, plants from nutrient impoverished areas may be enzyme limited and would be expected to exhibit an increased C_i/C_a. Sage and Pearcy (1987) show a decrease in C_i/C_a of 0.4 with increasing leaf nitrogen in *Amaranthus*, a C_4 plant. They did not see a similar relationship in *Chenopodium album*, a C_3 species. This implies a potentially significant effect on discrimination, although this has not been measured. However, Osmond (1987) cites examples of N nutrition effects on a C_4 grass that confirm this expectation. We do not know the magnitude of this potential nutrient effect in nature or its ecological significance. Since species tend to be sorted along nutrient gradients, any differences may be largely accounted for as life form differences (see next section). With regard to tracing the flow of carbon through food chains, impoverished plants generally would have low yields and be of low nutritive value to herbivores and therefore contribute relatively little to their diets.

At temperatures below the optimum for any species or plant, CO_2 uptake is reduced mainly by temperature effects on enzymatic reactions. This should result in an increase in C_i/C_a and a decrease in delta ^{13}C values. We should see the same response at high temperatures, assuming no additional water stress. One direct assessment with tomato showed that delta ^{13}C values were 3 $^{o}/oo$ more negative at 32° C than at 17° C (O'Leary and Treichel 1987) . This suggests plants at 32° C were well above their temperature optimum, although these comparisons have not been tested. In natural situations temperature effects may be minor, since plants show good genetic and often physiological adaptation to maintain temperature optima near ambient conditions. Only at very high latitudes or altitudes are some plants exposed to conditions where they may be below their optimum often. Preserved carbon in bogs or peat accumulation areas may preserve an as yet undefined temperature record. Recently, Stuiver and Braziunas (1987) have reported an increased fractionation (more negative delta ^{13}C) with latitude in a group of tree species. This supports the proposed temperature relationship, although changes in humidity and decreased irradiance also could contribute to the greater fractionation at high latitudes. As altitude increases the partial pressure of CO_2 decreases, resulting in a gradient ranging from ca. 350 μbar CO_2 near sea level to near 200 μbar CO_2 at the upper end of the biotic zone. Körner has shown an altitudinal increase in the efficiency of CO_2 uptake at the enzyme level. This results in a lower C_i/C_a at higher altitudes and is seen as less negative delta ^{13}C values (Körner et al. 1988). Their data show a significant relationship with altitude. The regression across all species and growth forms predicts a value around -29 $^{o}/oo$ at sea level and -26 $^{o}/oo$ around 5000 meters. Individual species varied as much as 2.7 $^{o}/oo$ with altitude. Obviously, this is of significance and needs to be verified.

Several studies of the relationships between discrimination and plant distributions have been completed recently. In East Africa (Tieszen and Boutton 1989), C_4 plants revealed genetic, species-based differences in delta ^{13}C values. For example, the delta ^{13}C values of C_4 sedges were consistently less negative than C_4 grasses. Within a species, delta ^{13}C values were very similar throughout an annual cycle, across a range of similar sites, and among plant parts. In a detailed survey of grasslands in southeastern South Dakota and Inner Mongolia (Tieszen and Bingyu, unpublished), plants occupying the driest habitats were about 1 $^{o}/oo$ less negative than species from the wettest habitats. It would be useful to do a thorough altitudinal assessment in the Great Basin.

Label Fidelity: Genetic Variation

It is not clear how much of the above variation is induced environmentally and how much is based on genetic differences. At the general level, growth form differences represent a long-term genetic response to various environments. But to what extent are intraspecific variations due to heritable characters? This question is of ecological, evolutionary, and agronomic interest (Johnson et al. 1990; O'Leary 1988; Hubick, et al. 1986; Farquhar and Richards 1984). O'Leary reports a genetic component in soybean and corn. Condon et al., (1987) show a range of >3.0 $^{o}/oo$ among 16 varieties of wheat in Australia. This is similar to the range (3.7 $^{o}/oo$) shown by Hubick et al. (1986) in field experiments with peanuts. Work with wheatgrasses show a similar genetic component and indicate that characters that determine C_i/C_a are heritable (Johnson et al. 1990).

Plant growth forms are associated highly with specific environments and resource (mainly water and nutrients) availability. Generally there is an inverse relationship between leaf longevity and photosynthetic rates, such that growth forms with short-lived leaves (annuals, some grasses, deciduous shrubs) possess higher rates of photosynthesis than growth forms with long-lived, evergreen leaves (Mooney and Gulmon 1982). Although the rate of photosynthesis is influenced by both stomatal and carboxylation components, consistent growth form patterns may exist and should be explored (Table 8.2).

Data on growth form effects on delta ^{13}C values are few and variously confounded with environmental factors such as altitude. De Lucia, et al. (1988) compared plants on different geological substrates. Their data indicate greater discrimination by shrubs than evergreen trees (as much as 5 $^{o}/oo$). An approximate classification of plants measured by Körner et al. (1988) into the life form categories of Table 8.2 suggests trees and grasses may show the most discrimination, followed by shrubs, forbs, cushion plants, and giant rosettes. When growth form delta ^{13}C was regressed against altitude, differences in both the intercepts and the slopes occurred. Data of Ehleringer and Cooper (1988) from a desert area indicate evergreen shrubs are slightly but consistently less negative than drought deciduous or "winter" deciduous shrub forms.

It is clear that C_3 and C_4 photosynthetic types bear a diagnostic label. It is also clear that within each group there is both genetic and environmentally imposed variation. This variation, on the order of a few $^{o}/oo$

Table 8.2. Predicted relationship between plant growth form and departure of C_i/C_a from a typical herbaceous plant growing under mesic conditions.

Plant Growth Form (Type)	Effect on C_i/C_a (Direction)	Effect on delta ^{13}C
Mean type, mesic	None	None
Short-lived annual	+	Decrease
Drought deciduous grass	+	Decrease
Drought enduring grass	-	Possibly increase
Drought deciduous dicots	+	Decrease
Deciduous shrubs	-	Increase or none
Deciduous trees	None	None
Short-lived evergreen	-	Increase
Long-lived evergreen	-	Increase
Cushion	-	Increase
Rosettes	-	Increase
Emergent macrophytes	+	Decrease

across strong gradients, is significant and must be recognized and accounted for when $^{13}C/^{12}C$ ratios are utilized as quantitative markers. In most cases, however, the magnitude and direction of these effects on delta ^{13}C values are predictable, and C_4 species will show substantially less variation than C_3 species.

Label Fidelity: Biochemical Variation

We have known for some time that delta ^{13}C values for certain biochemical fractions in plants deviate from that of bulk plant tissue. Vogel (1982) found seed proteins enriched in ^{13}C (3.8 $^{o}/oo$) relative to leaf and stem tissue (probably consisting mainly of cellulose). We have summarized some of our recent analyses (Tieszen and Boutton 1989) which are similar to those reported by Winkler et al. (1978). Our analyses showed lipids in corn, wheat, and wheatgrass leaves to be depleted in ^{13}C relative to bulk tissues by 3.5 to >5 $^{o}/oo$. Starches and celluloses were generally closer to bulk tissues. However, in most cases these were also the constituents that comprised a large proportion of the tissue mass analyzed. This illustrates a major difficulty with these analyses, the varying reference of the bulk tissue. DeNiro and Epstein (1978) formulated a mechanistic basis for the general depletion of ^{13}C seen in lipid fractions (Park and Epstein 1961; Whelan and Sackett 1970). Unfortunately, our quantitative understanding of these variations for other biochemical fractions is limited.

We clearly need a larger data base of isotopic values of biochemical components and we need to define the isotopic values for one component relative to all others under steady-state conditions and under conditions of various environmental stresses. Benner et al. (1987), for example, have shown that lignin in *Spartina* and several other species was depleted in ^{13}C by 2 to 6 $^{0}/oo$ relative to bulk tissue and 4 to 7 $^{0}/oo$ relative to cellulose. Since lignin is most certainly one of the determinants of both digestibility and decomposition rates, it is essential that we develop a clearer understanding of these isotopic values, especially since lignin often is believed to be the precursor of recalcitrant soil organic matter. It is difficult to reconcile, however, the apparent consequence of SOM delta ^{13}C values and bulk vegetation values with lignin. Either lignin is not the principle source for recalcitrant SOM or it becomes enriched in ^{13}C as it becomes converted. Although data are few, it does appear (Natelhoffer and Fry 1988) that SOM, which develops under some forests, is enriched by a few $^{0}/oo$. Generally, if this is confirmed, then it suggests differences in the formation of recalcitrant SOM in forests and grasslands. This would make direct interpretations of community changes more difficult.

Trophic Level Transfers

Soil, Soil Organic Matter, Carbonates, and Phytoliths

The reconstruction of vegetation assemblages requires that the proportional label is retained in some form that is preserved, and hence, available later for analysis. If the delta ^{13}C value of the material changes through time, the magnitude of depletion or enrichment must be estimated. There are but three or four materials that meet these requirements: 1) SOM and paleosols; 2) bone collagen; 3) soil carbonate; and 4) phytoliths. In addition to any transfer, depletion or enrichment, some estimate of mean residence time is necessary, since all these materials exist in a steady-state of addition and loss.

The isotopic composition of soil organic matter should reflect the relative abundance of C_3 and C_4 plants in a manner analogous to the transfer of carbon from plants to herbivores. The isotopic value should be identical to the contributing vegetation if there is no fractionation or differential decomposition as plant carbon is incorporated into SOM. In addition, a steady-state between SOM and vegetation will occur only if the vegetation is reasonably stable over the time period that it contributes to the SOM. Since plant biochemical components vary in their isotopic values, it is reasonable to expect that for a comparable biomass input, some species may contribute more to the isotopic signature in SOM than others. Furthermore, the extent of biochemical fractionation associated with microbial transformation is not established yet.

The fidelity of label transfer from living plant material to SOM appears high. It has been suggested that fractionation does not occur under anaerobic conditions (Stout et al. 1975; Stout et al. 1981). Under aerobic conditions, several studies indicate a slight enrichment of organic matter, (approximately 1 $^{o}/oo$) with increasing depth in the soil profile. This slight enrichment (Troughton et al. 1974; Stout et al. 1975; Goh et al. 1976; O'Brien and Stout 1978; Stout and Rafter 1978; Schleser and Bertram 1981; O'Brien et al. 1981; Stout et al. 1981) may reflect preferential microbial utilization of depleted biochemical fractions, e.g., lipids. A better understanding of the affects of selective microbial utilization of organic matter constituents will aid in interpreting SOM delta ^{13}C values. Benner et al., (1987) and Rodriguez and Tieszen (1988) found significant depletion (2 to 6 $^{o}/oo$) in ^{13}C of lignin. In vegetation where lignin is abundant, this can result in significant isotopic shifts upon decomposition and a steady-state SOM value more negative than that of the contributing vegetation. This is especially important in grasslands and woodlands because lignin will vary substantially and is an important determinant of the rate of decomposition (Parton et al. 1987; Melillo et al. 1982).

Although the potential for a departure of SOM delta ^{13}C from vegetation delta ^{13}C exists (Natelhoffer and Fry 1988), available data indicate the magnitude of discrepancy generally is small. As a result, major differences in soil delta ^{13}C should reflect, in a reasonably predictive manner, the vegetation that contributed to that organic material, thereby providing a very useful tool to examine vegetation changes over both short and long time periods.

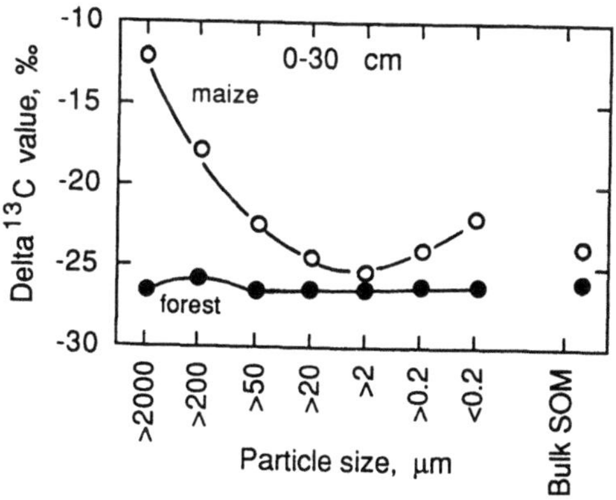

Figure 8.1. Relationship between soil particle size (μm) and delta ^{13}C values ($^o/_{oo}$) in soil from a stable forest system and in soil from that system after 23 years of continuous maize (C_4) cultivation (modified from Balesdent et al. 1987). Note the similar values for all size fractions in the steady-state system and the substantial change in the large size fractions after conversion to maize. The bulk SOM values are estimated and reflect the assumptions of slow exchange of the resistant fractions (0.2 to 20 μm) and the fact that these fractions also contain the bulk of the soil organic carbon. Patterns of delta ^{13}C value in size fractions of deeper soils on forest and converted sites were similar to those in the 0 to 30 cm zone, but showed less change in the direction of C_4.

In an early study in New Zealand, Hendy et al., (1972) observed that SOM delta ^{13}C values increased significantly approximately 10,000 years ago, indicative of a change in vegetation from C_3 (forest) to C_4 (subtropical grassland). A physiognomic shift such as this implies an increased temperature regime, aridity, or seasonality in the environment. Krishnamurthy and DeNiro (1982) sampled paleosols from Kashmir and established ranges of delta ^{13}C values between -16 and -25 $^o/_{oo}$, suggesting a change toward a greater C_3 domination with time. More recently, Dzurec et al. (1985) compared delta ^{13}C values for soil components along transects extending from within C_3-dominated winterfat (*Ceratoides lanata*) communities into C_4-dominated shadschale (*Atriplex confertifolia*) communities. The winterfat vegetation appeared to be in a steady-state with the SOM, whereas SOM of the shadschale community was more negative than could be accounted for by C_4 vegetation. Such results suggest shadschale importance has increased on this site.

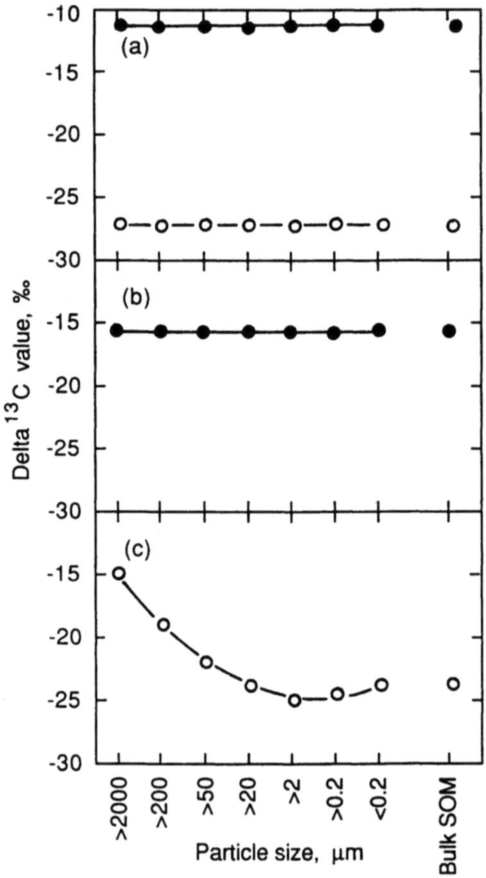

Figure 8.2. Diagrammatic representation of isotopic states of soil particle size fractions (μm) and bulk SOM under various conditions. In all cases the model assumes no enrichment or depletion from vegetation delta ^{13}C values. a, Steady-state values for pure C_4 (-11 $^{o}/_{oo}$) and C_3 (-27 $^{o}/_{oo}$) prairies; b, Conversion of pure C_3 prairie to 75% C_4 prairie, steady-state attained after a long time; c, Conversion of pure C_3 prairie to 75% C_4 prairie, steady-state not attained because conversion occurred relatively recently (perhaps 25 years ago). Note similarity of bulk SOM to original state because of resistant nature of 0.2 to 200 μm sizes.

A better resolution of the magnitude and rate of change in soil delta ^{13}C values could be provided by physical, chemical, or biological fractionation of SOM. Although the nature and chemical sources of fulvic acid, humic acids, and humins are unclear, these fractions possess different mean residence times and, therefore, could provide sources of material for delta ^{13}C analysis of various-aged fractions. Fulvic acid is usually more enriched in ^{13}C, and humic acid usually is depleted slightly (Dzurec et al. 1985). Humins are intermediate, as would be expected, when they are the principal contributor to bulk organic matter. However, it is our impression that this chemical fractionation does not provide substantial insight into our understanding of changes in SOM because these constituents are chemically heterogenous, variable, and are not necessarily sequential components in a degradative pathway.

Physical fractionation of SOM holds the promise of contributing significantly to our ability to document historical and paleovegetation changes (Schwartz et al. 1986). Balesdent et al. (1987, 1988) quantitatively demonstrate replacement of SOM in various size fractions in a C_3 forest system after 23 years of continuous maize (C_4) cultivation (Figure 8.1). Large fractions turn over fastest, whereas fractions between 0.2 and 20 μm are the most resistant to exchange. Volkoff and Cerri (1987) have shown similar differences between forest and grassland humus in the A horizon, but indicate the isotopic signals from the two types converge in the B1 and lower horizons of soil from subtropical Brazil. This is in contrast to our observations of profile patterns in grasslands from South Dakota, Texas, and East Africa and that of Balesdent et al. (1987, 1988) in Europe and a tallgrass prairie in Missouri.

As modelled in Figure 8.2, measurements of bulk SOM can indicate a change in vegetation. However, an analysis of the changes in delta ^{13}C of particle size fractions may be necessary to determine whether or not a new steady-state has been attained. The particle size analysis therefore has the potential to provide high sensitivity and resolution. Figure 8.2a and b illustrates the steady-state condition of SOM size fractions that should occur following change in vegetation from pure C_3 to increasingly C_4 types. This new steady-state would likely take a long time (up to 1000 years?) to attain because of the substantial proportion of SOM bound in highly resistant size fractions. An intermediate state of SOM change is illustrated in Figure 8.2c. Notice that we expect the largest, and to a lesser extent the smallest, particle sizes to change relatively quickly. Bulk SOM values, however, would underestimate the extent of C_3-C_4 replacement and obscure the initial state of the SOM.

Soil carbonates are formed from the CO_2 in the soil column and because this is derived largely from microbial and root respiration (Marion et al. 1985), this carbon also should reflect C_3 and C_4 sources (Cerling 1984; Gardner 1984; Schlesinger 1985). Cerling's preliminary data show a reasonable relationship between the calculated C_4 composition and measured delta ^{13}C values that show an expected enrichment around 10.4 $^o/oo$ resulting from isotopic fractionation between CO_2 and carbonate. Cerling (1984) also indicates that the contribution by atmospheric CO_2 could range from near 0 to 4.4 $^o/oo$. More recently, he has shown that soil carbonate and SOM systematically differ by 14 to 16 $^o/oo$ in native grasslands (Cerling et al. 1989). This difference was also apparent in paleosols.

Woodland and Grassland Conversions

The broad generalizations concerning the control of the distribution of C_3 and C_4 species clearly define latitudinal and altitudinal clines of increasing C_3 presence as temperatures decrease with either latitude or altitude (Terri and Stowe 1976; Tieszen et al. 1979). At low altitudes in Kenya, for example, all grasses in open savannas are C_4 and nearly all shrubs and all trees are C_3. Thus, the opportunity to evaluate community type conversions with distinct isotopic signals is readily apparent. Furthermore, the change from C_4 dominance at low altitudes to C_3 dominance at high altitudes by grasses in open grasslands provides the possibility to assess climatic changes that would be apparent in the signals of organic matter from former grasslands. In North America analogous parallels exist. C_4 grasses are replaced with C_3 species as latitude or altitude increase (Boutton et al. 1980). In the south and southwest, open grasslands, which are predominantly C_4, also are interspersed with woodland systems, which are predominantly C_3. As a result, we should be able to assess community changes in these types of systems both within recent times and in older systems where organic matter is available for analysis (eg., Hendy et al. 1972; Krisnamurthy and DeNiro 1982; Dzurec et al. 1985; Balesdent et al. 1988).

Case Study from a Subtropical Woodland-Savanna System

The potential natural vegetation of southern Texas and northern Mexico has been classified as *Prosopis-Acacia-Andropogon-Setaria* savanna (Kuchler 1964). However, much of the present vegetation is subtropical thorn woodland (Blair 1950). A component of the Tamaulipan Biotic Province (Dice 1943), the thorny shrubs and small trees in this region have counterparts throughout much of the world's tropical and subtropical zones (Brown 1982). In many instances it is believed these vegetation types have replaced former grasslands (e.g., Johnston 1963; Schofield and Bucher 1986). However, the basis for this contention is primarily from indirect, historical sources, many of which are conflicting (Malin 1953). Could thornwoodlands have replaced savannas in recent history? To build a case for such a conversion, successional processes must be identified, their rates quantified and supportive evidence gathered.

Indirect assessments of vegetation change. Recent investigations on the Texas A&M La Copita Research Area 65 km west of Corpus Christi, Texas (27° 40′N; 98° 12′W) have focused on the stability of grass-woody mosaics in subtropical regions of southern Texas. The site, which is dominated by a savanna parkland vegetation complex, has no history of anthropogenic disturbance except cattle grazing since the mid-1800s. On sandy loam upland portions of the landscape, discrete clusters of woody plants are scattered throughout an herbaceous matrix of C_4 grasses. Ephemeral drainages and playas are dominated by closed-canopy woodlands. Initial investigations focused on quantifying the structure of the woody clusters and examining the herbaceous zone for seedlings and saplings of woody plants. Field data (Archer et al. 1988) indicate:

1. The majority of woody plants in the herbaceous zones, in terms of density and frequency, are of one species, *Prosopis glandulosa*.
2. Woody clusters range in size from 1 to 40 m in diameter.
3. Typically there is one *Prosopis* plant centrally located in each cluster, where it is the largest plant in terms of height, basal diameter, and canopy area.
4. The number of subordinate woody species in clusters ranges from 1 to 15 and is correlated positively with the size of the *Prosopis* plant (R^2 = 0.86).
5. With the exception of *Prosopis*, most species in clusters are bird-dispersed.
6. Species composition and relative abundance in large clusters is comparable to that of closed-canopy woodlands in neighboring ephemeral drainages.

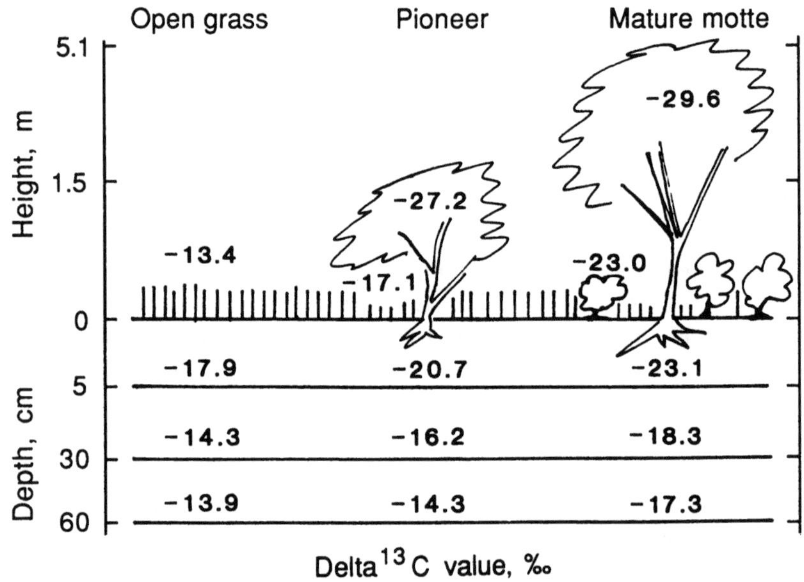

Figure 8.3. Delta ^{13}C values in components of vegetation and soil on developing woody clusters on sandy loam uplands in southern Texas. Coefficients of variation were <0.06 for means of all data shown. The grass below mature motte is especially noteworthy. Archer et al. (1988) hypothesized that *Prosopis glandulosa* invades herbaceous zones and adds vertical structure, which attracts birds. These disseminate seeds of other woody plants. *Prosopis* also modifies soils and microclimate to facilitate germination and/or establishment of subordinate woody species. The first woody species appear 10-15 yr after *Prosopis* establishment and within 60 yr, clusters contain 12 woody species, on average (Archer 1989).

These data suggest *Prosopis* invades grass-dominated sites and serves as the nucleus of cluster organization, apparently facilitating the ingress and/or establishment of additional woody species otherwise restricted to other habitats (Figure 8.3). If *Prosopis* plants continue to establish and develop in the herbaceous zones, new clusters will form. At the same time, existing clusters will enlarge as new species are added and canopies of established plants develop. The present savanna parkland may thus represent an intermediate stage in the conversion of a grassland to woodland. Closed-canopy woodlands in the region appear to represent sites where this process has already occurred.

Direct assessments of vegetation change. The scenario presented in the previous section was developed from inferences derived from a static space-for-time substitution study of vegetation structure (van der Maarel and Werger 1978). Inferences from this static approach can be misleading (Austin 1977; Shugart et al. 1981). However, direct assessments of changes in grass-woody composition of these landscapes using aerial photographs and stable carbon isotopes support the contention that woody plants have displaced grasses in recent history. Woody species in the clusters possess the C_3 photosynthetic pathway (delta ^{13}C value range = -27 to -32 $^{o}/oo$), whereas vegetation of grass-dominated zones between clusters is characterized by plants with the C_4 pathway (delta ^{13}C range = -13 to -17 $^{o}/oo$) (Figure 8.3). If shrubs have been a long-term constituent of these landscapes, the delta ^{13}C signature of soils beneath them should reflect this and fall in the -27 to -32 $^{o}/oo$ range. However, if C_3 shrubs had displaced C_4 grasses: (1) The soil delta ^{13}C value would be less negative than -27 to -32 $^{o}/oo$; (2) The degree of departure from the expected ratio would decrease as time of site habitation by shrubs increases; and (3) Soil delta ^{13}C values would become less negative with depth along the chronosequence. An analysis of soil organic carbon delta ^{13}C confirms these predictions. The organic carbon of soils beneath herbaceous zones was strongly C_4 and generally reflected the composition of the current vegetation throughout the profile (delta ^{13}C = -14 to -18 $^{o}/oo$). In contrast, the mean delta ^{13}C value in the upper horizon of soils beneath clusters at early and late stages of development was -21 and -23 $^{o}/oo$, respectively. The decrease in delta ^{13}C from -21 to -23 $^{o}/oo$ appears to reflect the additional input of C_3 carbon associated with the passage of time required for *Prosopis* plants and clusters to develop. Among soils supporting woody vegetation, the contribution of C_3-derived carbon decreased with depth to 60 cm, converging on the values observed for the herbaceous zones.

These data constitute direct evidence that C_3 shrubs have displaced C_4 grasses on the site. But when did these changes take place? Estimates of cluster growth rates under different precipitation regimes (Archer, et al. 1988) were used to model growth and determine size/age relationships for *Prosopis* plants and the woody clusters that form beneath them. Results indicate the age-class distribution of *Prosopis* plants and clusters on the site was that of a young population (mostly $<$90 yr of age) expanding geometrically (Archer 1989). This reconstruction concurs with historical accounts from the late 1800s, which describe landscapes in the region as savannas with "mottes" (i.e., small patches of

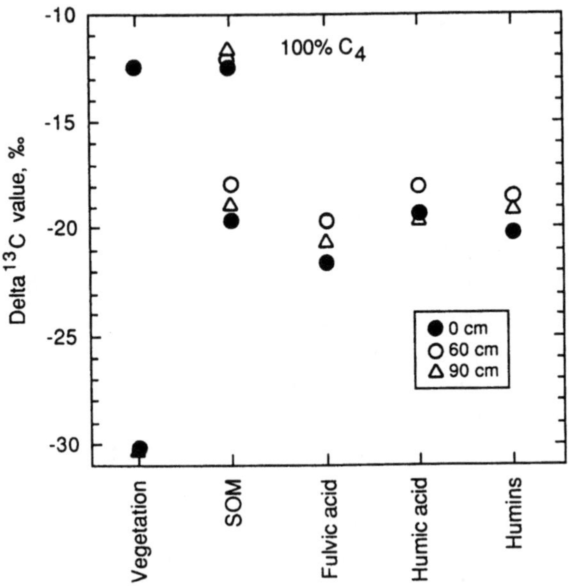

Figure 8.4. Comparison of delta ^{13}C values in vegetation, SOM, and soil fractions in a closed, canopy bur oak gallery forest in the tallgrass prairie region near Sioux Falls, S.D. Values represent means of three replicate samples with ranges of $<1\ ^o/oo$. The C_4 vegetation and SOM plotted for comparison are from a pure C_4 grassland in Kenya, East Africa.

woods in a prairie land) (Crosswhite 1980). Predicted stand age and temporal development on this site are comparable to those reported for encroachment of *Pinus ponderosa* (Madany and West 1983), *Larrea tridentata* (Chew and Chew 1965), *P. monophylla* and *Juniperus osteosperma* (Blackburn and Tueller 1970), and *J. occidentalis* (Young and Evans 1981) in other arid and semi-arid grasslands of western North America.

Case Study From a Temperate Grassland-Forest System

We also have examined the SOM status of small, oak-dominated gallery forests in Eastern South Dakota. These forest stands occur within grassland formations ranging from true prairies to mixed prairies having a significant representation by C_4 species (Ode, Tieszen, and Lerman 1980). Areas of grassland that are near rivers, in steep ravines, or on fire-protected sides of lakes occasionally support closed forests dominated by bur oak (*Quercus macrocarpa*), hackberry (*Celtis occidentalis*), ironwood (*Tilia americana*), and some elm (*Ulmus americana*). These wooded inclusions range from closed-canopy forests with little understory to woodland and savanna systems. We recently analyzed the SOM in a closed forest stand in southeastern South Dakota, near Sioux Falls.

As is illustrated in Figure 8.4, all compartments of living and identifiable dead material possesses a clear C_3 signal. The weighted vegetation value is around -30 $^0/oo$. The C_3 grasses in the understory have a considerably more negative delta ^{13}C signature than the C_3 grasses in the open grassland, probably reflecting carbon recycling and low irradiance effects. Roots and other material are around the -27 to -30 $^0/oo$ range. Notice, however, that all the SOM components are distinctly more positive, between -18 and -21 $^0/oo$ generally. Our interpretation is that the forest vegetation in this particular location is relatively recent and has displaced grassland vegetation with a significant C_4 component. The delta ^{13}C values of grassland of known 100% C_4 composition is also plotted to indicate the other potential equilibrium. It is interesting that in all three samples from this site, the delta ^{13}C values for fulvic and humic acid fractions are similar to that of the more abundant humin materials. This may indicate that replacement of grassland by forest has been recent, with insufficient time for "new" fulvic acid and humic acid fractions to turn over. A particle size analysis of these soils and dating would constitute additional tests of this hypothesis.

Case Study from a Temperate Grassland System

The similarity of fulvic acid, humic acid, and the humin fractions in the prairie-forest reconstruction contrasts markedly with that of an eastern Montana mixed-grass prairie. We have examined two sites: onc dominated by the C_3 western wheatgrass (*Agropyron smithii*), and the other dominated by the C_4 grass, blue grama (*Bouteloua gracilis*). On both sites there is a striking difference between vegetation values and bulk SOM and humin materials (Figure 8.5). The composite vegetation of the wheatgrass site possesses a delta ^{13}C value around -27 $^0/oo$, indicating nearly pure dominance by C_3 species. This is in accordance with foliar cover data. Yet the SOM at all depths is less negative and becomes increasingly less negative with depth. Humins are still more positive than the bulk SOM and show the same pattern with depth. In contrast, delta ^{13}C values of fulvic and humic acid fractions depart

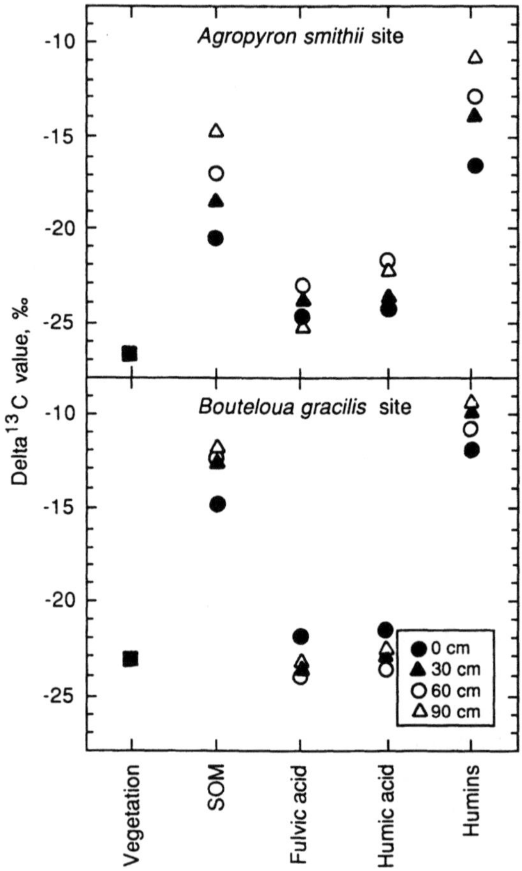

Figure 8.5. Comparison of delta ^{13}C values in vegetation, SOM, and soil fractions in two grassland sites in a mixed-grass prairie in eastern Montana. Values represent means of two separate pooled samples.

markedly from that of humins and approach the wheatgrass vegetation values. The logical interpretation is that *Agropyron* presently dominates a site previously occupied by C_4 species whose signature remains in the bulk SOM and is expressed most strongly in the humins. As has been suggested by others, this interpretation further suggests that fulvic and humic acids turn over faster than do humins.

The *Bouteloua* site is equally interesting. Note that the vegetation at this site consists of at least 50% C_3 plants. Nevertheless, the vegetation values are also substantially more negative than the -10 $^{o}/oo$ of the SOM and humin values and show the same depth distribution as the *Agropyron* site. As was the case on the *Agropyron* site, fulvic and humic acids largely reflect the present vegetation. However, humin and bulk SOM delta ^{13}C values suggest this site was at one time nearly 100% C_4, and is becoming increasingly dominated by more C_3 species, as reflected in the fulvic and humic acid SOM fractions. Again, dating for evidence of mean residence times would be very useful. It is interesting that in all three case studies (subtropical savanna in Texas, temperate prairie and forest in South Dakota, and two temperate grassland sites in Montana) there is a consistent trend in the isotopic values of SOM and soil fractions with depth that suggest a general trend toward more C_3 vegetation in recent times.

Conclusions

To a significant extent, the ratio of ^{13}C to ^{12}C incorporated by a plant is determined by the photosynthetic pathway utilized during CO_2 uptake and reduction. Within each photosynthetic group, however, species and individuals may show variation from the modal values for these groups. In C_3 plants this variation is determined largely by the C_i/C_a ratio, which we now know has both genetic and environmental determinants. Furthermore, plant biochemical fractions differ in their delta ^{13}C values, with lipids and lignin often more negative than bulk plant material. Since these sources of stable isotopic variation can confound interpretations of the quantitative transfer of organic matter between trophic levels, care must be exercised in the use of stable isotopes as ecosystem tracers.

The carbon isotopic value of soil organic matter, however, closely reflects that of the vegetation with which it has been in a steady state for an extended period. Thus, a C_3 forest results in a characteristic C_3 delta ^{13}C value for the SOM; and C_4 grasslands, as in tropical East Africa, develop signatures around -12 to -13 $^{o}/oo$, reflecting that of the vegetation. The SOM associated with any system, therefore, will record carbon signals from earlier vegetation; and where vegetation types differ in C_3 and C_4 composition, it should be possible to reconstruct vegetation history from SOM delta ^{13}C values. The slow turnover of bulk SOM, however, results in rather poor time resolution.

Particle size analysis reveals that turnover is a clear function of SOM associated with various size fractions. Some fractions turn over with a half-life measured in years, others with a half-life measured in centuries. This approach makes it possible to both estimate the direction of vegetation change and to make inferences about the rate of that change. Particle size analysis holds promise for the reconstruction of past vegetation in the Great Basin and Great Plains. The approach requires further validation and a better understanding of the rate of biochemical inputs, microbial degradation, soil texture, and other factors that affect label integrity and rates of SOM turnover through time.

References

Ambrose SH and DeNiro MJ (1986) The isotopic ecology of East African mammals. Oecologia 69:395-406.

Archer S (1989) Have southern Texas savannas been converted to woodlands in recent history? Amer. Nat. 134:545-561.

Archer S, Scifres CJ, Bassham CR, and Maggio R (1988) Autogenic succession in a subtropical savanna: rates, dynamics and processes in the conversion of a grassland to a thorn woodland. Ecol. Monogr. 58:111-127.

Austin MP (1977) Use of ordination and other multivariate descriptive methods to study succession. Vegetatio 35:165-175.

Balesdent J, Mariotti A, and Guillet B (1987) Natural ^{13}C abundance as a tracer for studies of soil organic matter dynamics. Soil Biol. Biochem. 19:25-30.

Balesdent J, Wagner GH and Mariotti A (1988) Soil organic matter turnover in long-term field experiments as revealed by the ^{13}C natural abundance tracer technique. Soil Sci. Soc. Am. J. 52:118-124.

Bender MM (1968) Mass spectrometric studies of carbon 13 variations in corn and other grasses. Amer. J. Sci. Radiocarbon. Suppl. 10:468-472.

Benner R, Fogel ML, Sprague EK, and Hodson RE (1987) Depletion of ^{13}C in lignin and its implications for stable carbon isotope studies. Nature 329:708-710.

Blackburn WH and Tueller PT (1970) Pinyon and juniper invasion in black sagebrush communities in east-central Nevada. Ecology 51:841-848.

Blair WF (1950) The biotic provinces of Texas. Texas J. Sci. 2:93-117 .

Boutton TW, Harrison AT, and Smith BN (1980) Distribution of biomass of species differeing in photosynthetic pathway along an altitudinal transect in Southeastern Wyoming grassland. Oecologia 45:287-298.

Brown JR (ed) (1982) Biotic communities of the American southwest-United States and Mexico. Desert Plants 4:101-106.

Cerling TE (1984) The stable isotopic composition of modern soil carbonate and its relationship to climate. Earth Planet. Sci. Lett. 71:229-240.

Cerling TE, J Quade, Y Wang, and JR Bowman (1989) Carbon isotopes in soils and paleosols as ecology and palaeoecology indicators. Nature 341:138-139.

Chew RM, and AE Chew (1965) The primary productivity of a desert shrub (*Larrea tridentata*) community. Ecol. Monogr. 35:355-375.

Condon AG, RA Richards, and GD Farquhar (1987) Carbon isotope discrimination is positively correlated with grain yield and dry matter production in field-grown wheat. Crop Sci. 27:996-1001.

Crosswhite FS (1980) Dry country plants of the South Texas Plains. Desert Plants 2:141-179.

DeLucia EH, WH Schlesinger, and WD Billings (1988) Water relations and the maintenance of Sierran conifers on hydrothermally altered rock. Ecology 69:303-311.

DeNiro MJ (1987) Stable isotopy and archaeology. Am. Sci. 75:198-91.

DeNiro MJ and Epstein S (1978) Influence of diet on the distribution of carbon isotope ratios in animals. Geochim. Cosmochim. Acta 42:495-506.

Dice LR (1943) The Biotic Provinces of North America. Univ. Michigan Press, Ann Arbor.

Dzurec RS, Boutton TW, Caldwell MM, and Smith BN (1985) Carbon isotope ratios of soil organic matter and their use in assessing community composition changes in Curlew Valley, Utah. Oecologia 66:17-24.

Ehleringer JR and Cooper TA (1988) Correlations between carbon isotope ratio and microhabitat in desert plants. Oecologia 76:562-566.

Ehleringer JR, Field CB, Lin Z, and Kuo C (1986) Leaf carbon isotope and mineral composition in subtropical plants along an irradiance cline. Oecologia 70:520-526.

Emanuel WR, Shugart HH, and Stevenson M (1985a) Climatic change and the broad-scale distribution of terrestrial ecosystem complexes. Climatic Change 7:29-43.

Emanuel WR, Shugart HH, and Stevenson M (1985b) Response to comment: climatic change and the broad-scale distribution of terrestrial ecosystem complexes. Climatic Change 7:457-460.

Farquhar GD, Hubick KT, Condon A, and Richards RA (1988) Carbon isotope fractionation and plant water-use efficiency. pp. 21-40. In Rundel PW, Ehleringer JR, and Nagy KA (editors) Stable Isotopes in Ecological Research, Ecological Studies Series, Springer-Verlag, Heidelberg.

Farquhar GD and Richards RA (1984) Isotopic composition of plant carbon correlates with water-use efficiency of wheat genotypes. Aust. J. Plant Physiol. 11:539-552.

Fry B and Sherr EB (1984) Delta ^{13}C measurements as indicators of carbon flow in marine and freshwater ecosystems. Contrib. Mar. Sci. 27:13-47.

Gardner JR (1984) Carbon and oxygen isotope composition of pedogenic $CaCO_3$ from soil profiles in Nevada and New Mexico, USA. Isotope Geoscience 2:55-73.

Goh KM, Rafter TA, Stout JD, and Walker TW (1976) The accumulation of soil organic matter and its carbon isotope content in a chronosequence of soils developed on aeolian sand in New Zealand. J. Soil Sci. 27:89-100.

Guy RD, Reid DM, and Krouse HR (1980) Shifts in carbon isotope ratios of two C_3 halophytes under natural and artificial conditions. Oecologia 44:241-247.

Hendy CH, Rafter TA, and MacIntosh NWG (1972) The formation of carbonate nodules in the soils of the Darling Downs, Queensland, Australia, and the dating of the Talgai craniu. pp. D106-D126. In Rafter TA, Grant-Taylor T (editors) Proc. 8th Int. Radiocarbon Dating Conference, Lower Hutt, New Zealand: Royal Society of New Zealand.

Hubick KT, Farquhar GD, and Shorter R (1986) Correlation between water-use efficiency and carbon isotope discrimination in diverse peanut (*Arachis*) Germplasm. Aust. J. Plant Physiol. 13:803-816.

Johnson DA, Asay KH, Tieszen LL, Ehleringer JR, and Jefferson PG (1990) Carbon isotope discrimination: potential in screening cool season grasses for water-limited environments. Crop Science 30:338-343.

Johnston MC (1963) Past and present grasslands of southern Texas and northeastern Mexico. Ecology 44:456-466 .

Körner Ch, Farquhar GD, and Roksandic Z (1988) A global survey of carbon isotope discrimination in plants from high altitude. Oecologia 74:623-632.

Krishnamurthy RV and DeNiro MJ (1982) Isotope evidence for Pleistocene climatic changes in Kashmir, India. Nature 298:640-644.

Kuchler AW (1964) The potential natural vegetation of the conterminous United States. Amer. Geographical Soc., NY.

Madany MH and West NE (1983) Livestock grazing-fire regime interactions within montane forest of Zion National Park, Utah. Ecology 64:661-667 .

Malin JC (1953) Soil, animal, and plant relations of the grassland, historically recorded. Scientific Monthly 76:207-220.

Marion GM, Schlesinger WH and Fonteyn PJ (1985) CALDEP: A regional model for soil $CaCO_3$ (Caliche) formation in Southwestern Deserts. Soil Science 139:468-481.

Medina E, Montes G, Cuevas E, and Rokzandic Z (1986) Profiles of CO_2 concentration and delta ^{13}C values in tropical rain forests of the upper Rio Negro Basin, Venezuela. J. of Tropical Ecology 2:207-217.

Melillo JM, Aber JD, and Muratore JF (1982) Nitrogen and lignin control of hardwood leaf litter decomposition dynamics. Ecology 63:621-626.

Mook WG (1986) Delta ^{13}C in atmospheric CO_2. Netherlands J. Sea Research. 20:211-223.

Mooney HA and Gulmon SL (1982) Constraints on leaf structure and function in reference to herbivory. BioScience 32:198-206.

Mulkey SS (1986) Photosynthetic acclimation and water use efficiency of three species of understory herbaceous bamboo (Gramineae) in Panama. Oecologia 70:514-519.

Natelhoffer KJ and Fry B (1988) Controls on nitrogen-15 and carbon-13 abundances in forest soil organic matter. Soil Sci. Soc. Amer. J. 52:1653-1640.

Neales TF, Fraser MS, and Roksandic Z (1983) Carbon isotope composition of the halophyte *Disphyma clavellatum* (Haw.) Chinnock (Aizoaceae), as affected by salinity. Aust. J. Plant Physiol. 10:437-444.

Nier AV and Gulbransen EA (1939) Variations in the relative abundance of the carbon isotopes. J. Am. Chem. Soc. 61:697-698.

O'Brien BJ and Stout JD (1978) Movement and turnover of soil organic matter as indicated by carbon isotope measurements. Soil Biol. Biochem. 10:309-317.

O'Brien BJ, Stout JD, and Goh KM (1981) The use of carbon isotope measurements to examine the movement of labial and refractory carbon in the soil. pp. 46-74. In Flux of Organic Carbon by Rivers to the Ocean. Carbon Dioxide Effects Research and Assessment Program. U.S. Department of Energy CONF-8009140, UC-11. Washington, DC.

Ode D, Tieszen LL, and Lerman JC (1980) The seasonal contribution of C_3 and C_4 plant species to primary production in a mixed prairie. Ecology 61:1304-1311.

O'Leary MH (1988) Carbon isotopes in photosynthesis. BioScience 38:328-335.

Osmond CB (1987) Photosynthesis and the carbon economy of plants. New Phytol. 106(suppl.):161-175.

Park R and Epstein S (1961) Metabolic fractionation of $^{13}C/^{12}C$ in plants. Plant Physiol. 36:133-137.

Parton WJ, Schimel DS, Cole CV, and Ojima DS (1987) Analysis of factors controlling soil organic matter levels in Great Plains grasslands. Soil Sci. Soc. Amer. J. 51:1173-1179.

Peterson BJ and Fry B (1987) Stable isotopes in ecosystem studies. Ann. Rev. Ecol. Syst. 18:293-320.

Rodriguez S and Tieszen LL (1988) A comparison of ^{13}C abundance in plant biochemical fractions (in prep).

Rundel PW, Ehleringer JR, and Nagy KA (1989) Stable Isotopes in Ecological Research, Ecological Studies Series, Vol. 68. Springer-Verlag, Heidelberg.

Sage RF and Pearcy RW (1987) The nitrogen use efficiency of C_3 and C_4 plants. Plant Physiol. 84:959-963.

Schell DM and Ziemann PJ (1989) Natural carbon isotope tracers in arctic aquatic foodwebs. pp. 230-251. In Stable Isotopes in Ecological Research. Rundel PW, Ehleringer JR, and Nagy KA (editors) Ecological Studies Series, Vol. 68. Springer-Verlag, Heidelberg.

Schleser GH and Bertram HG (1981) Investigation of the organic carbon and delta ^{13}C profile in a forest soil. pp. 201-204. In Frigerio A (editor) Recent Developments in Mass Spectrometry in Biochemistry, Medicine, and Environmental Research, Vol. 7. Elsevier, Amsterdam.

Schlesinger WH (1985) The formation of caliche in soils of the Mojave Desert, California. Geochim. Cosmochim. Acta 49:57-66.

Schofield CJ and Bucher EH (1986) Industrial contributions to desertification in South America. Tree 1:78-80.

Schwartz HP, Melbye J, Katzenberg MA, and Knyf M (1985) Stable isotopes in human skeletons of southern Ontario: reconstructing palaeodiet. J. Archaeol. Sci. 12:187-206.

Schwartz D, Mariotti A, Lanfranchi R, and Guillet B (1986) $^{13}C/^{12}C$ ratios of soil organic matter as indicators of vegetation changes in the Congo. Geoderma 39:97-103.

Shugart HH, West DC, and Emanuel WR (1981) Patterns and dynamics of forests: an application of simulation models. In West DC, Shugart HH, and Botkin DB (editors) Forest Succession: Concepts and Applications. Springer-Verlag, Heidelberg.

Smith BN and Epstein S (1971) Two categories of $^{13}C/^{12}C$ ratios of higher plants. Plant Physiol. 47:380-384.

Stout JD and Rafter TA (1978) The $^{13}C/^{12}C$ isotopic ratios of some New Zealand tussock grassland soils. pp. 75-83. In Robinson BW (editor) Stable isotopes in the earth sciences. DSIR Bull. 220. Wellington, New Zealand.

Stout JD, Goh KM, and Rafter TA (1981) Chemistry and turnover of naturally occurring resistant organic compounds in soil. pp. 1-73. In Paul EA and Ladd JN (editors) Soil Biochemistry, Vol 5. Marcel Dekker, NY.

Stout J, Rafter T, and Troughton JH (1975) The possible significance of isotopic ratios in paleoecology. pp. 279-286. In Suggate R and Cresswell M (editors) Quarternary Studies. Royal Soc. New Zealand, Wellington.

Stuiver M and Braziunas TF (1987) Tree cellulose $^{13}C/^{12}C$ isotope ratios and climatic change. Nature 328:58-60.

Terri JA and Stowe LG (1976) Climatic patterns and distribution of C_4 grasses in North America. Oecologia 23:1-12.

Tieszen LL, Senyimba MM, Imbamba SK, and Troughton JH (1979) The distribution of C_3 and C_4 grasses and carbon isotope discrimination along an altitudinal and moisture gradient in Kenya. Oecologia 37: 337-350.

Tieszen LL and Imbamba SK (1980) Photosynthetic systems, carbon isotope discrimination, and herbivore selectivity in Kenya. African J. Ecology 18:237-242.

Tieszen LL and Boutton TW (1989) Stable carbon isotopes in terrestrial ecosystem research. pp. 167-195. In Rundel PW, Ehleringer JR, and Nagy KA (editors). Stable Isotopes in Ecological Research, Ecological Studies Series, Vol. 68. Springer-Verlag, Heidleberg.

Troughton JH, Wells PV, and Mooney HA (1974) Photosynthetic mechanisms and paleoecology from carbon isotope ratios in ancient specimens of C_4 and CAM plants. Science 185:610-612.

van der Maarel E and Werger MJA (1978) On the treatment of succession data. Phytocoenosis 7:257-278.

Vogel JC (1982) Koolstofisotoopsamestelling van plantproteine. S. Afr. Tyd. Naturwet. Tergnol. 1:7-8.

Volkoff B and Cerri CC (1987) Carbon isotopic fractionation in subtropical Brazilian grassland soils: Comparison with tropical forest soils. Plant and Soil 102:27-31.

Whelan T and Sackett WM (1970) Carbon isotope discrimination in a plant possessing the C4 dicarboxylic acid pathway. Biochem. Biophys. Res. Commun. 41:1205-1210.

Winkler FJ, Wirth E, Latzko E, Schmidt HL, Hoppe W, and Wimmer P (1978) Influence of growth conditions and development on $\delta^{13}C$ values in different organs and constituents of wheat, oat, and maize. Z. Pflanzenphysiol. 87:255-263.

Young JA and Evans RA (1981) Demography and fire history of a western juniper stand. J. Range Manage. 34:501-506.

Chapter 9

The North American Great Basin: A Sensitive Indicator of Climatic Change

Robert A. Wharton, Peter E. Wigand, Martin R. Rose, Richard L. Reinhardt, David A. Mouat, Harold E. Klieforth, Neil L. Ingraham, Jonathan O. Davis, Carl A. Fox, and J. Timothy Ball

Introduction

Climatic change has become a major scientific and political issue during the past decade. Articles concerning global warming due to the greenhouse effect and ozone depletion from industrial chemicals are common in the news media and scientific literature. As stated by Schneider (1989), "the intense heat, forest fires, and drought of the summer of 1988 and the observation that the 1980s are the warmest decade on record have ignited an explosion of media, public, and governmental concern that the long-debated global warming has arrived."

Within the scientific community there is considerable controversy over climatic change. Are the recently observed climatic events simply short-term fluctuations or true indications of a rapidly changing climate? Continued observations over time ultimately will provide a more definitive understanding of global warming and ozone depletion. In the meantime, however, much can be done to increase our understanding of these and other global processes, as well as their effects on global, regional, and local environments.

The Great Basin of North America has been identified as a region highly sensitive to climatic change (Maggs 1989). The unique climatological, geological, and hydrological characteristics of the Great Basin have provided and preserved an unparalleled record of climate over the past 50,000 years. Studies of past climatic conditions using paleoenvironmental proxy data have shown that this desert ecosystem is responsive to short- and long-term fluctuations in temperature, precipitation, solar radiation, wind, and other environmental variables. This chapter will focus primarily on the reconstruction of past climates using calibrations between paleoenvironmental proxy data and instrumental climatic data, and integrations of reconstructions based on paleoenvironmental data that cannot be calibrated easily. This chapter will document the responsiveness of the Great Basin to past climatic change and demonstrate the Great Basin's utility in improving our understanding of contemporary global climatic change.

Definitions

Mitchell et al. (1966) noted that many words have been introduced in the literature and used more or less interchangeably as synonyms for "climatic change." It is common to find references to climatic change, fluctuation, variation, trend, and so on; an assortment of terms that can introduce considerable confusion in discussions of climate and weather. It is conceivable that each term might be associated with a particular numerical characteristic, physical origin, or time scale. For this reason, the following definitions will be used (Mitchell et al. 1966):

Climatic change: a general term encompassing all forms of climatic inconstancy, disregarding its statistical nature or physical cause.

Climatic variation: a fluctuation that occurs over a length of time, such that it results in an appreciable inconstancy of successive 30-year averages of the variable under investigation.

Effects of Climatic Change on Vegetation

Climatic change can be manifested in a number of ways, including both natural and cultural effects. Clearly, one of the major natural effects of climatic change is on vegetation. An ideal area for the study of climatic change is one which straddles the transition zone between biomes. Biome structure and function are predicated on climate. Biome boundaries may consist of relatively narrow or wide ecotones, which have characteristics of the adjacent biomes. Wide ecotones reflect broad transitional gradients between climatic regimes. Where ecotones are narrow the transition between climatic regimes is typically more abrupt. Areas especially sensitive to climatic change typically have narrow ecotones accompanied by sharp climatic zone boundaries. (Climatic zones are areas with similar temperature, precipitation, wind regimes, etc., that have a zonal arrangement, that is not strictly latitudinal.) Conceptually, those areas which straddle major biomes and are influenced by highly divergent climatic zones are ideal for the study of climatic change, because when the movement of these boundaries occurs the vegetational response usually will be noticeable.

Although examples of the above can be found on most continents, the resulting ecotones usually tend to be too broad for effective study. Occasionally, terrain factors exist which serve to narrow ecotones. For example, changes in elevation with minimal changes in slope angle and aspect often result in relatively narrow ecotones. In the Great Basin of western North America, elevational changes associated with the "island" mountains are accompanied by extremely sharp changes in vegetation (Billings 1951; Chapter 3). Some of these changes could be attributed to slope, exposure, and rock (soil) type, but they primarily result from the climatic effect associated with elevation. An example of a recent change in Great Basin biomes is the boundary between shrubland (dominated by *Artemisia tridentata*) and woodland (*Pinus* spp. and *Juniperus* spp.). Varying vegetation boundaries between these two basic biomes reflect climatic change during the past few hundred years (Hastings and Turner 1965; Turner and Karpiscak 1980; Rogers 1982; Mehringer and Wigand 1987; Wigand 1987).

The North American Great Basin

Environmental Factors

The boundaries of the Great Basin are the crests of the Sierra Nevada and Cascade Mountains on the west, the Snake River Plain on the north, the Wasatch Range on the east, and the Colorado River drainage basin on the south. Within the Great Basin are numerous mountain ranges, trending north-south and separated by basins or valleys. The Great Basin is characterized by an interior drainage resulting in hydrologically closed systems (Houghton 1969). At present, the characteristic climate of the Great Basin is semiarid to arid (Houghton 1969;

Houghton et al. 1975; Chapter 2). Except on a few mountain summits, precipitation averages <50 cm yr^{-1}, and in about three quarters of the province, 25 cm yr^{-1}. In a transect from west to east across the central Great Basin, average precipitation ranges from less than 13 cm yr^{-1} in the rainshadow of the Sierra Nevada to about 25 cm yr^{-1} near the Wasatch Front. In the northern Great Basin, winters are cold (-12 to 4° C) and summers are warm (10 to 27° C). The southern Great Basin has milder winters (4 to 10° C) and hotter summers (27 to 32° C).

The Great Basin lies at the convergence of Pacific, Continental and Gulf atmospheric circulation systems (Houghton et al. 1975; Chapter 2). Movement of these systems and their impact on local and regional climate is reflected in changes in hydrology, erosional and depositional processes, and vegetation. Displacement of winter and summer storm-tracks and penetration of the summer monsoon all are affected by the positions of these pressure systems. Decreased monsoonal penetration in the Great Basin caused by southeastward movement of the Gulf of Mexico-dominated Bermuda High often results in decreased summer precipitation. However, this effect is sometimes offset by a southward displacement of summertime storms originating in the Pacific. Expansion of the Continental pressure systems during the winter often results in increased winter precipitation in the Great Basin by southward displacement of the Pacific storm-track (Houghton et al. 1975).

Topographic diversity in the Great Basin serves as a trigger for orographic rainfall (Houghton 1969). The resulting disparity in the distribution of precipitation between mountaintops and valley floors has resulted in the distribution of a multiplicity of vegetation types within close proximity to each other. Elevational changes in plant distributions through time can be dramatic as storm-track shifts combine with the orographic effect of the north-south trending mountain ranges to radically change the distribution of precipitation on the landscape (Houghton 1969; Wigand 1987).

The arid nature of the Great Basin enhances its usefulness as an indicator of climatic change through the relatively immediate and direct effect of climate on landscape and vegetation, as well as by the preservation of paleoenvironmental indicators. Despite sparse vegetation cover, relatively low rainfall and high evaporation rates with resulting low "effective" precipitation generally lead to lower slope erosion and slower basin bottom deposition rates (Wigand 1987). Large increases in effective precipitation, because of sparse vegetation cover, will be marked in the sedimentological record by increases in slope erosion rates and basin bottom deposition rates. With continued higher effective precipitation rates, vegetation density will increase and stabilizing soils will form, thereby reducing the rate of sediment movement (Mehringer and Wigand 1987). A return to lower effective precipitation for a time will sustain low slope-erosion rates. However, in the basin bottoms, aeolian erosion and deposition will be initiated as lakes dry, exposing playa bottoms to wind action (Currey and James 1982; Davis 1982; Mehringer 1986).

Low effective precipitation favors the preservation of certain kinds of paleoenvironmental indicators. Plant microfossils preserved in dry cave deposits and woodrat middens can provide intermittent records of local vegetation spanning tens of thousands of years (Spaulding et al. 1983). Wood from long-dead trees, preserved by the relatively dry conditions that characterize upper and lower tree-lines in the Great Basin, provides continuous long-term evidence of climatic change that may extend the record already obtained from living trees perhaps to 10,000 years B.P. (Fritts 1976; Hughes et al. 1982; Graybill 1985, 1987). Although lakes with lifetimes spanning the Holocene are rare in the Great Basin, especially in the southern portion, desert springs such as those found in Ash Meadows (Mehringer and Warren 1976) and at Tule Springs (Mehringer 1967) often provide a record of past vegetational change. These springs typically have relatively high deposition rates, which hold the potential for the generation of high frequency regional pollen and local aquatic plant microfossil records. These records can allow for comparisons with tree-ring data resulting in the generation of long, detailed and mutually corroborative records of regional climate.

Annual production of plant biomass is extremely sensitive to effective precipitation (Walter 1954; Whittaker 1970; Chapters 5-7). Therefore, the arid climate of the Great Basin results in a vegetation highly responsive to even minor changes in precipitation. Such changes in vegetation can be measured directly today through satellite imagery, historically through repeat photography, and prehistorically through tree-ring analysis and fluctuating pollen production reflected in pollen influx.

The rapid response of Great Basin vegetation to changes in climate is primarily the result of both the physiological and morphological characteristics of the individual plant species that have evolved as a result of millions of years of adaptation to arid and semiarid conditions (Brown 1987). Many of the plants growing in the Great Basin are opportunistic

(Odum 1971), and they can take advantage of increased precipitation rather quickly. Yearly fluctuations of precipitation can be reflected in increased pollen production (Mehringer and Wigand 1990) and even single rainfalls can initiate new foliage production. Also, many of these species have wind-dispersed pollen and seeds, which enhances rapid dispersal of propagules when conditions are favorable. The seeds of other plants such as juniper and pinyon are ingested by animals which may carry the seeds great distances (Wells 1983). Long-term seed and plant dormancy until adequate precipitation has occurred is also characteristic of many of the plants found in the Great Basin (Went 1957). The year-to-year vegetative response of individual plants is also dramatic. As a result, subtle changes in the patterns of plant distribution can occur within one to two years. Thus, the Great Basin is responsive to the full spectrum of climatic change, not only to long-term trends and rather abrupt discontinuities, but also to higher frequency fluctuations and variability.

Paleoclimatic and Modern Climatic Data Sources

Many locations worldwide have paleoenvironmental proxy data of one form or another that yield information on long-term climatic change. However, few have as many different types of paleoenvironmental proxies as the Great Basin, especially with the length of available records and spatial coverage. The uniqueness of the Great Basin results in part from the arid conditions that allow for outstanding preservation of paleoenvironmental information. The paleoenvironmental reconstructions possible with such a constellation of data sets resolve changes occurring from seasons and years to decades, hundreds, and thousands of years. Furthermore, the different data sets frequently cover the same points in time and space and allow for mutual corroboration, or they complement one another in the sense that each can be used to reconstruct a different aspect of climatic change.

There are many data sets pertinent to the analysis of climatic variations in the Great Basin over the last 50,000 years. These sets are comprised of pollen data, woodrat midden data, tree-ring data, and lake level. Modern observations including stream flow, isotopic data, aircraft and satellite imagery, and meteorological data obtained from the National Cooperative Observers Network, Soil Conservation Service (SNOTEL), Bureau of Land Management and Forest Service (Remote Automatic Weather Stations - RAWS), are discussed below.

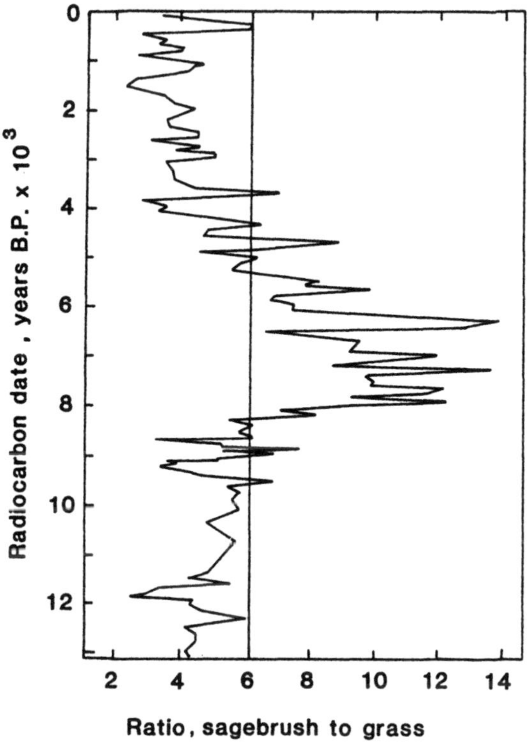

Figure 9.1. Ratio of sagebrush to grass pollen for the past 13,000 years at Fish Lake in Steens Mountain of south-central Oregon. Increasing sagebrush corresponds with decreasing effective precipitation. The ratio is plotted around the mean of the past 9,700 radiocarbon years (modified from Mehringer and Wigand 1987).

Pollen. Studies of pollen from stratified deposits in lakes, marshes, and alluvium, and from plant remains preserved in ancient woodrat nests (middens) are important for reconstructing the general trends of past vegetation and climate in the Great Basin (Mehringer 1977, 1985, 1986). Pollen analysis can provide a continuous record of both regional terrestrial and local aquatic vegetation. In addition, sediments containing pollen often hold seeds of aquatic plants that reflect a history of fluctuating water tables (Wigand 1987). To reconstruct past plant communities, fossil pollen spectra are correlated to pollen samples collected in modern plant communities (Davis 1984; Thompson 1984; Overpeck et al. 1985). The precipitation and temperature parameters of the modern analogue plant communities are then assigned as the climatic parameters of the correlating fossil plant community. The longest, best preserved pollen records within the Great Basin (Fish Lake and Wildhorse Lakes in Steens Mountain, Figure 9.1, and Lake Cleveland in the Albion Range) are found in the north. The ratio of sagebrush to grass pollen in the fossil pollen at Fish and Wildhorse Lakes suggests changing effective precipitation during the last 13,000 years in the northern Great Basin. In the south, many localities consist primarily of Pleistocene and/or late Holocene records (e.g., Tule Springs in the Las Vegas Valley and Cofer Spring northeast of Beatty, Nevada) that lack a mid-Holocene component. In addition, pollen records from the northern Great Basin come from elevations ranging from mountaintop to valley bottom. High elevation records usually have been obtained from cirque basins that were glaciated during the last ice age and therefore span only the time since then. Due to the absence of major Pleistocene glaciation in the southern Great Basin, high elevation pollen localities are lacking. Glaciation produces depositional environments conducive to pollen accumulation.

Low elevation sites, even in the northern Great Basin, usually lack the pollen record covering the warm, dry mid-Holocene. However, mid-to-late Holocene marsh records abound. In the south, Pleistocene and mid-to-late Holocene spring and marsh records occur rarely (e.g., Tule Springs; Mehringer 1967).

Woodrat middens. The record of past vegetation, often to species level, derived from fossil woodrat middens provides uniquely detailed information about past climate and soil conditions at a particular site. Woodrats forage plant remains (twigs, flowers, seeds, etc.) from within a 50 m radius. Therefore, their nests (preserved due to aridity or protection in caves and overhangs) are excellent repositories of ancient vegetation that provide site-specific information as to the exact plants that grew within a very restricted locality at a particular point in time (Spaulding 1985). Because the distribution of plants is determined by specific moisture and temperature conditions, their occurrence in the past at elevations above or below today's provides clues to changing temperature and precipitation (Spaulding et al. 1983; Spaulding 1985). For example, in the Eleana Range southeast of Pahute Mesa on the Department of Energy's Nevada Test Site, Spaulding (1985) found that between 18,000 and 13,000 years ago subalpine woodland occurred up to 900 m lower than it does today. He suggests that a combination of decreased temperature and increased precipitation was responsible for this marked depression in the number of species that characterize this plant community today.

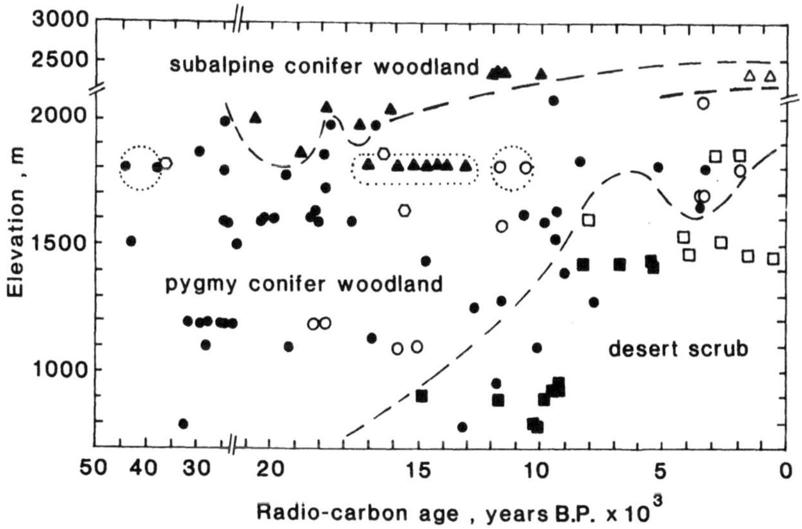

Figure 9.2. The elevational distribution of dated late Quaternary woodrat midden samples from the northern Mojave Desert. Inferred position of major vegetation boundaries are shown with broken line and extrapolated to modern vegetation elevations. Samples dominated by subalpine conifers (▲); with mountain conifers (△); containing abundant Utah juniper (●); with abundant juniper and pinyon pine (○); with juniper and limbar pine (○); with anomalous desert scrub (■) and with present day desert scrub affinity (□). Middens from sites on volcanic substrate are encircled with dotted lines; all other sites are on limestone. Modified from Spaulding 1985.

Woodrat middens are abundant throughout the Great Basin. However, certain rock types, such as limestone and welded tuff, are more resistant to weathering than other rock types and are therefore excellent preservers of woodrat nests. Volcanic rock substrates, such as basalt and ash-flow tuffs, are highly susceptible to weathering and usually tend to contain younger middens. Dominance of basalts, ash-flow tuffs, and andesites in the northern Great Basin may be responsible partially for the fact that few middens in excess of four to five thousand years in age have been found (Webb 1986). In addition, the effect of more mesic conditions in the northern Great Basin may play a role in limiting the longevity of woodrat middens. In the south, numerous middens in excess of 15,000 years B.P. reveal much more information on vegetational change during the late Pleistocene. Using this longer record, Spaulding (1985) has traced the movement of the boundaries between major biomes during the past 50,000 years (Figure 9.2). Lower treeline, which was below 800 m during the late Pleistocene, has risen since about

13,000 years ago. The same pattern is reflected in the higher elevation communities. Depressions of the subalpine conifer woodland occurred about 20,000 years ago and between 17,000 and 13,000 years ago. A slight depression of the lower juniper treeline about 4,000 years ago may record a period of significant increased effective precipitation.

Dendroclimatology. Climatic influence on the growth of trees (i.e., dendroclimatology) has been a major focus of dendrochronology (Douglass 1914, 1919). However, only during the past 20 years, with the application of statistical methods and the calibration and verification of tree-ring series with modern rainfall, temperature and other meteorological records, has dendroclimatology become an integral part of the biological data set used to reconstruct past climate (Fritts 1976; Hughes et al. 1982). Most of this pioneering research has been carried out in the arid Southwest where slight variations in precipitation can result in major variations in the growth rings of trees. Tree-ring records covering the period of historical climate records can be correlated to derive transfer functions that are used to extrapolate temperature and precipitation to the beginning of the tree-ring record (Fritts 1976). A large body of tree-ring data has been developed over the past quarter century that can be used to reconstruct climatic change in western North America over the past 400 years (Stokes et al. 1973). These chronologies have variable beginning dates and most end in the early 1960s. The quality of the chronologies in this data set varies primarily as a function of the sample size in the earliest years of each chronology. A second data set, which includes the above information, is computerized at the Laboratory of Tree-Ring Research at the University of Arizona. It also incorporates an extensive network of tree-ring chronologies recently developed for California, Oregon, and the northern Great Basin (Holmes et al. 1986).

An important set of tree-ring chronologies for studying long-term climatic change and variability on an annual basis for the past several thousand years is partially described by Graybill (1987). This set was derived from bristlecone and limber pine growing at relatively high elevations and distributed from the Front Range of the Rocky Mountains to the Nevada-California border, and from northern Nevada and Utah to north central Arizona and New Mexico. It was developed in conjunction with the U.S. Department of Energy program to study the influence of elevated CO_2 on vegetation but has yet to be used for paleoclimatic reconstructions. The potential exists to expand this network north to the Canadian border, using limber, whitebark, and western white pine. The outstanding feature of these tree-ring series is their great age.

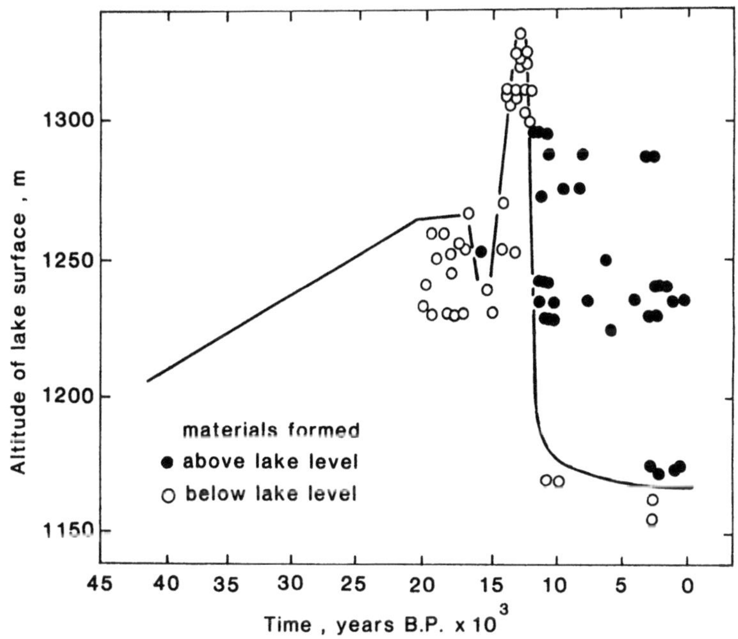

Figure 9.3. Proposed late Pleistocene lake-level chronology of pluvial Lake Lahontan. Radiocarbon dated materials formed above the lake level primarily consist of fossil woodrat middens, and materials formed below lake level primarily consist of tufa. Modified from Benson and Thompson (1987).

Living trees can be found that are 1,000 to 2,000 yr in age (and even older in some locations), and at many of the sites standing snags and deadwood remnants offer the possibility to extend tree-ring records backward in time several thousand more years.

Paleohydrology. Another data set that is important in documenting past climatic changes is lake-level fluctuations (Russell 1895; Hubs and Miller 1948, 1973; Morrison 1964; Davis 1978; Mifflin and Wheat 1979). As pointed out by Smith and Street-Perrott (1983), the value of fluctuations in lake levels as an indicator of climate has been recognized for more than 250 years. The paleoclimatic records derived from pluvial lake basins provide inferential methods of dating or correlating sequences of climate-related geological phenomena such as glaciation or soil and terrace development. For example, Benson and Thompson (1987) used selected radiocarbon data on surficial materials from the Lahontan Basin of Nevada and California to develop a chronology of lake-level variation for the past 50,000 yr. They suggest that the shape of the lake-level curve (Figure 9.3) indicates that past changes in the hydrologic balance (and therefore climate) of the Lahontan Basin were large in magnitude and took place in a rapid step-like fashion. They hypothesize that the rapid changes in lake level resulted from changes in the mean position of the jet stream, as it was forced north or south by the changing size of the North American continental ice sheet.

Streamflow estimates derived from historical measurements, and reconstructed from tree-ring chronologies, are also important in assessing recent changes in regional climate, particularly precipitation (Meko and Stockton 1984). For example, the Truckee River drains a large area of the east side of the Sierra Nevada, as well as a part of the western Great Basin. Due to the rapid decrease in precipitation from west to east across the area, most of the flow of the Truckee is derived from precipitation (viz. winter snow) in the Sierra Nevada. The present terminus of the Truckee system is Pyramid Lake.

Watson (1977) conducted a dendrohydrology study on the Truckee River and Pyramid Lake for the period from 1745 to 1904. Three tree-ring chronologies obtained from the Reno and Truckee area were screened for their correlations with precipitation and runoff. One chronology had a correlation of 0.72 with 44 years (1929 to 1972) of Truckee River discharge. Probably, the correlations between streamflow and tree-rings would be even better if it were not for a persistent bias during high streamflow years. This is due to the fact that during wet years tree growth is limited primarily by the tree's genetic potential. In dry years, the physiological processes controlling tree growth are limited primarily by water availability.

Isotopes. The isotopic composition of meteoric water (i.e., protium H, deuterium D, oxygen-16 ^{16}O, and oxygen-18 ^{18}O) is another useful source of information for studying climatic change in the Great Basin. The two primary factors that govern the isotopic composition of meteoric water at any location are the isotopic composition of the parent condensing vapor and the temperature of condensation. Warm precipitation generally has a heavier isotopic composition (more D and ^{18}O), whereas cold precipitation generally has a lighter composition, therefore the stable isotopic composition of meteoric water is a function of its atmospheric history, including storm-path trajectory as well as local meteorological conditions.

Ingraham and Taylor (1986) have suggested that the Great Basin approximates an isotopically closed system that contains groundwater in isotopic equilibrium with the atmospheric circulation patterns upwind. The relatively light isotopic composition (delta D = ca. -135 $^{o}/oo$ relative to SMOW standard) of meteoric water in the Great Basin is the result of isotopically advanced storm systems that evolve from continued hydrologic processes upwind. A collary to this observation is that the stable isotopic composition of meteoric water in the Great Basin should be highly sensitive to climatic change.

Claassen (1986) used stable isotopic ratios along with ^{14}C dating of ancient water to determine paleoclimatic conditions of the Amargosa Desert in Nevada. Unfortunately, uncontaminated ancient water samples with credible age determinations are difficult to obtain. In lieu of ancient water samples, several other techniques are available to determine paleoclimatic conditions. For example, the hydrogen isotopic composition of ancient cellulose has been used to reconstruct paleoclimate (Libby et al. 1976; Epstein et al. 1977; Yapp and Epstein 1982; Burk and Stuiver 1981). Also, cave speleothems preserve the record of long-term isotopic fluctuations in recharge waters. Harmon et al. (1979) studied the isotopic composition of ancient water from fluid inclusions in Pleistocene speleothems. These data can be interpreted in terms of both the paleotemperature variations within the cave and the stable isotopic composition of meteoric water (Schwarcz 1986).

Another source of climatic information is contained in volcanic glasses (Friedman et al. 1985). These glasses are known to absorb meteoric water once they have cooled. Hydrated tephra glasses from the Great Basin yield ancient meteoric water samples for isotopic analysis (N. L. Ingraham, unpublished data). This type of study is simplified by the abundance of described tephra layers in the Great Basin and known ages of these layers (determined by the radiocarbon dating of wood and other materials stratigraphically associated with the tephra).

Winograd et al. (1985) reconstructed a two million year record of the deuterium content of precipitation in the Great Basin by analyzing fluid inclusions from vein fillings. More recently, Winograd et al. (1988) reported on a 250,000-year climatic record obtained from a uranium-series dated calcite vein in the southern Great Basin. This latter study provided a continuous record of oxygen-18 variations in the continental hydrosphere during the middle-to-late Pleistocene. They show that the configuration of their curve for delta ^{18}O versus time, developed for the

Great Basin, closely resembles those developed for marine and Antarctic (Vostok) environments. However, the U-Th dates obtained for the Great Basin indicate that the last interglacial stage began before 147,000 years B.P., at least 17,000 years earlier than indicated by the marine record and 7,000 years earlier than by the Antarctic record. They suggest that this discrepancy, and other differences in the timing of key climatic events, indicate that the indirectly-dated marine chronology may need revision and that orbital forcing (i.e. Milankovitch Theory) may not be the principal cause of the Pleistocene ice ages.

Remote sensing. Remote sensing affords an opportunity to examine a number of relatively more recent factors associated with climatic change in the Great Basin. For example, the synoptic and repetitive capabilities provided by the Landsat satellite together with the spatial and spectral characteristics of the Multispectral Scanner (MSS) and Thematic Mapper (TM) sensors are ideally suited for spatial and temporal analyses.

Many environmental factors are associated closely with climatic change and can be identified, analyzed, and monitored by the Landsat system. Vegetation type boundaries and surface hydrology, especially lake-level measurements, are two environmental factors that are especially suited to Landsat analyses. Although the spectral, spatial, and radiometric resolution properties of the TM sensor are more appropriate for discriminating environmental phenomena, the MSS sensor is nevertheless useful. In fact, temporal comparisons which involve the use of MSS data from before 1982 (the year in which the TM was deployed) and TM data after that date can be especially useful for discriminating both vegetation and lake-level changes. Even though the Landsat satellite series only began operation in 1972, the capability of using such a system for climatic change research cannot be understated. In addition, the capability for defining vegetation compositional units (or types) as well as for identifying lacustrine properties associated with the climatic regime in place at the time of the satellite overpass provides a discrete spatial and temporal record which can be used to evaluate the correlation of lake level and vegetation information for past climatic regimes.

Landsat MSS and TM imagery can be used to identify vegetation compositional units and delineate boundaries. For example, Mouat and Johnson (1981) discussed the use of MSS imagery for the delineation of vegetation types within a hierarchical vegetation classification system for Grand Canyon National Park. They successfully identified and delineated vegetation types at the association (and finer) level. The chief problem in delineating vegetation types at the biome level lies in the characterization of ecotones.

As well, Landsat MSS and TM imagery has been used successfully in the discrimination of lake surface extent in semiarid terrain (Currey 1987). A comparison of terrain in western Nevada imaged by the Landsat MSS sensor in the summer of 1977 (a drought year) and also imaged by the Landsat TM sensor in the summers of 1984 and 1985 (wet years), is particularly instructive. The pine and juniper covered ranges east of the Sierra Nevada are clearly visible in each image, as are the remnants of Pleistocene lakes. During the drought year of 1977, the water bodies shrank to small vestiges of their size during moist years. Although the boundaries of the water bodies are very precise and distinct, the higher and lower elevation boundaries of the tree-covered mountain ranges are less distinct. With careful analysis, however, these boundaries, especially the lower boundaries, can be identified. These boundaries represent the integration of a considerable period of climatic fluctuation, perhaps extending back several hundred years. They may be considered a contemporaneous baseline against which past vegetation assemblages and their concomitant climates may be compared.

Climatic data. A considerable amount of 19th-century temperature and precipitation data also exists for the Great Basin (Chapter 2). Most of these data are derived from detailed meteorological records that were kept by military surgeons at outposts in the region. After 1871, the records were kept by the Army signal corps until 1892, when the U.S. Weather Bureau was established. The State of Nevada alone has records prior to 1890 from 21 different stations. Many of these stations have continuous records that were maintained from the 19th century through to the present. Historical Climate Network data from 1895 to the present are available from the Department of Energy, Oak Ridge National Laboratory. These data have been subjected to extensive evaluation and corrections to insure that the information is of the highest quality. Variables include monthly minimum, maximum, and mean temperature, total monthly precipitation, and the Palmer Drought Severity Index.

Convergence of data. Review of the site locations (Figure 9.1) at which pollen, woodrat midden, tree-ring, temperature and precipitation data were obtained reveals that there are several areas within the Great Basin that already have a significant spatial and temporal data overlap. These data are sufficient to develop a paleoclimatic reconstruction for many parts of the Great Basin in Nevada. Data sets are also available for southeastern Nevada (P.E. Wigand, unpublished data). The integration of these data sets and problems associated with integration are discussed below.

Climatic Reconstructions

Paleoclimatic proxy data are actually paleoenvironmental indicators used to reconstruct past climate. The term "proxy" is used because the measurement of climatic parameters is indirect through biological, geological, or isotopic data and not by direct measurement of precipitation or temperature. For example, instrumental records of annual precipitation in the Great Basin are only 50-100 years long. A study of long-term change in annual precipitation requires the use of paleoenvironmental records (i.e., proxy data), such as tree-rings or pollen distribution, that are sensitive to spatial-temporal changes in many aspects of the moisture regime and that can be used as a surrogate. The appropriateness of a particular proxy is determined by establishing its covariance with the target climatic variable during a period of overlap between the two records.

Integration of Paleoclimatic Proxy Data

Although fossil pollen and plant remains have been used extensively for reconstructing past vegetation and climatic history in the arid west, they have rarely been used in conjunction with each other or with tree-ring analyses to provide more accurate estimates of precipitation and temperature. The use of regional environmental pollen data in combination with woodrat midden data has been carried out primarily in the northern Great Basin where both kinds of data are available (Mehringer and Wigand 1987; Wigand 1987; Thompson 1984). The feasibility of integrating these data with tree-ring data has been illustrated only recently (Mehringer and Wigand 1990). The limitation to integration of these data has been the problem of difference in sampling intervals. In most cases, the sampling intervals of pollen data have not been frequent enough to allow comparisons with tree-ring records. With continuous or near continuous pollen sampling intervals, as in the upper part of the Diamond Pond record from south-central Oregon, much closer correspondence with the tree-ring record has been achieved (Petersen 1988; Mehringer and Wigand 1990). The integration of

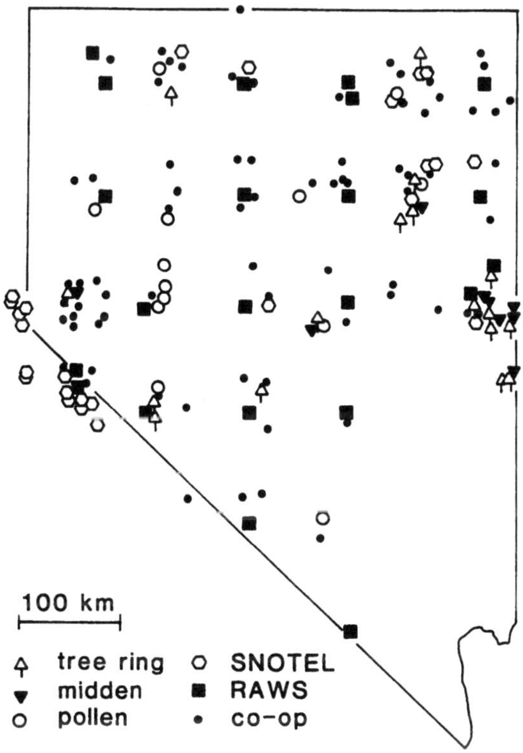

Figure 9.4. Locations of study sites within Nevada and eastern California. SNOTEL, snow telemetry data; RAWS, remote automatic weather station; co-op, National Cooperative Observers Network.

isotopic information with pollen records for paleoclimatic reconstruction is somewhat limited. Some analyses of the isotopic content of plants recovered from woodrat middens have been attempted (Long et al. 1990). Isotopic studies of sediments containing pollen have been conducted elsewhere (e.g., Heusser and Shackleton 1979). In the Great Basin, ostracode valves recovered from lake sediments have been analyzed to reconstruct an ^{18}O isotope curve to correlate with other paleoenvironmental data including pollen (Benson 1988). Changes in ^{18}O and deuterium (Libby et al. 1976) and carbon isotopes in recent trees (Pearman et al. 1976) correlate well with measured temperature. Additional isotopic analyses of newly obtained tree-ring cores would add much to our knowledge of past temperature variation.

The use of other paleolimnological data in the Great Basin is very limited. At Walker Lake in west central Nevada, microfossil data help to unravel the complex record of fluctuating lake levels based upon a combination of climatic and geomorphic factors (Benson 1988; Bradbury 1987; Bradbury et al. 1989). These data indicate that from 16,000 to 5,000 years ago Walker Lake became saline and may have dried up due to diversion of the Walker River (which feeds the lake) into the Carson River. A low water episode between 2,500 and 2,000 years ago probably reflects climatic change rather than river diversion.

Ostracode data together with pollen data are being used currently to detail the fluctuating Pleistocene lake level of Pluvial Lake Chewaucan in south central Oregon by A. Cohen and J.O. Davis (Pippin and Davis 1986). Although the results of this research are still incomplete, it appears that during periods when pine dominates the pollen rain, the ostracode record indicates relatively deep water. Conversely, sediments with lower levels of pine pollen include salt tolerant ostracodes, and therefore suggest lower lake levels.

Important Factors Affecting Data Interpretation

Temporal and spatial comparisons and integrations of various measured climatic parameters with paleoenvironmental indicators may be complicated by many factors, including: 1) sampling frequency, 2) period of record, 3) environmental signal strength, 4) response time, 5) data incongruencies, and 6) spatial distribution.

Sampling frequencies vary between the different types of paleoclimatic proxy data. For example, tree-rings are dated accurately to the year in which they developed (Fritts 1976), and at the highest frequency they yield climatic information on a yearly and seasonal basis. Also they potentially respond to climatic changes occurring over the entire frequency range, from years to decades, to hundreds of years. In contrast, single pollen or sediment samples obtained in geological or archaeological contexts may comprise the data from a few to many years. However, potentially both pollen and sediment samples can be obtained from locations with relatively rapid sedimentation rates such that each sample represents a smaller time interval.

To some extent, the temporal and spatial control exhibited by different climatic records and paleoclimatic proxy data is a function of the number of studies and the research interests of individual scientists. Therefore, a distinction can be made in both time and space between potential and actual resolution of a paleoclimatic reconstruction. Potential resolution is empirically based and grounded only in the formation processes of real data. Actual resolution is constrained by the detail of data records already published, and by practical operational considerations and funding limitations imposed on future research.

The highest frequency data, for example weather observations, are made throughout the day at meteorological stations all over the Great Basin (Bradley et al. 1982; NOAA 1988). However, the locations of these stations are related primarily to the long-term socioeconomic histories and cultural geography of the area. Even during the most recent period of instrumentally recorded meteorological data, geographical biases (particularly elevational biases) are present in the data. For example, lower elevations in the vicinity of water supplies and agricultural land are reasonably well represented, whereas higher elevations are sparsely covered. As a consequence, longer climatic records come from the lower elevation sites, whereas the few records available from higher elevation sites are usually only a few years in length.

The overall length of a paleoclimatic proxy record, in conjunction with sampling frequency, exerts a strong influence on the types of interpretations that can be made. The total length of a record ultimately determines the lowest frequency of change that is resolvable. The highest frequency resolvable is one divided by twice the sampling interval. The sampling interval for tree-ring series is one year, therefore, the highest frequency resolvable is 0.5 cycles yr^{-1}, or an event occurring with a periodicity of one cycle every two years. Tree-ring series are used as an example here because the sampling interval does not change over the period of record. If a paleoenvironmental indicator is analyzed that has a variable sampling interval (e.g., pollen distribution in sediments), or for which a regular sampling interval cannot be obtained through interpolation, the types of conclusions that can be drawn regarding the regularity of certain events is even more restricted.

Another factor to consider when evaluating a paleoclimatic proxy record is the strength of the targeted climatic signal contained in it. For example, different tree species frequently have different climatic responses. That is, some species may contain a strong winter and spring precipitation signal while others are more sensitive to late spring and summer moisture (Fritts 1974). Similarly, climatic change may be detected in one pollen record, but not another from the same region. For example, a pollen record obtained from a pinyon-juniper woodland might yield a sensitive paleoclimatic reconstruction while one from a spruce-fir forest would not. Pinyon and juniper trees respond more dramatically to slight changes in effective precipitation when compared to spruce and fir trees. In addition, records obtained from ecotonal areas might indicate greater sensitivity than those obtained well within the boundaries of a biome. Pollen sites in ecotonal areas will readily show the movement of the biome boundary while localities deep within the biome will register only slight deviations in abundance of the pollen type characterizing the biome's major plant species. The strength of a climatic signal can be determined by a temporal or spatial calibration (predictive numerical relationship) between the paleoenvironmental indicator (e.g., tree-ring index series) and the actual climatic variable, such as annual precipitation.

Temporal calibrations are established when a large number of temporally sequential samples of a paleoenvironmental indicator correspond to the same number of observations of the targeted climatic variable. This is basically a statistical problem in that there must be a large enough set of sample observations to establish a probabilistically valid transfer function. Transfer functions can range from bivariate regression between a paleoenvironmental proxy (independent variable) and a climatic variable (dependent variable) to equations expressing the covariation between sets of paleoclimatic proxy data and climatic data. The sample on which the transfer function is based should encompass the full range of variability expected in the dependent climatic variable(s) during the period of interest.

A spatial calibration is required when the temporal overlap between the paleoenvironmental indicator and historical or instrumental climatic data is not represented by a large set of tightly-dated observations. In this situation, a transfer function can be established between a large spatial set of observations of climatic data (dependent variable) and paleoenvironmental data (independent variable). Two examples of spatial calibration are mean annual or seasonal precipitation and temperature values from an array of weather stations and a spatial array of surface pollen samples. A transfer function can be generated to express the relationship between the pollen data (independent variables) and the climatic data (dependent variables). If a verifiable transfer function can be established between the two data sets, then its coefficients can be applied to the temporal pollen series to obtain estimates of past climate.

In the absence of statistically valid calibrations between paleoenvironmental indicators and climatic data during periods of spatial or temporal overlap, observed and repetitive covariance through time among the same type of indicator at different locations suggests that some common macroclimatic signal is acting in unison at all locations. A macroclimatic signal is a climatic factor that is present over large areas. For example, there is a paucity of upper elevation (>3,000 m) climatic data with any length (>40 yr) that can be calibrated with tree-ring series developed from high elevation conifers in the western United States. However, principal components analyses performed on sets of upper elevation tree-ring chronologies from the western U.S. result in 62-72% of the variance explained by the first eigenvector, depending on the particular data set employed. A common macroclimatic signal is the only conceivable factor that could cause such a large amount of common variability in tree-ring series in an area ranging from the Front Range of the Rocky Mountains to the Sierra Nevada, and from the central Great Basin and central Rocky Mountains into north central Arizona and New Mexico.

An additional factor important when integrating paleoenvironmental indicators with climatic data is response time. Response time is a measure of how quickly a paleoenvironmental indicator responds to changes in environmental conditions. Tree-rings respond quickly during the growing season to variability in precipitation, temperature, and soil moisture conditions (Fritts 1976). In comparison, response time in pollen records may vary considerably because the different species in a plant community will respond to climatic change in two ways. There will be the immediate response of increased pollen production and the slower response of species migration to variation in climate. When conditions are favorable, grasses, sedges, forbs, and most aquatic plants will spread rapidly, being constrained only by competition from other species. Trees and shrubs require relatively more time to mature and produce pollen, and therefore may not reflect climatic variation for years. Thus, it is conceivable that the same climatic inputs over a period of time, whether it be a year or a century, can yield vastly different but complementary responses from different paleoenvironmental indicators.

Incongruencies or inconsistencies in the data are also important when integrating paleoenvironmental indicators with climatic parameters.

Almost all paleoenvironmental data sources have the potential of presenting a biased picture of climatic change because of sampling problems. The question that must be asked of each paleoenvironmental indicator is to what extent the samples permit generalizations to a larger data set. More specifically, to what extent do samples obtained from a particular environmental stratum allow inferences to be made to a larger or nearby area, different elevational zones, or different ecological zones (biomes). For example, tree-ring chronologies over most of the Colorado Plateau normally are correlated highly with one another. Therefore, the general aspects of a paleoclimatic reconstruction based on tree-rings at one location may reasonably represent variability over a much larger area (Fritts 1976; Rose et al. 1982).

Alternatively, an attempt to reconstruct climatic variability from pollen obtained from archaeological sites may be hindered by the effects of increased disturbance, the manner in which local architectural barriers such as walls of dwellings, and human removal of local forests for firewood and agriculture, affect pollen dispersal. Woodrat middens also have potential sample bias because they disintegrate in areas of high rainfall, are preserved better in less easily weathered substrates (e.g., limestone) than others (e.g., basalts and andesites), and are restricted to rocky environments (Betancourt 1987). Additionally, different species of woodrats have different food preferences so that plant species represented in the midden may not reflect "real" relative plant community composition.

In presenting a paleoenvironmental integration of the Great Basin, it is important to address the strengths and weaknesses of the different paleoclimatic proxy data sets with respect to the factors discussed above.

Paleoclimatic Reconstruction of Great Basin Climate Using Tree-Ring Data

Paleoclimatic analyses using tree-ring chronologies offer techniques for reconstructing annual, or in some cases seasonal, temperature, precipitation, streamflow, and other factors associated with climate. In the following example, recent variability in instrumental climatic data from the northwestern sector of Nevada is analyzed from 1895 to 1983. Monthly temperature and precipitation data are analyzed statistically to characterize the nature of central tendency and variability over seasons. Time series plots of total annual precipitation and mean temperature are used to portray recent changes. The instrumental record of the Palmer Drought Severity Index (PDSI) since 1895 is used to present a more integrated view of moisture availability through time than is available from either temperature or precipitation. The PDSI is a hydrological accounting scheme that employs monthly temperature and precipitation data in its computation, in addition to robust information about soil moisture retention characteristics, evapotranspiration, and runoff.

Instrumental Climatic Data. An understanding of recent climatic variability is required as a baseline to which paleoclimatic reconstructions can be compared and by which hypothesized scenarios of climatic changes in the near future can be evaluated. Instrumental climatic data also serve as the dependent variable data set for calibration with tree-ring index chronologies.

In this example, the statistical characteristics of mean monthly temperature, total monthly precipitation, and the Palmer Drought Severity Index for the northwestern quadrant of Nevada are described, based on the years 1895 to 1983. This area is designated Division 1 (northwestern Nevada) in the spatial delineation scheme used by the National Oceanic Atmospheric Administration (NOAA). The raw data for this time period were obtained on magnetic tape from the Historical Climate Network data set of the Department of Energy, Oak Ridge National Laboratory (Karl et al. 1987). Mean monthly precipitation and temperature data are presented in Figure 9.5. For the period of record, average annual precipitation was 22.1 cm, ranging from a minimum of 9.9 cm to a maximum of 41.6 cm (Figure 9.6, upper). The coefficient of variation for precipitation indicates that the most variable (in a relative sense) season is summer and the least variable is spring. Least variable in a statistical sense implies high predictability, whereas high variability indicates lower predictability.

Total annual precipitation (1895 to 1983) for northwestern Nevada is shown in Figure 9.6. Above-average values were recorded near the beginning of the century, the mid-thirties to mid-forties, and the mid-sixties to mid-seventies. The highest annual value during the period of record occurred in 1983. An extended dry period is indicated from the early twenties to the mid-thirties. Other dry intervals occurred during the late forties to the early sixties. A time series plot of annual temperature is also presented in Figure 9.6. An obvious feature is the extended period of low temperature from the 1890s through the early twenties. The mid-twenties through the early forties are an interval when values

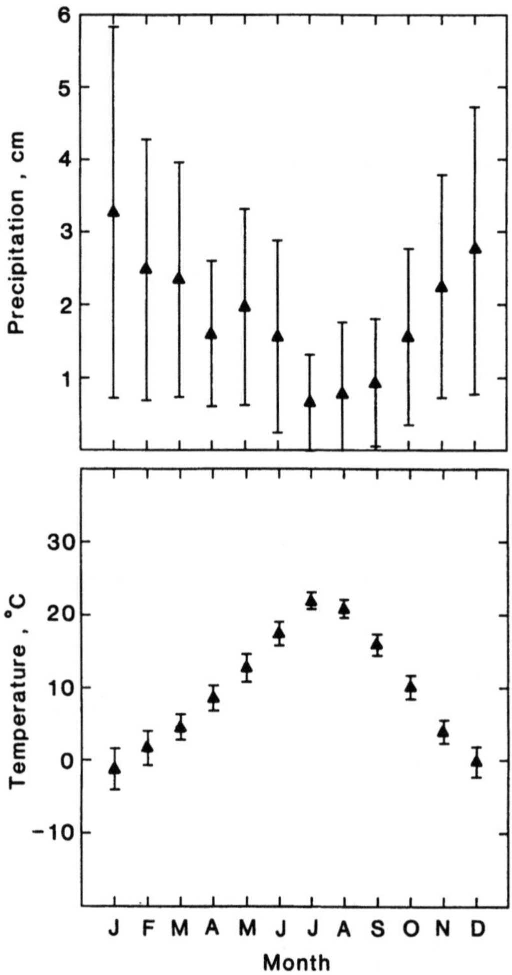

Figure 9.5. (upper) Average total monthly precipitation in centimeters (± 1.0 SD), Nevada division 1, 1895 to 1983. (lower) Mean monthly temperature in degrees C (± 1.0 SD), Nevada division 1, 1895 to 1983.

were frequently above the long-term mean (about 10° C). During the period from the late forties to the present there have been several instances of consecutive years with temperature below the long-term average.

The precipitation and temperature descriptions for the period of instrumental record noted above can be integrated on an annual basis

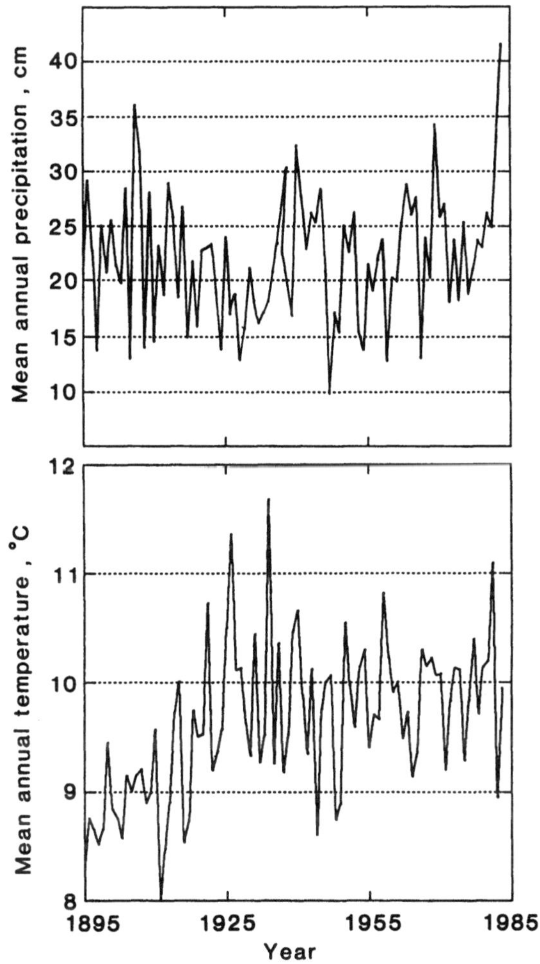

Figure 9.6. (upper) Total annual precipitation (cm.), Nevada division 1, 1895 to 1983. (lower) Mean annual temperature (° C), Nevada division 1, 1895 to 1983.

with the Palmer Drought Severity Index (PDSI). Monthly temperature and precipitation, in addition to soil moisture retention information, are used in its computation; it therefore offers a reasonable and integrated measure of moisture availability (effective precipitation). For these reasons, it frequently exhibits a higher covariation than temperature or precipitation alone with tree-ring series, since a tree responds to the integrated effects of temperature and precipitation through its interface with the soil and the atmosphere.

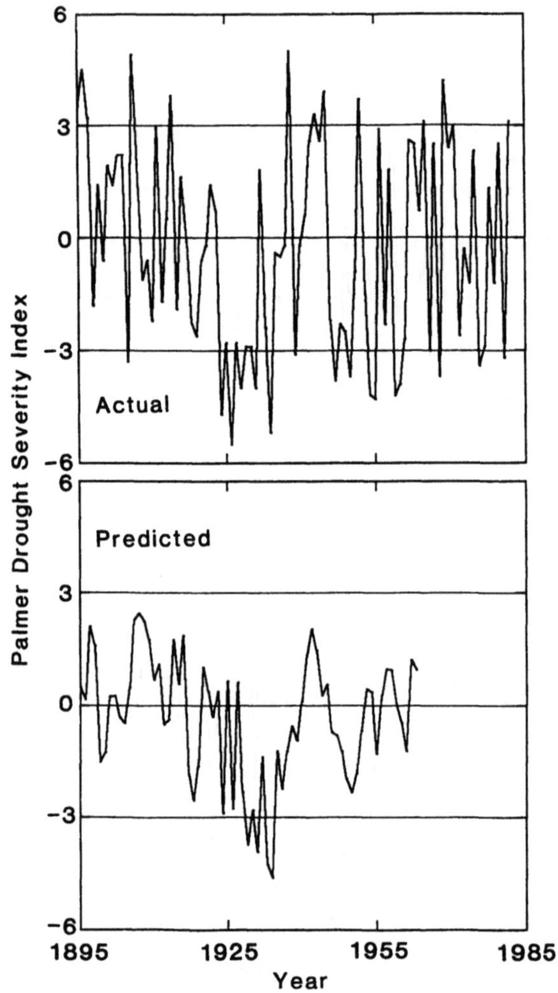

Figure 9.7. (upper) Actual Palmer Drought Severity Index for July, Nevada division 1, for the years 1895 to 1983. (lower) Predicted Palmer Drought Severity Index for July, Nevada division 1, 1895 to 1962.

A time series plot of July PDSI is presented in Figure 9.7. Values near zero indicate "normal" meteorological conditions, while with increasing distance from zero, positive values indicate increasingly mesic conditions and negative values indicate increasingly severe drought-conditions. Prominent in Figure 9.7 is the extended dry period from the early twenties to the mid-thirties. The "dust bowl" year of 1934 is especially evident, but the frequency of such years during this extended dry interval is also noteworthy. Figure 9.7 also shows that consecutive drought years occur quite frequently, even during this relatively short period of instrumental record.

Dendroclimatic Reconstruction. Tree-ring chronologies, developed from locations where climate limits the physiological processes controlling tree growth are paleoenvironmental indicators that can be used to reconstruct climatic conditions over the past several hundred to several thousand years. The chronologies are developed from measured widths of accurately dated annual growth rings of trees from selected localities. They can be constructed solely from long-lived trees or by extending living tree chronologies back in time with dead wood from geological or archaeological contexts. Now we will numerically calibrate some of these tree-ring chronologies with the July PDSI for Nevada's Division 1. The linkage between tree growth, as the independent variable set, and PDSI, as the dependent variable, is established during the years of common overlap between the two data sets.

The PDSI data set begins in 1895 and ends in 1983. The tree-ring chronology data set, beginning in A.D. 1600 and ending in 1962, contains 104 chronologies from the western United States. Although there has been more recent dendrochronological research in the region (Holmes et al. 1986), the 104 chronology grid was chosen because of its greater spatial coverage. A description and analysis of this data set has been presented by Rose and Fritts (unpublished report). The overlap between the climate and tree-ring data sets extends from 1895 to 1962. Therefore, the statistical model expressing the relationship between the two is based on this interval.

July PDSI was selected as the dependent variable because it is the time of year when most historically recorded droughts have peaked, when water storage becomes depleted severely, and when plant growth is affected adversely. As discussed previously, July is the month of maximum temperature and minimum precipitation.

Many of the tree-ring chronologies in the 104 series data set exhibit significant correlations with one another. In tree-ring research, principal components analysis is frequently used as a data reduction technique and orthogonalization device (Fritts 1976). A principal components analysis was performed on the 104 chronology data set to reduce the number of potential independent predictor variables and to create

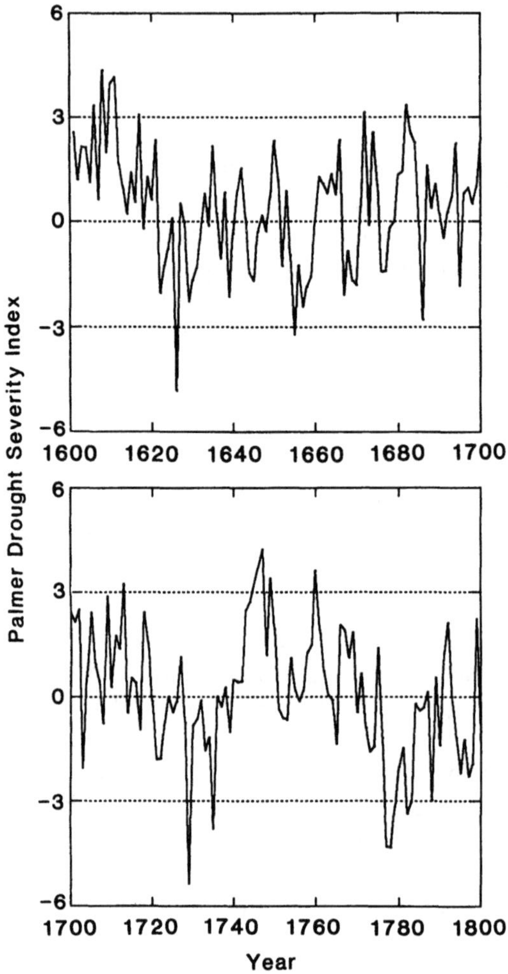

Figure 9.8. (above and facing page) Palmer Drought Severity Index values for the years 1601 to 1962 in Nevada division 1, reconstructed from tree-ring data as described in the text.

linear combinations of the original variables which are uncorrelated with one another. The principal components also have the property that the first component accounts for the maximum amount of variance in the data set, the second accounts for the next largest amount while being uncorrelated with the first, the third accounts for the next largest amount while being uncorrelated with the first and second, and so on. Although there are as many components as original variables, only a

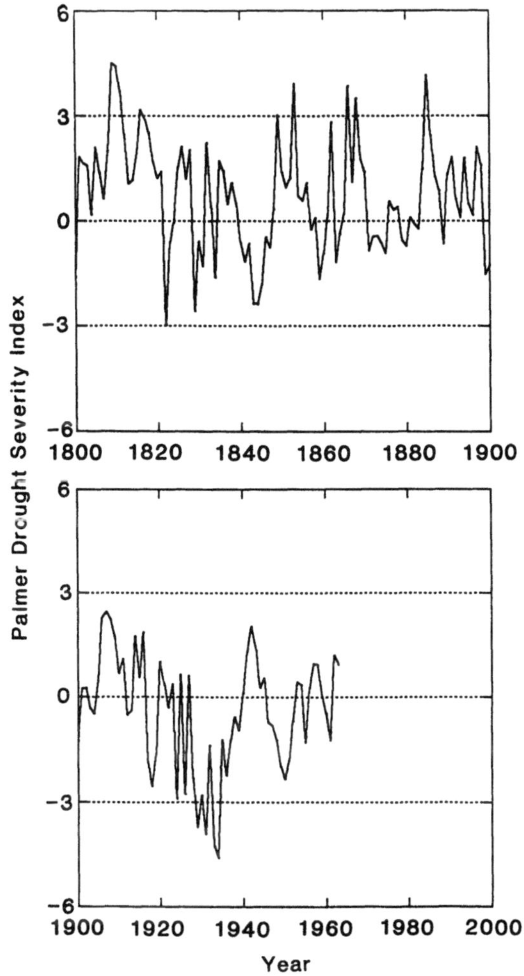

subset of the components needs to be used because they account for a large amount of the variance in the data set. In the analysis of the 104 chronologies, the first component explains 17.3% of the total variance in the data set, the second explains 12.7%, the third 4.5%, and so on through all 104 components. Approximately 75% of the variance in the tree-ring chronology data set is explained by the first twenty principal components, so they were selected as potential independent variables. A multiple regression equation was constructed using the 1895 to 1962 PDSI and selected tree-ring data (i.e., the scores on principal components of the latter). The correlation between the PDSI and the several components that entered into the equation as predictors for this period was 0.76, indicating that over half of the variability in the PDSI can be explained by the tree-ring data. A plot of predicted PDSI is shown under the actual value in Figure 9.7. To test the time stability of this relationship the data were divided into halves, from (period A) 1895 to 1928 and (period B) 1929 to 1962. A regression equation developed for the latter half of the data (B) and applied to the tree-ring data from the earlier half (A) determines how well it can predict the PDSI. The reverse operation, where an equation developed using the first data half (A) is applied to the tree-ring data from the second half (B), determines how well the PDSI for that interval is predicted. The correlation between the tree-ring data and the PDSI for the second data half (B) is 0.77, and the correlation between the actual and predicted PDSI for the first data half (A) is 0.72. There is no significant difference of means (.05 significance level) between the actual and predicted series for the verification comparisons. Difference of means is not tested for the model calibration period because they are identical by definition.

The same operation using the first half of the data set (A) for model calibration yields a correlation of 0.63, and the correlation between the actual and predicted PDSI for the second data half (B) is 0.67. There is also no significant difference of means between the actual and predicted PDSI for the latter data half. The decrease in correlation in the calibration equation between the most recent period (B) and the first half of the data (A) is probably due to the poorer quality of the meteorological data during the earliest years of record, especially since the regional averages used to compute the PDSI are based on fewer stations. Both of the calibration relationships between the PDSI and tree-ring data are highly significant and indicate the relationship is reasonably stable over time. For this reason, the equation developed using all of the data is used to produce a reconstruction of July PDSI for the period 1600 to 1962. The equation developed for the 1895 to 1962 period is applied to the complete length of the tree-ring data set in order to produce the reconstruction shown in Figure 9.8.

The recent instrumental record (post-1900) does not fully represent the nature of dry and wet years over the past 400 years. If the frequencies of wet and dry years were determined solely from the post-1900 data they would not adequately represent the complete record from 1600 to the present. Also, the types of year-to-year changes present in the complete record are not always accurately portrayed by the period of modern record. Watson's (1977) reconstruction for flow in the Truckee

River overlaps with the modern instrumental record for the years 1929 to 1972. During this interval the dry years shown in the Palmer Drought Serverity Index (PDSI) plot correspond to periods of reduced flow on the Truckee. Over the period from 1740 to the present the variance in reconstructed Truckee discharge remains relatively constant, but there were periods in the late 1700s, 1820s, 1830s and late 1850s when flow was considerably below normal.

Thus the long record of drought reconstructed from the tree-ring series, can be used to make more accurate statements about the characteristics of future conditions. For example, with what frequency do drought conditions occur? If a drought of a given severity is experienced, what is the chance that it will be followed by more moderate or more severe conditions the following year? The record of climate reconstructed from long tree-ring series offers a unique foundation on which such assessments can be based.

Conclusions and Recommendations

In this chapter, we have discussed the use of the North American Great Basin as a sensitive indicator of climatic change. We have shown that there are several paleoclimatological records that overlap spatially and temporally. These records are better preserved and encompass a much longer time span than the majority of similar records available elsewhere. Therefore, we believe that intensified study should be undertaken that will utilize these paleoclimatological records, and in conjunction with modern climatological data, develop correlations to consider the range of variability in climate that might reasonably be expected to occur given changes in global climate.

The increasing pressure on water resources in arid and semiarid regions from urban, industrial, and agricultural users makes it necessary to develop a better understanding of the relationships between climatic variation and water availability. The Great Basin has been identified as the hydrological region in the United States most vulnerable to global warming (Maggs 1989). The hydrological cycle that determines distributions of water resources does appear to be quite sensitive to short-term climatic change. An increase in mean annual temperature of a few $^{\circ}$C could amplify seasonal effects, for example, evapotranspiration is expected to be higher under warmer conditions and summer droughts more common. Understanding the potential limits climate may impose on water resources is helpful in making informed planning decisions to balance water needs and availability, both effectively and realistically.

It is not possible at the present time to accurately predict climatic trends or variability on the scale of a few years, let alone decades. Therefore, we believe it is necessary to examine past records of climate, water supply, and plant distributions within the Great Basin in order to consider the range of variability that might reasonably be expected to occur in the future. In addition, the Great Basin serves as a natural laboratory with which to study changes that may be manifestations of climatic variability on a global scale.

References

Benson LV (1988) Preliminary Paleolimnologic Data for the Walker Lake Subbasin, California and Nevada. Water-Resources Investigations Report 87-4258. U.S. Geological Survey, Denver, CO.

Benson LV and Thompson RS (1987) Lake level variation in the Lahontan Basin for the past 50,000 years. Quatern. Res. 28:69-85.

Betancourt JL (1987) Paleoecology of pinyon-juniper woodlands: summary. pp 129-139 In Everett RL (editor) Proceedings of the pinyon-juniper conference, USDA Technical Report INT-215, Reno, NV.

Billings WD (1951) Vegetational zonation in the Great Basin of western North America. pp. 101-122. In Compt. Rend. du Colloque sur les Bases Ecologique de la Regeneration de la Vegetation des Zones Arides. IUBS, Paris.

Bradbury JP (1987) Late Holocene diatom paleolimnology of Walker Lake, Nevada. Archiv. Hydrobiol. 79:1-27.

Bradbury JP, Forester RM, and Thompson RS (1989) Late quaternary paleolimnology of Walker Lake, Nevada. J. Paleolimn. 1:249-267.

Bradley RS, Barry RG, and Kildas G (1982) Climatic fluctuations of the western United States during the period of instrumental record. Contribution 42, Department of Geology and Geography, Univ. Massachusetts, Amherst, MA.

Brown RW (1987) Ecophysiology and water relations research in the Pinyon-Juniper Vegetation Type. pp 398-405 In Everett RL (editor) Proceedings of the Pinyon-Juniper Conference, USDA Technical Report INT-215, Reno, NV.

Burk RL and Stuiver M (1981) Oxygen isotope ratios in trees reflect mean annual temperature and humidity. Science 211:1417-1419.

Claassen HC (1986) Late-Wisconsin paleohydrology of the westcentral Amargosa Desert, Nevada. Chem. Geol. (Isotope Geoscience Section) 58:311-323.

Currey, DR (1987) A reconstruction of high levels of Great Salt Lake, Utah, in late Holocene time. pp 45-47. In Proceedings of the Third Annual LANDSAT Workshop, Univ. California, Santa Barbara, CA.

Currey DR, Atwood G, and Mabey DR (1983) Major levels of Great Salt Lake and Lake Bonneville. Utah Geological and Mineral Survey, Utah Department of Natural Resources, Map 73.

Currey DR and James SR (1982) Paleoenvironments of the northeastern Great Basin and northeastern Basin rim region: a review of geological and biological evidence. pp. 27-52. In Madsen DB and O'Connell JF (editors), Man and Environment in the Great Basin, Society for American Archaeology Papers Vol. 2.

Davis JO (1978) Quaternary tephrochronology of the Lake Lahontan area, Nevada and California. Univ. Nevada Archaeological Survey Research Paper 7.

Davis JO (1982) Bits and pieces: the last 35,000 years in the Lahontan area. pp. 53-75. In Madsen DB, O'Connell JF (editors) Man and Environment in the Great Basin. Society for American Archaeology Papers Vol. 2.

Davis OK (1984) Pollen frequencies reflect vegetation patterns in a Great Basin (U.S.A.) mountain range. Review of Palaeobotany and Palynology 40:295-315.

Douglass AE (1914) A method of estimating rainfall by the growth of trees. In Huntington E (editor) The Climatic Factor, Carnegie Institution of Washington, Washington, DC.

Douglass AE (1919) Climatic cycles and tree-growth: a study of the annual rings in trees in relation to climate and solar activity. Carnegie Institution of Washington Publication 289, Washington, DC.

Epstein S, Thompson P, and Yapp CJ (1977) Oxygen and hydrogen isotopic ratios in plant cellulose. Science 198:1209-1215.

Flaschka IM (1984) Climatic change and water supply in the Great Basin. Unpublished thesis, Univ. Arizona, Tucson.

Friedman I, Izett GA, and Gleason JD (1985) Isotopic paleoclimate from hydrated volcanic ash. Geological Society of America, Abstracts with Programs, 17:587.

Fritts HC (1974) Relationships of ring widths in arid-site conifers to variations in monthly temperature and precipitation. Ecol. Monogr. 44:411-440.

Fritts HC (1976) Tree-rings and Climate. Academic Press.

Graybill DA (1985) Western U.S. tree-ring index chronology data for detection of arboreal response to increasing carbon dioxide. pp. 1-63. Laboratory of Tree-Ring Research, Series 26. Univ. Arizona, Tucson, AZ.

Graybill DA (1987) A network of high elevation conifers in the western U.S. for detection of tree-ring growth response to increasing atmospheric carbon dioxide. pp. 463-474. In Jacoby GC, Hornbeck JW (editors) Proceedings of the International Symposium on Ecological Aspects of Tree-ring Analyses. Department of Energy Conference-8608144, Washington, DC.

Harmon RS, Schwarcz HP, and O'Neil JR (1979) D/H ratios in speleothem fluid inclusions: a guide to variation in the isotopic compositions of meteoric precipitation. Earth Planetary Sci. Letts. 42:254-266.

Hastings JR and Turner RM (1965) The Changing Mile. Univ. Arizona Press, Tucson, AZ.

Heusser LE and Shackleton NJ (1979) Direct marine-continental correlation: 150,000-year oxygen isotope-pollen record from the North Pacific. Science 204:837-839.

Holmes RL, Adams RK, and Fritts HC (1986) Tree-ring chronologies of western North America: California, eastern Oregon and northern Great Basin, with procedures used in the chronology development work, including users manuals for computer programs COFECHA and ARSTAN. Univ. Arizona, Laboratory of Tree-Ring Research, Chronology Series VI. Tucson, AZ.

Houghton JG (1969) Characteristics of Rainfall in the Great Basin. Univ. Nevada, Desert Research Institute. Reno, NV.

Houghton JG, Sakamoto CM, and Gifford RO (1975) Nevada's weather and climate. Nevada Bureau of Mines and Geology, Special Publication 2, Reno, NV.

Hubbs CL and Miller RR (1948) The zoological evidence/correlation between fish distribution and hydrographic history in the desert basins of the western United States. pp. 17-166. In The Great Basin, with emphasis on glacial and post glacial times. Bulletin of the Univ. Utah Vol. 38.

Hubbs CL, Miller RR, and Hubbs LC (1974) Hydrographic history and relict fishes of the north-central Great Basin. Memoirs of the California Academy of Sciences 7.

Hughes MK, Kelly PM, Pilcher JR, and LaMarche VC (1982) Climate from Tree-rings. Cambridge Univ. Press, Cambridge.

Ingraham NL and Taylor BE (1986) Hydrogen isotope study of large-scale meteoric water transport in northern California and Nevada. J. Hydrol. 85:183-197.

Karl TR, Metcalf LK, Nicodemus ML, and Quayle RG (1987) Historical Climatological Series, National Climate Data Center, Asheville, NC.

Libby ML, Pandolphi LJ, Payton PH, Marshall J III, Becker B, and Giertz-Siebenlist V (1976) Isotopic tree thermometers. Nature 261:284-288.

Long A, Warnecke L, and Betancourt JL (1990) Stable isotopes, fossil packrat middens and paleoclimate. In Martin PS, Betancourt J, Van Devender JR (editors) Fossil Packrat Middens: The Last 40,000 Years of Biotic Change, Univ. Arizona Press (in press).

Maggs, WW (1989) Warming will alter water resources. EOS 70:67-74.

Mehringer PJ Jr. (1967) Pollen analysis of the Tule Springs site area, Nevada. pp. 129-200. In Wormington HM and Ellis D (editors) Nevada State Museum Anthropological Papers Vol 13, Carson City, NV.

Mehringer PJ Jr. (1977) Great Basin late Quaternary environments and chronology. pp. 113-167. In Fowler, DD (editor), Models and Great Basin Prehistory: A Symposium. Desert Research Institute Publications in the Social Sciences 12, Reno, NV.

Mehringer PJ Jr. (1985) Late-Quaternary pollen records from the interior Pacific Northwest and northern Great Basin of the United States. pp. 165-187. In Bryant VA, Holloway RG (editors) Pollen Records of Late-Quaternary North American Sediments. American Association of Stratigraphic Palynologists, Dallas, TX.

Mehringer PJ Jr. (1986) Prehistoric environments. pp. 31-50. In D'Azevedo WL (editor) Handbook on North American Indians. Volume 11: Great Basin. Sturtevant WC (editor), Smithsonian Institution, Washington, DC.

Mehringer PJ Jr. and Warren CN (1976) Marsh, dune and archaeological chronology, Ash Meadows, Amargosa Desert, Nevada. pp. 120-150. In Elston R (editor) Holocene Environmental Change in the Great Basin. Nevada Archeological Survey Research Papers Vol. 6.

Mehringer PJ Jr. (1986) Holocene history of Skull Creek dunes, Catlow Valley, Oregon, U.S.A. J. Arid. Environ. 11:117-138.

Mehringer PJ Jr. and Wigand PE (1987) Western Juniper in the Holocene. pp. 109-119. In Proceedings of the Pinyon-Juniper Conference, USDA Technical Report INT-215, Reno.

Mehringer PJ Jr. and Wigand PE (1990) Comparison of Late Holocene environments from woodrat middens and pollen, Diamond Craters, Oregon. In Martin PS, Betancourt J, and Van Devender TR (editors) Fossil Packrat Middens: The Last 40,000 Years of Biotic Change. Univ. Arizona Press. (in press).

Meko DM and Stockton CW (1984) Secular variations in streamflow in the western United States. J. Clim. Appl. Meteor. 23:889-897.

Mifflin MD and Wheat MM (1979) Pluvial lakes and estimated pluvial climates of Nevada. Nevada Bureau of Mines and Geology, Bulletin 94, Reno, NV.

Mitchell JM, Dzerdzeevskii B, Flohn H, Hofmeyr WL, Lamb HH, Rao KN, and Wallen CC (1966) Climatic change, Technical Note 79, World Meteorological Organization, Geneva.

Morrison RB (1964) Lake Lahontan: geology of southern Carson Desert, Nevada. U.S. Geological Survey Professional Paper 401.

Morrison RB and JC Frye (1965) Correlation of the middle and late Quaternary successions of the Lake Lahontan, Lake Bonneville, Rocky Mountain (Wasatch Range), southern Great Plains, and eastern Midwest areas. Nevada Bureau of Mines and Geology Report 9, Reno, NV.

Mouat DA and Johnson RR (1981) An inventory and assessment of wildlife habitat in Grand Canyon National Park using remote sensing techniques. pp. 105-113. In Proc. 4th Pecors Symposium on Remote Sensing, Sioux Falls, Iowa.

NOAA (1988) Climatological data annual summary. National Climate Data Center, Asheville, 102(13).

Odum EP (1971) Fundamentals of Ecology. Third Edition. W. B. Saunders Company, Philadelphia, PA.

Overpeck JT, Webb T III, and Prentice IC (1985) Quantitative interpretation of fossil pollen spectra: dissimilarity coefficients and the method of modern analogs. Quatern. Res. 23:87-108.

Pearman GI, Francey RJ, and Fraser PJB (1976) Climatic implications of stable carbon isotopes in tree rings. Nature 260:771-773.

Petersen KL (1988) Climate and the Dolores River Anasazi. Univ. Utah Anthropological Papers 113.

Pippin LC and Davis JO (1986) Long pollen record from pluvial lake sediments at Summer Lake, south central Oregon. Geological Society of America, Abstracts with Programs 18:719.

Rogers GF (1982) Then and Now: A Photographic History of Vegetation Change in the Central Great Basin Desert. Univ. Utah Press, Salt Lake City, UT.

Rose MR, Robinson WJ, and Dean JS (1982) Dendroclimatic reconstruction for the southeastern Colorado Plateau. Laboratory of Tree-ring Research, Univ. Arizona, Tuscon, AZ.

Russell IC (1895) Present and extinct lakes of Nevada. National Geographic Society Monographs 1:101-136.

Schneider SH (1989) The greenhouse effect: science and policy. Science 243:771-781.

Schwarcz HP (1986) Geochronology and isotopic geochemistry of speleothems. In Fritz P and Fontes JCh (editors) Handbook of Terrestrial Geochemistry, vol 2, The Terrestrial Environment. Elsevier, Amsterdam.

Siegal RD (1983) Paleoclimatic significance of D/H and $^{13}C/^{12}C$ ratios in Pleistocene and Holocene wood. Unpublished M.S. thesis, Univ. Arizona, Tucson, AZ.

Smith GI and Street-Perrott FA (1983) Pluvial lakes of the western United States. pp. 190-191. In Porter SC (editor) Late-quaternary Environments of the United States, Vol. 1 The Late Pleistocene. Univ. Minnesota Press, Minneapolis, MN.

Spaulding WG (1985) Vegetation and climates of the last 45,000 years in the vicinity of the Nevada Test Site, south-central Nevada. U.S. Geological Survey Professional Paper 1329.

Spaulding WG, Leopold EB, and Van Devender TR (1983) Late Wisconsin paleoecology of the American Southwest. pp. 259-293. In Wright HE Jr and Porter SC (editors), Late Quaternary Environments of the United States, Vol. 1, The Late Pleistocene. Univ. Minnesota Press, Minneapolis, MN.

Stokes MA, Drew LG, and Stockton CW (1973) Tree-ring chronologies of western North America, I. Selected tree-ring stations. Laboratory of Tree-ring Research Chronology Series 1, Univ. Arizona, Tucson, AZ.

Thompson RS (1984) Late Pleistocene and Holocene environments in the Great Basin. Unpublished PhD dissertation, Univ. Arizona, Tucson, AZ.

Turner RM and Karpiscak MM (1980) Recent vegetation changes along the Colorado River between Glen Canyon Dam and Lake Mead, Arizona. U.S. Geological Survey Professional Paper 1132.

Walter H (1954) Le facteur eau dans les regions arides et sa signification pour l'organisation de la vegetation dans les contrees sub-tropicales. pp. 27-39. In Les Divisions Ecologiques du Monde. CNRS, Paris.

Watson PC (1977) Dendrochronologic Reconstruction of Water Levels for Pyramid Lake, Nevada, 1745 to 1904 A.D., unpublished M.S. Thesis, Univ. Nevada, Reno, NV.

Webb RH (1986) Spatial and temporal distribution of radiocarbon ages on rodent middens from the southwestern United States. Radiocarbon 28:1-8.

Wells PV (1983) Paleobiogeography of montane islands in the Great Basin since the last glaciopluvial. Ecol. Monogr. 53: 341-382.

Went FW (1957) The Experimental Control of Plant Growth. Chronica Botanica, Waltham, MA.

Whittaker RH (1970) Communities and Ecosystems. Macmillian, NY.

Wigand PE (1987) Diamond Pond, Harney County, Oregon: vegetation history and water table in the eastern Oregon Desert. Great Basin Naturalist 47:427-458.

Winograd IJ, Szabo BJ, Coplen TB, Riggs AC, and Kolesar PT (1985) Two-million year record of deuterium in Great Basin ground waters. Science 227:519-522.

Winograd IJ, Szabo BJ, Coplen TB, and Riggs AC (1988) A 250,000-year climatic record from Great Basin calcite: implications for Milankovitch Theory. Science 242:1275-1280.

Yapp CJ and Epstein S (1982) A reexamination of cellulose carbon-bound hydrogen δD measurements and some factors affecting plant-water D/H relationships. Geochim. Cosmochim. Acta 46:955-965.

TAXONOMIC INDEX

INDEX

MIX
Papier aus verantwortungsvollen Quellen
Paper from responsible sources
FSC® C105338

If you have any concerns about our products,
you can contact us on
ProductSafety@springernature.com

In case Publisher is established outside the EU,
the EU authorized representative is:
Springer Nature Customer Service Center GmbH
Europaplatz 3, 69115 Heidelberg, Germany

Printed by Libri Plureos GmbH
in Hamburg, Germany